Ecological Studies, Vol. 89

Analysis and Synthesis

Ecological Studies

Volume 68
Stable Isotopes in Ecological Research
Edited by P.W. Rundel, J.R. Ehleringer, and K.A. Nagy (1989)

Volume 69
Vertebrates in Complex Tropical Systems
Edited by M.L. Harmelin-Vivien and F. Bourlière (1989)

Volume 70
The Northern Forest Border in Canada and Alaska
Edited by J.A. Larsen (1989)

Volume 71
Tidal Flat Estuaries: Simulation and Analysis of the Ems Estuary
Edited by J. Baretta and P. Ruardij (1988)

Volume 72
Acidic Deposition and Forest Soils
By D. Binkley, C.T. Driscoll, H.L. Allen, P. Schoeneberger, and D. McAvoy (1989)

Volume 73
Toxic Organic Chemicals in Porous Media
Edited by Z. Gerstl, Y. Chen, U. Mingelgrin, and B. Yaron (1989)

Volume 74
Inorganic Contaminants in the Vadose Zone
Edited by B. Bar-Yosef, N.J. Barnow, and J. Goldshmid (1989)

Volume 75
The Grazing Land Ecosystems of the African Sahel
Edited By H.N. Le Houérou (1989)

Volume 76
Vascular Plants as Epiphytes: Evolution and Ecophysiology
Edited by U. Lüttge (1989)

Volume 77
Air Pollution and Forest Decline: A Study of Spruce (*Picea abies*) on Acid Soils
Edited by E.-D. Schulze, O.L. Lange, and R. Oren (1989)

Volume 78
Agroecology: Researching the Ecological Basis for Sustainable Agriculture
Edited by S.R. Gliessman (1990)

Volume 79
Remote Sensing of Biosphere Functioning
Edited by R.J. Hobbs and H.A. Mooney (1990)

Volume 80
Plant Biology of the Basin and Range (1990)
Edited by B. Osmond, G.M. Hidy, and L. Pitelka (1990)

Volume 81
Nitrogen in Terrestrial Ecosystem: Questions of Productivity, Vegetational Changes, and Ecosystem Stability
By C.O. Tamm (1991)

Volume 82
Quantitative Methods in Landscape Ecology: The Analysis and Interpretation of Landscape Heterogeneity
Edited by M.G. Turner and R.H. Gardner (1990)

Volume 83
The Rivers of Florida
Edited by R.J. Livingston (1990)

Volume 84
Fire in the Tropical Biota: Ecosystem Processes and Global Challenges
Edited by J.G. Goldammer (1990)

Volume 85
The Mosaic-Cycle Concept of Ecosystems
Edited by H. Remmert (1991)

Volume 86
Ecological Heterogeneity
Edited by J. Kolasa and S.T.A. Pickett (1991)

Volume 87
Horses and Grasses
Edited by P. Duncan (1991)

Volume 88
Pinnipeds and El Niño: Responses to Environmental Stress
Edited by F. Trillmich and K.A. Ono (1991)

P. J. C. Kuiper · M. Bos (Eds.)

Plantago: A Multidisciplinary Study

With 136 Figures

Springer-Verlag
Berlin Heidelberg New York
London Paris Tokyo Hong Kong
Barcelona Budapest

Prof. Dr. Pieter Jan Cornelis Kuiper
Dr. Marten Bos
Biological Centre
University of Groningen
P.O. Box 14
NL-9750 AA Haren
The Netherlands

ISBN-13: 978-3-642-76394-6 e-ISBN-13: 978-3-642-76392-2
DOI: 10.1007/978-3-642-76392-2

Library of Congress Cataloging-in-Publication Data. Plantago: a multidisciplinary study / P. J. C. Kuiper, M. Bos, eds. p. cm. – (Ecological studies; v. 89) Includes bibliographical references and index.
1. Plantago. I. Kuiper, P. J. C. (Pieter Jan Cornelis) II. Bos, M. III. Series. QK495.P714P53 1991 583'.89–dc20
91-24529 CIP

Softcover reprint of the hardcover 1st edition 1992

Production Editor: Herta Böning, Heidelberg
Reproduction of the figures: Gustav Dreher GmbH, Stuttgart
Typesetting: K+V Fotosatz GmbH, Beerfelden
31/3145-543210 – Printed on acid-free paper

Preface

The results, published in this book, present the outcome of a cooperative research between plant ecologists, physiologists and population geneticists. The project received generous financial support from the Foundation for Biological Research, which is subsidized by the Netherlands Organization for Scientific Research. During the course of the Project the value of the integrative approach on a limited number of *Plantago* species and populations became evident and it was decided that the results should be published as a book.

The start of the preparation of the "plantago" book was done by Dr. P. J. M. van der Aart and Dr. H. Lambers but due to their appointment as full professor of the University of Utrecht they were both unable to continue. We are greatly indebted to Van der Aart and Lambers for the groundwork they have laid.

The book presents a sample of results obtained over a period of more than 10 years. Research on *Plantago* is still continuing, as evident from a number of publications and doctoral theses. We want to thank the authors for their patience and assistance to complete the job. The editorial assistance of Mrs. I. Cameron-Doornbos was extremely valuable, as well as the help of Mr. E. Leeuwinga und Mrs. N. Tolmeijer with the drawings and the help of Mrs. T. E. Stuit with the list of references.

Haren, August 1991

P. J. C. KUIPER
M. BOS

Contents

List of Contributors

BLACQUIÈRE, T.
Proefstation voor de Bloemisterij
Linnaeuslaan 2a
NL-1431 JV Aalsmeer
The Netherlands

BLOM, C. W. P. M.
Botanisch Laboratorium
Toernooiveld
NL-6525 ED Nijmegen
The Netherlands

BOS, M.
Biological Centre
University of Groningen
P.O. Box 14
NL-9750 AA Haren
The Netherlands

DE NOOIJ, M. P.
Institute of Terrestrial Ecology
I. O. O.
P.O. Box 40
NL-6666 ZG Heteren
The Netherlands

FREIJSEN, A. H. J.
Institute of Terrestrial Ecology
I. O. O.
P.O. Box 40
NL-6666 ZG Heteren
The Netherlands

HAECK, J.
Institute of Terrestrial Ecology,
I. O. O.
P.O. Box 40
NL-6666 ZG Heteren
The Netherlands

KUIPER, D.
CABO
P.O. Box 14
NL-6700 AA Wageningen
The Netherlands

KUIPER, P. J. C.
Biological Centre
University of Groningen
P.O. Box 14
NL-9750 AA Haren
The Netherlands

LOTZ, L. A. P.
CABO
P.O. Box 14
NL-6700 AA Wageningen
The Netherlands

MØLGAARD, P.
Department of Botany and
Pharmacognosy
Royal Danish School of Pharmacy
Universitetsparken 2
DK-2100 Copenhagen
Denmark

Mook, J.H.
Institute of Terrestrial Ecology
I.O.O.
P.O. Box 40
NL-6666 ZG Heteren
The Netherlands

Pons, T.L.
Botanical Ecology
University of Utrecht
P.O. Box 80084
NL-3508 TB Utrecht
The Netherlands

Smit, A.J.
Institute of Terrestrial Ecology
I.O.O.
P.O. Box 40
NL-6666 ZG Heteren
The Netherlands

Soekarjo, R.
Botanical Ecology
University of Utrecht
P.O. Box 80084
NL-3508 TB Utrecht
The Netherlands

Stoutjeskijk, P.
Institute of Terrestrial Ecology
I.O.O.
P.O. Box 40
NL-6666 ZG Heteren
The Netherlands

Stulen, G.
Biological Centre
University of Groningen
P.O. Box 14
NL-9750 AA Haren
The Netherlands

Tànczos, O.G.
Biological Centre
University of Groningen
P.O. Box 14
NL-9750 AA Haren
The Netherlands

Troelstra, S.R.
Institute of Terrestrial Ecology
I.O.O.
P.O. Box 40
NL-6666 ZG Heteren
The Netherlands

Van Damme, J.M.M.
Institute of Terrestrial Ecology
I.O.O.
P.O. Box 40
NL-6666 ZG Heteren
The Netherlands

Van Delden, W.
Biological Centre
University of Groningen
P.O. Box 14
NL-9750 AA Haren
The Netherlands

Van Dijk, H.
Lab. de Génétique et Evolution
des Populations Végétales
Universite de Lille
URA, CNRS 1185, Bat. SN 2
F-59655 Villeneuve d'Asq Cedex
France

Van Dijk, P.J.
Institute of Terrestrial Ecology
I.O.O.
P.O. Box 40
NL-6666 ZG Heteren
The Netherlands

Van Hasselt, P.R.
Biological Centre
University of Groningen
P.O. Box 14
NL-9750 AA Haren
The Netherlands

VAN TIENDEREN, P.H.
Institute of Terrestrial Ecology
I.O.O.
P.O. Box 40
NL-6666 ZG Heteren
The Netherlands

VAN DER AART, P.J.M.
Botanical Laboratory
University of Utrecht
Lange Nieuwstraat 106
NL-3215 PN Utrecht
The Netherlands

VAN DER TOORN, J.
Institute of Terrestrial Ecology
I.O.O.
P.O. Box 40
NL-6666 ZG Heteren
The Netherlands

VULTO, J.C.
Institute of Terrestrial Ecology
I.O.O.
P.O. Box 40
NL-6666 ZG Heteren
The Netherlands

WOLDENDORP, J.W.
Institute of Terrestrial Ecology
I.O.O.
P.O. Box 40
NL-6666 ZG Heteren
The Netherlands

WOLFF, K.
TNO-CIVO
P.O. Box 360
NL-3700 AJ Zeist
The Netherlands

1 Introduction

In the Netherlands in the period after World War II, plant ecology concentrated mainly on classifying the vegetation according to the methods of Braun Blanquet. This period came to an end in the 1970s and influenced by John Harper research into the level of the population took over. This shift in approach was considerably facilitated by support from animal population biology, which in the Netherlands is of a longstanding tradition and at that time was flourishing. The joint project of which the results are published in this book started at that time as a co-operative effort between plant ecologists, plant physiologists and animal population geneticists. Most of the contributors were associates of the Institute for Ecological Research of the Royal Netherlands Academy of Arts and Sciences and the Departments of Plant Physiology and Population Genetics of the University of Groningen. This multidisciplinary project, in which also soil chemists, micrometeorlogists and microbiologists participated, began in 1976 under the title *Comparative research on the demographic, physiological and genetic properties of plant species in relation to the properties of their location in grasslands.* The original idea was that the properties of phylogenetically related species would be nearly alike but that their differences would be adaptations to their specific environments. The genus *Plantago* was chosen for this comparative research. This genus is represented in the Netherlands by five species, i.e. *P. lanceolata, P. major, P. media, P. maritima* and *P. coronopus.* No hybridization between these species is known and they occur in different habitats. These species were chosen because they are common, occur in different, non-vulnerable vegetation types and are easy to grow from seed in the laboratory.

In the multidisciplinary research project, that received considerable financial support from the Foundation for Biological Research (BION/NWO), a large number of senior scientists, PhD students, undergraduates and technicians collaborated. During this period, there were 17 theses and 200 primary publications, as well as a number of reviews. A newsletter to keep the participants up to date with the project's progress, as well as with the new research proposals and preliminary results, was issued four times a year. There were more than 50 issues as well as many general and specialized meetings, enabling the participants to integrate their results with those of others, which gave their publications a wider scope.

In the initial stages of the project, data on the demographic and genetic structure of populations of the species were collected, whereas ecophysiological data were gathered in the laboratory and also, to some extent, in the field. In addition, an inventory was made of soil chemical and micrometeorological characteristics of the sites of the various populations. During this phase, attention was primarily focussed on a comparison of the five species. Subsequently, when the results could be related to each other, the comparison became mainly restricted to populations of *P. major* and *P. lanceolata.* These two species, the first of which is primarily an inbreeding species and the second an obligatory outbreeding species, were thought to exhibit different strategies. From the period in which the populations were compared, several lines of research emerged. Much attention was paid to the production of nitrogenous compounds in the soil and their uptake and assimilation by the plant, the ecological significance of the alternative respiratory chain, the internal mechanisms behind the adaptation to sea salt, the breeding system (in particular the occurrence of gynodioecy within populations of *P. lanceolata*), the genetic structure within populations and the local adaptation of populations. *P. lanceolata* as well as *P. major* showed considerable local adaptation that occurred in the morphological characteristics of the plants. Reciprocal transplant experiments appeared to be a mighty tool to evaluate the fitness value of these adaptations. Selection of ecotypes took place primarily in the adult phase and not, as was originally thought, in the juvenile phase. The contrasting strategies of *P. major* (local adaptation of genetically different populations composed of non-plastic genotypes) and *P. lanceolata* (no genetic variation between populations that are composed of one general-purpose genotype) were not as pronounced as was originally thought.

In the final period of the project, attention was focussed on the variation within populations and on the impact of biotic factors such as mycorrhiza and plant-pathogenic fungi. The presence within populations of genetic variation that was related to fitness could be shown in *P. major* as well as within *P. lanceolata*, but the differences between genotypes were small. An important result was that in the latter species genetic variation for morphological properties was present in all populations but, due to selection, clusters of morphological properties occurred, which resulted in pasture and hayfield ecotypes.

Looking back on the project, we conclude that we achieved a satisfying synthesis between population ecology, population genetics and plant physiology. With regard to environmental factors, the distribution of the species is more influenced by factors such as how the grasslands are used and water availability than by soil chemical factors, although sea salt is, of course, an important discriminating factor. Although we found considerable differences in physiological properties between the five species, we often were not able to explain them in terms of fitness. A possibly more effective approach could be to establish physiological differences between ecotypes of which the genetically determined differences in fitness parameters are known.

The present volume is a synthesis of the results of the previous years. Over the years we have collected a great deal of data about plantains and so the species remain our favourite subject by which we study general ecological problems. New horizons are emerging in the study of the genetic structure of populations with the use of new molecular-biological techniques such as fingerprinting and the use of specific DNA probes. The studies on the utilization of nitrate in the rhizosphere are also continuing, as well as those on the tripartite co-adaptation between *P. lanceolata*, a specific pathogenic fungus and a weevil species that is the vector of the fungus. The presence of electrogenic pumps at the symplast-xylem interface is being studied in greater detail to try to explain the mechanisms of salt tolerance. The previous studies are an impetus to similar studies with species other than plantains, that are 'at times' more suitable objects for researching specific aspects. This book does not pretend to give all the answers, but we hope that it provokes further questions at the interface between ecology, genetics and physiology that will need to be answered by a new generation of students.

2 General Biology of *Plantago*

2.1 Evolutionary Status

The Plantaginaceae are thought to be an ancient family (Day and Stebbins 1966; Cronquist 1968). They are only remotely related to modern families and are considered a separate order, the Plantaginales (Pilger 1937). The primitive forms from which the Plantaginaceae have evolved are unknown. The Scrophulariaceae show a phylogenetic relation with the Plantaginaceae (Souèges 1921, 1923; Li 1946; Takhtajan 1953; Dietrich 1975). A relation with the Primulales was suggested by Endlicher (1836) and Lindley (1836); with the Rubiales by Mattei (1909). The family includes mostly annual and perennial herbs and some halfshrubs.

Plantago is the only genus of the family rich in species (Pilger 1937): about 240. Besides *Plantago*, the Plantaginaceae consist of the genus *Littorella* with three and the genus *Bougueria* with one single species. Due to its great antiquity and the wealth of species, the genus *Plantago* is well suited for evolutionary studies (Day and Stebbins 1966) by morphological-anatomical methods and by allozyme techniques. The genus provides an excellent opportunity to compare and test both methods. Particularly the stem-forming *Plantago* species (section Palaeopsyllium) seem to be of great age. Dietrich (1975 a) considers the high stem-forming species of the section Palaeopsyllium as a separate genus and renames the rest of the section *Palaeopsyllium* to *Dendropsyllium*. The more recent stem-forming *Plantago* species belong to the subgenus *Psyllium*. The rosette-forming *Plantago* species are gathered in the subgenus *Plantago*.

Morphological traits like flower morphology (strict tetramery of calyx, petals and stamens), form of inflorescence, rosette form, venation and, to a lesser extent, root morphology appear to be conservative. In contrast, ecological characteristics such as seed number, seed weight, vegetative propagation, time of reproduction, and seed dormancy appear to react strongly to selection pressure (Gregor 1938, 1939; Gregor and Lang 1950). Even with a single species like *P. major*, ecotypes may represent r or K types (Sects. 6.2 and 6.3). These ecotypes differ in many traits, which appear to be closely interlinked; intermediates, exhibiting both r and K traits, seldom occur.

2.2 Biogeography and Human Effects

2.2.1 Biogeography

The genus *Plantago* is widely distributed in the temperate regions and higher locations in the tropics (Pilger 1937). It is lacking in the Antarctic continent and in the greater part of the Arctic. The section *Palaeopsyllium* is endemic on oceanic islands. The Mediterranean area is thought to be the centre of the more recent forms of the erect-growing subgenus *Psyllium*. Sections of the subgenus *Plantago*, such as, e.g., the section *Coronopus*, also have their centre of origin in the Mediterranean region. The rosette-forming sections *Leucopsyllium, Novorbis, Oreophytum* and *Oliganthos* are mainly distributed in South America, and *Microcalyx, Mesembrynia and Oliganthos* occur in the Antarctic-Australian region. The biogeographical distributions suggest that stem-forming and rosette-forming *Plantago* species already existed at the end of the Palaeozoicum.

The weeds *P. major, P. lanceolata* and *P. media* have a nearly cosmopolitan distribution. These species, and to a lesser degree *P. coronopus*, have been distributed by man all over the world: the name "Plantago" refers to the anthropogenic distribution of the weed species (planta: sole of the foot; agere: to stride).

2.2.2 Pollen Grains as Indicatives for Human Settlement

Pollen grains of *Plantago lanceolata* and, to a lesser extent, of *P. major* and *P. media*, are regarded as indicators of human settlements (Müller 1947; Erdtman et al. 1961). In settlements from the neolithicum onwards, pollen grains of cereals and *Plantago lanceolata* are often found together. In settlements from before the neolithicum, pollen grains of *P. lanceolata* are rarely found. Pollen grains of the *P. major/P. media* type are not seldom found in older sediments (late glacial and early and middle post glacial; Iversen 1958). They are limited to specific places such as a washing zone ashore. From the late glacial onwards, all places influenced by man offered new habitats for the weedy *Plantago* species.

Pollen grains of *P. coronopus* and *P. maritima* are restricted to coastal areas or salty inland places (Müller 1947; Zoller 1968). The disjunct distribution in Europe is regarded as a relict distribution, following glacial times.

2.2.3 Economic Value

The Plantaginaceae are economically of minor importance. *P. major, P. lanceolata* (Hawthorn 1974; Cavers et al. 1980) and *P. media* are weeds in pastures and arable land. *P. alpina* is a good pasture plant. *P. coronopus* and *P. maritima* are locally grown as vegetables.

P. psyllium, P. indica and *P. ovata* have economic importance for the production of semen psyllii (flea-seeds; Chandler 1954). The seeds of these species have a high content of mucilage. The mucilage serves as a dressing agent for silks and muslin (Ranachandraw and Venkataraman 1968), and is also applied for certain diseases of the respiratory organs (Woodhouse 1954) and diseases of the intestine (Bianchini et al. 1975). Pollen of *Plantago lanceolata* is one of the major dicotyledonous allergenic components of hay fever in temperate region (Baldo et al. 1982). Since ancient times, leaves of P. major have been used for their wound-healing properties. Leaves of *P. lanceolata* in Europe and of *P. asiatica* in Asia (Tanaka et al. 1982) are in use as *herba plantaginis* for its urine production-stimulating properties and as a major constituent of cough mixture (Bianchini et al. 1975).

2.3 General Ecology

The subgenous *Plantago* with its prostrate rosettes is restricted to comparatively open vegetations that provide sufficient light on the ground level. Arable fields and grazed or trampled sites provide such open vegetations on relatively nutrient-rich and moist soils. Under more natural conditions these species occur in open vegetations of low temperature, low nutrient status, high salt status or heavy metal contamination. On a global scale, *Plantago major* and *P. lanceolata* belong to the top ten noxious weed species. *P. lanceolata* (Hawthorn 1974; Cavers et al. 1980) and *P. major* (Sager and Harper 1964) behave like annual weeds in various crops. Their nutritional value for cattle in grassland is comparatively low. *P. major* is completely avoided by horses. The nutritional value of this species for slugs was found to be low.

In the course of a vegetation succession, developing towards a more closed canopy structure, rosette-forming *Plantago* species occur in the open pioneer phase. Like most other rosette-forming species, they are poor competitors for light and in a dense vegetation they are replaced by tall erect growing species.

In this book attention is focussed on five species of contrasting ecology: two halophytes, *P. maritima* and *P. coronopus*; a species from a Ca-rich environment, *P. media*; a species tolerant to a wide pH range, *P. lanceolata*; and a species from relatively nutrient-rich environments, *P. major*. Such a choice yields ample possibilities to investigate the physiological and genetical components of adaptation of the individual plant to its habitat.

2.4 General Morphology in *Plantago*

Plantago is a genus of annual and perennial herbs and semi-shrubs. The plants are herbaceous; in some cases a persistent woody stem gives rise to woody semi-shrubs. Most of the species form basal rosettes with a compressed stalk and a leafless flower stalk; other species are erect herbs with a multi-branched stalk.

In the Netherlands the *Plantago* species form true rosettes. The inflorescences develop out of axillary buds and the apical meristem remains vegetative; this is in contrast to semi-rosette plants as *Sempervivum, Digitalis* and *Verbascum*, where the apical meristem develops into an elongated stem with flowers. In *Plantago* species with elongated stems, the inflorescences develop in the leaf axils and the apical meristem remains vegetative throughout the plant's life cycle. The stem part of the *Plantago* species found in the Netherlands is mostly inconspicuous: a thickend stem base (caudex) with hardly any elongation of the internodes, as is the case in many other true rosette plants (e.g. *Taraxacum*). Stem internodes may elongate under natural conditions, e.g. when the plant is covered with sand as shown for *P. major* and *P. lanceolata* (Wiesner 1891), *P. coronopus* (Dodds 1953) and *P. media* and *P. maritima* (Soekarjo, unpublished experiments). When the mother rosette remains active, a thick subterranean stem is formed by elongation of the internodes of the mother axis itself. When the mother rosette dies, the internodes of the newly developing daughter rosettes also show internode elongation. This elongation takes place until the soil surface is reached (Fig. 2.1). Treatment with gibberellic acid does not result in elongation of the internodes (Lona 1959). *Plantago* leaves are narrowly to widely lanceolate and never composite. In general they are undivided, but in some species various degrees of incision, lobation and division may occur (pinnatifid as in *P. coronopus*; Gravis 1932; Gorenflot 1964) with a basal sheath and parallel veins, as in monocotyledonous plants. Stipules are lacking. The undivided leaf type is thought to be derived from the pinnate leaf type by reduction of the margin and by broadening of the rhachis (von Gaisberg 1922). The veins of *P. major* and *P. lanceolata* are well protected and extremely tough, due to an expanded midrib. The thick-walled epidermis cells above the veins are smaller and more rounded than those above the mesophyll. Below these epidermis cells there is a thick layer of collenchyma cells with varying cell wall thickening. The veins are protected by a sclerenchyma fibre bundle on the upper and lower sides, or enclosed by a sclerenchyma sheath (Pilger 1898; Troll 1939). In *P. major* the vein itself is not easily damaged mechanically. Large parts of the lamina may be lost by mechanical damage and the veins may be separated, but distal parts of the lamina still remain turgescent. There are genetic differences: *P. major* ssp. *major*, occurring on tracks, and *P. major* ssp. *pleiosperma*, occurring on river banks and in agricultural fields, can be distinguished by the degree of elasticity of the leaf sclerenchyma, the latter ssp. having less elastic veins.

With the exception of *P. maritima*, the leaves of Dutch species are bifacial, with stomata occurring on both sides. The structure of the leaves differs in relation to the growing site: *P. lanceolata* plants, growing in an open site, have dorsiventral leaves with palisade parenchyma on the upper side only; plants in a closed vegetation show an equal development of the palisade parenchyma, oriented towards the incident light (Pick 1882). Ectodesmata are present in the epidermis of *P. major*, and function in absorption of water and solutes and in exudation (Franke 1957, 1960, 1964a, b).

In some perennial species the weak main root dies soon and is replaced by numerous adventitious roots (*P. major*). Most species have a perennial main

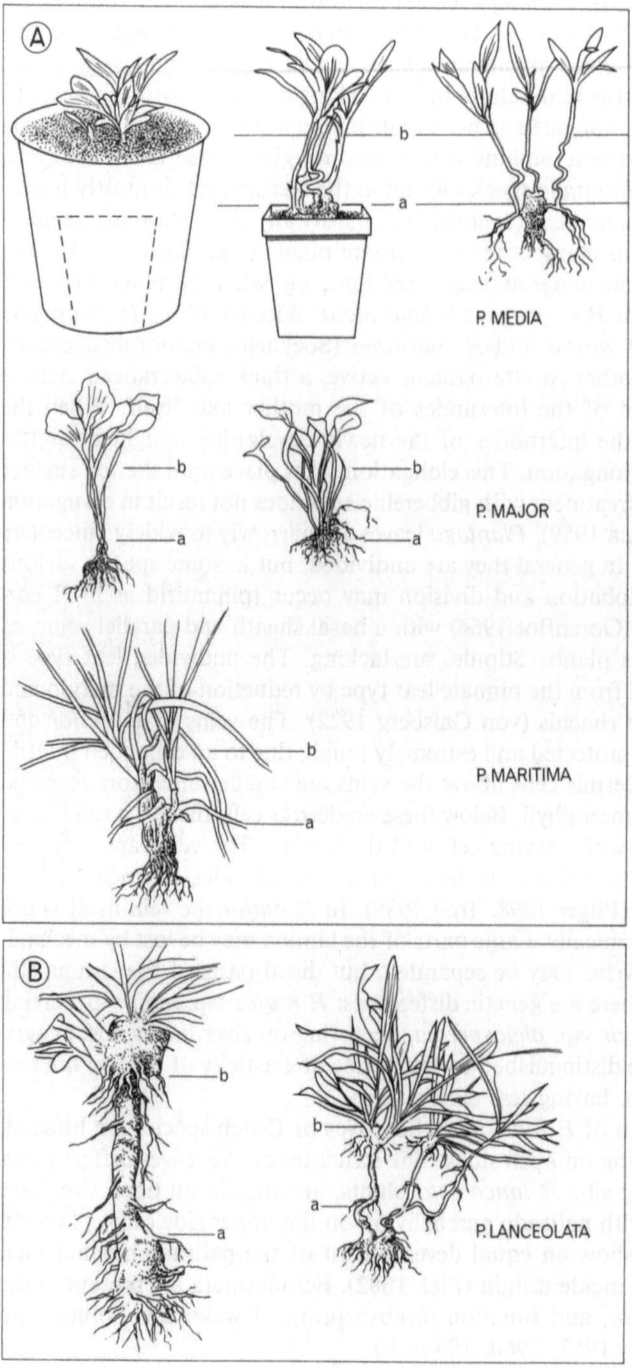
A
b
a
P. MEDIA
b
b
P. MAJOR
a
a
b
P. MARITIMA
a
B
b
a
b
a
P. LANCEOLATA

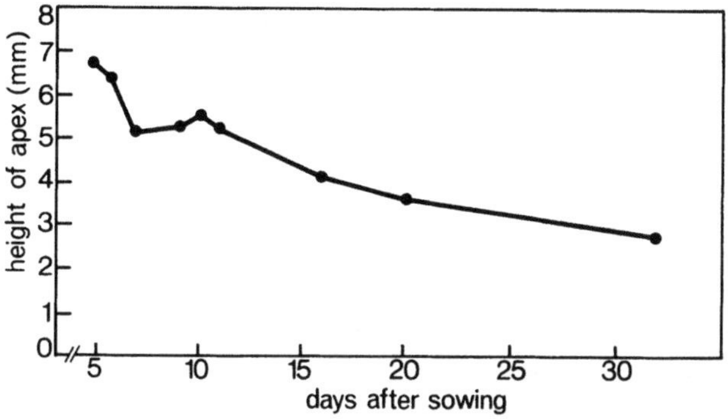

Fig. 2.2. Retraction of the seedlings of *Plantago coronopus* by root contraction. The height of the apex above the surface of the sand decrease with time. The hypocotyl may still show some growth, as reflected in the increase in height around 10 days after sowing. Plants were grown in short-day conditions (8 h light) and low light intensity at 20 °C. Mean values of nine plants are given

root (*P. lanceolata, P. media*). Extended branching of the rhizoma leads to turf formation (*P. maritima*). A feature related to resistance to treading is the occurrence of root contraction (*P. major, P. media* and *P. lanceolata*; Rimbach 1898, 1899, 1929). Contraction of roots is strong in *P. major* and *P. media* and weak in *P. lanceolata*. We also observed root contraction in *P. coronopus* and *P. maritima*; contraction may occur within a short period (Fig. 2.2). Differences in root contraction (and in withdrawal of the apical meristem into the ground) cause varying vulnerability of the shoot to treading. In experiments in which plants of *P. major, P. media, P. coronopus* and *P. lanceolata* were subjected to a daily rhythm of mechanical pressure, Blom (1979) found *P. lanceolata* – the species with the lowest root contraction – to be the one most susceptible to mechanical pressure. In open habitats old plants of *P. major* have the leaves closely pressed to the soil, suggesting contraction of adventitious roots (Sagar and Harper 1964), as no main root is present.

The small inconspicious flowers usually are bisexual (hermaphroditic; see also Sect. 2.6.5 and Fig. 2.3) and arranged in axillary spikes. The spikes project from the plants and are adapted to wind pollination. The calyx is nearly always four-lobed. The corolla is membranaceous and regular with a four-lobed short tube. There are four stamens, alternating with the corolla lobes (Eichler 1875) and often far exerted from the corolla tube. Before flowering, the corolla lobes cover style and stamens. In the protogynous flower the style appears first; hereafter the corolla lobes will reflex and the anthers emerge (Bos et al. 1985;

◄

Fig. 2.1. A Elongation of the internodes of *P. major, P. maritima* and *P. media* when the plants are covered with sand (5 cm); after 6 months. **B** The shape of the central axis in *P. lanceolata*, under the same conditions; after 1 year. *Left* the mother axis continued to grow; *right* the development of axillary-born daughter rosettes. Distance between bars *a* and *b* is 5 cm. Experiments done in greenhouse conditions. (Plants drawn after photographs)

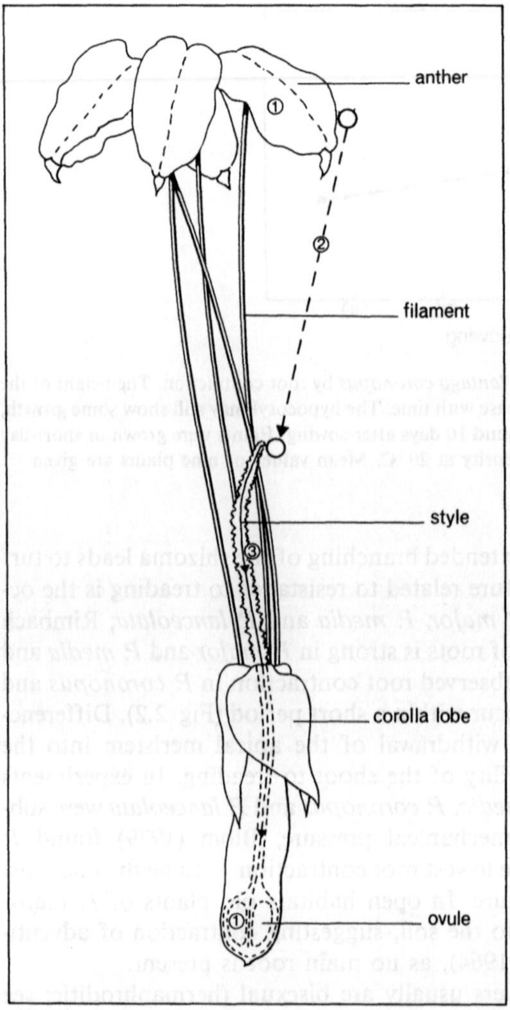

Fig. 2.3. A single flower of *Plantago lanceolata* illustrating three stages in sexual reproduction: *1* the production of gametes; *2* transport of pollen; *3* pollen growth and fertilization

Fig. 2.3). The style is solitary and elongated; the stigma is simple, from papillose to nearly hairy.

The ovary is superior. The two fused carpels form two locules, each with one or more ovules on a central placenta (Cooper 1942). The capsule contains one to numerous (50) seeds. In a given species, the number of seeds per capsule is variable. In *Plantago* the fruit is a capsule with a circumscissile dehiscence (opening with an operculum; Fig. 2.4). The mature capsule splits along a circu-

Fig. 2.4. Fruits of *Plantago coronopus, P. lanceolata, P. major* and *P. media*, showing the position of the seeds. The smaller seed of *P. coronopus* is located in the operculum. (Drawn by A. Du Bois-van Heezik)

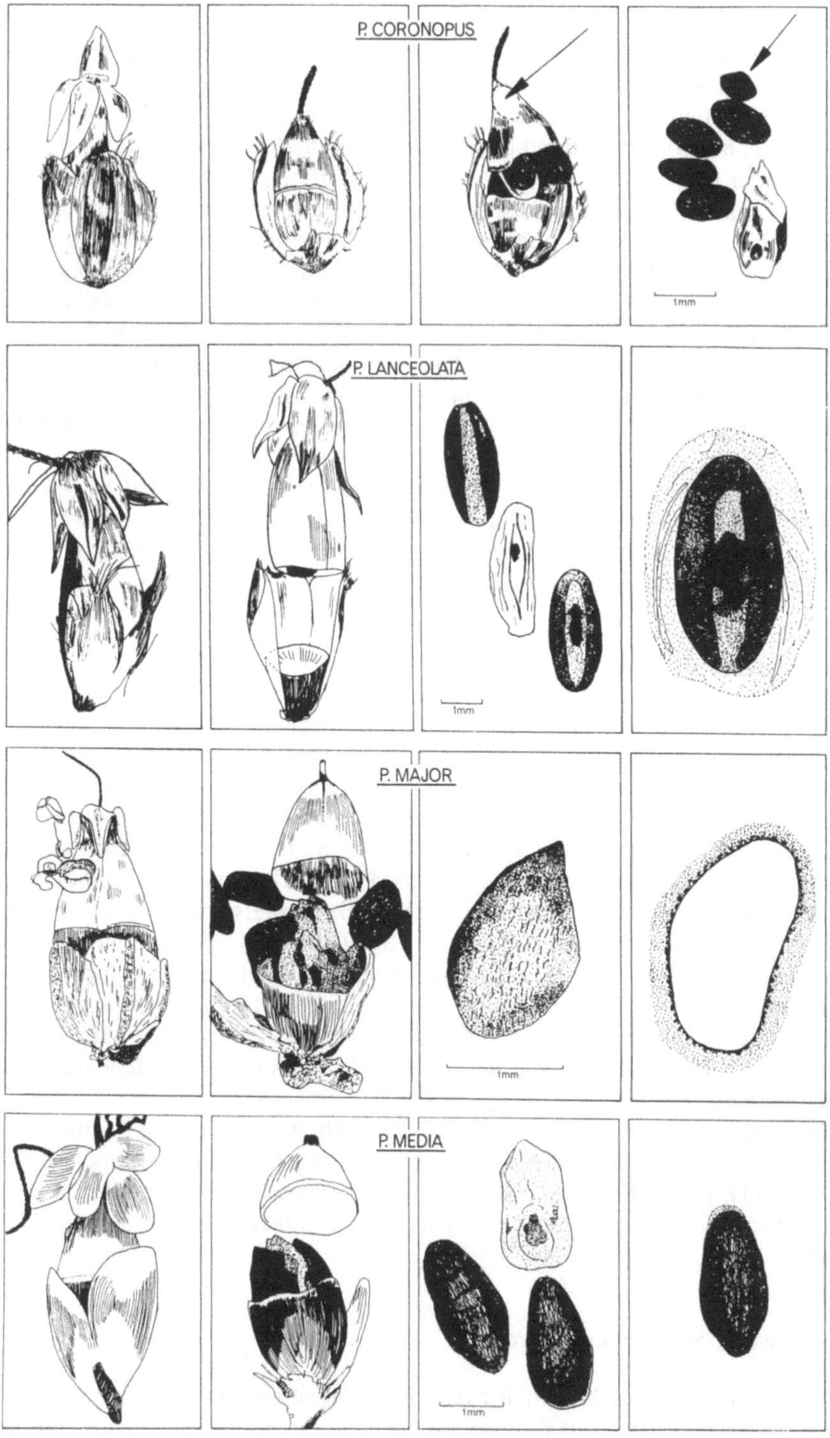

P. CORONOPUS
1mm
P. LANCEOLATA
1mm
P. MAJOR
1mm
P. MEDIA
1mm

lar line in or under the middle, the upper part (operculum) falling down. The seed is mucilaginous and is easily transported by cattle and man and his equipment. The thickness of the mucilagineous layer differs in the five species. The endosperm is fleshy. The species show large differences in seed size, which is largest in *P. maritima*, followed by *P. lanceolata, media, major* and *coronopus* in that order (Fig. 2.4). The seeds are thick and often glossy with a central attachment place. *P. coronopus* displays seed dimorphism: one of the up to five seeds in the capsule, i. e. the one attached to the operculum, is smaller than the others (Dowling 1933; Schat 1982). The smaller seed does not form a conspicuous mucilaginous layer when in contact with water and is not as easily shed as are the larger ones. As it stays in the floating operculum, this may function as an additional means of dispersal. The germination of the small seeds is similar to that of the large seeds and they are genetically not different (Bos, unpublished results). The embryo is of the capsella-type: straight or slightly curved.

The induction of flowering in *P. lanceolata* has been analyzed by Snyder (1948): it is a long-day plant. Maximal production of inflorescences was obtained in continuous light (Van Damme 1983 a). The time of flowering of other species suggests that they also are long-day plants.

2.5 Flowering Biology

The flowers of *Plantago* are typical wind-flowers; nectar, colour, and scant are mostly lacking. The small inconspicuous flowers are densely packed in oval or oblong spikes. The scapes are long, easily moved by wind and projected into the open air. The four long delicate anther filaments project far out of the calyx. The rough (long papillose to featherly) stigma is well equipped for the reception of the small spherical pollen grains (Pöhl 1929; Cavers et al. 1980). The flower is protogynous. In most species the spike flowers from below upward; each part goes first through the female stage (Bos et al. 1985). Some species show entomophilous adaptation. *P. lanceolata*, with its yellow-coloured anthers, attracts pollen grain-collecting insects (Clifford 1962). *P. media*, with its fine scent and coloured anthers and filaments, is visited by bees, bumblebees and syrphids (Knuth 1905; Van der Goot and Grabandt 1970; Stelleman and Meeuse 1976).

Generally, the flowers are hermaphroditic. The rosettes of the Andine species *P. rigida* and *P. tubulosa* have unisexual flowers (dioecious species; Weberbauer 1911). In some species occasionally flowers with only stamens or a stigma occur (gynodioecy, androdioecy) as with *P. lanceolata* (Ross 1970; Van Damme 1983).

2.6 Breeding Systems in *Plantago*

The plant kingdom is known for its wide diversity in reproductive systems, which varies between plant families. The Plantaginaceae hold a rather unique position and the modes of reproduction in plantains will be discussed in this section. Although plantain species can form side rosettes, they do not reproduce vegetatively and vegetative reproduction will therefore not be discussed here. Sexual reproduction comprises the production of haploid gametes and their fusion into (diploid) zygotes that will grow up to form the new generation. Three phases of this process are of interest here because they show variation in plantains (Fig. 2.3).

2.6.1 The Production of Male and Female Gametes

The predominant sex type among the members of the Plantaginaceae is the hermaphrodite, in which each flower produces both male and female gametes through pollen and ovules, respectively. Other sexual systems also occur in this family. ***Bougeria***, consisting of a single species, probably is also hermaphrodite, but the species of ***Littorella*** are monoecious, i.e. individuals have male and female flowers on the same plant (Pilger 1937). The large genus ***Plantago*** contains, besides hermaphrodite species, relatively many gynodioecious species. In this last system plants have either female or hermaphrodite flowers. The female sex type is usually designated as male sterile since it is thought to have evolved as a male sterility mutation in a hermaphrodite plant (Lewis 1941). In Table 2.1 all literature reports of gynodioecy in European ***Plantago***

Table 2.1. Distribution of self-incompatibility (SI) and gynodioecy in Europe among sections and species of the genus *Plantago* based on literature reports and unpublished results of the author (indicated with *). The classification used is according to Dietrich (1975a) except for *Psyllium*, which is represented here as a section instead of a genus

Section	Species	SI	Gynodioecy
Plantago	*major*	–	+ (loc.)[a]
Lamprosantha	*media*	+	+
Montana	*atrata*	+ *	MS[b]
	nivalis	–	–
	ovata	–	+
Lancifolia	*lagopus*	+ *	+
	lanceolata	+	+
Maritima	*maritima*	+	+
	sulbulata	+	–
Coronopus	*coronopus*	–	+
	crassifolia	–	–
	macrorhiza	+ *	–
	serraria	+	MS[b]
Psyllium	*afra*	–	–
	arenaria	–	–
	sempervirens	+ *	–

[a] loc. = at least locally.

[b] MS = male sterility observed but gynodioecious population have not been established.

species are compiled (Pilger 1937; Atal 1958; Fröst 1959; Mital and Issar 1970; Ross 1970 and references therein; Delannay 1978). With at least seven gynodioecious species in Europe (out of 35 tested species; Tukin et al. 1964–1980, Vol. 4), the Plantaginaceae exhibit a relatively high frequency of sexual dimorphism (Delany 1978; Van Damme 1985). Dioecy does not occur in European species, but Pilger (1937) has reported two South American *Plantago* species (*P. regia* and *P. tubulosa*) to be dioecious.

2.6.2 The Transport of Pollen to the Stigmatic Surface

Most *Plantago* species are regarded as wind-pollinated and their flower characteristics seem to be well adapted to wind pollination (Pilger 1937; Hammer 1978; Primack 1978). Plantain flowers have relatively long, papillate styles and versatile stamens with long filaments and anthers, containing numerous pollen. The inflorescences are usually well exposed to wind.

Insects visiting plants of *P. lanceolata* and *P. media* have been recorded (Knuth 1906; Clifford 1962; Stelleman 1981). In spite of attempts with *P. media* (Leereveld et al. 1976; Meeuse 1984), experimental evidence that insect pollination actually occurs was obtained only with *P. lanceolata* (Stelleman and Meeuse 1976). Stelleman (1984a) showed for wind-sheltered populations of the latter species that insect pollination significantly contributed to seed set. Local adaptation in *P. lanceolata* to insect pollination was suggested by differences in degree of pollen clumping between plants from sheltered and from open sites. No pollen vector is required in cleistogamous *Plantago* species, where selfing is enforced before the flower opens. Selfing is induced by early opening of anthers and by their close position to the stigmatic surface in the flower bud. Cleistogamy does occur in North and South American species. They belong to three sections (sensu Dietrich 1975a) of the genus that have no species represented in Europe (Pilger 1937).

The timing of stigma receptivity and pollen presentation influences the transport of pollen from the stamen to the stigmatic surface. Protogyny, when flowers have their stigmas receptive before the anthers release their pollen, is widespread in *Plantago* (Pilger 1937; Rahn 1957 and others). According to Dietrich (1975a) protogyny, an adaptation to outcrossing, occurs even in cleistogamous species. *Plantago* species differ in degree of protogyny (Van Damme unpubl. data; Bos et al. 1985). A similar difference in timing is found in *Littorella*, where the female flowers are receptive before the male flowers of the same plant present their pollen (Pilger 1937).

2.6.3 The Growth of the Male Gametophyte Until Fertilization

Variation among plantain species is found in the ability to self-fertilize. Numerous *Plantago* species cannot self-fertilize due to a self-incompatibility system. Table 2.1 shows the distribution of self-incompatibility over 16 (out of 35;

Tutin et al. 1964–1980, Vol. 4) European species (data from Ross 1970 and references therein; Van Damme unpubl.). More than half of the species examined are self-incompatible and the system occurs in almost all sections in spite of the few tested species per section. *P. maritima* is noteworthy, since it is reported as self-incompatible in Europe and self-compatible in North America (Ross 1970). The genetic basis of self-incompatibility has been examined in *P. lanceolata* (Ross 1973). It concerns a gametophytic system, determined by one locus with multiple alleles. Since the type of self-incompatibility tends to be similar within a family (Brewbaker 1967), the above system may well apply to the other self-incompatible species of Table 2.1. Support comes from the fact that in gametophytic systems location of the self-incompatibility reaction is in the style instead of on the stigmatic surface (De Nettancourt 1977): *P. atrata, P. lapogus, P. subulata* and *P. sempervirens* also show inhibition of pollen tube growth in the style (examined by fluorescence microscopy; Van Damme unpubl. data).

2.6.4 Outcrossing Rate

The self-incompatible species *P. lanceolata* and *P. maritima* most likely are completely outcrossing. Partial failure of the self-incompatibility system has been observed in *P. lanceolata* in the greenhouse under a high temperature regime of 32 °C during several days (Van Damme unpubl. data). Although the exact conditions for selfing are unknown, it seems unlikely that in the cool climate in the Netherlands any selfing will occur. As to *P. media*, which has been claimed to be self-incompatible (Rahn 1957; Sagar and Harper 1964), there is some doubt whether the system fully excludes selfing. In a small experiment in which spikes of seven plants were bagged individually to prevent outcrossing, five plants set some seeds and the percentage of ovules that had become seed was low, 3.2 (Van Damme unpubl.). H. van Dijk (pers. commun.) also found evidence for occasional selfing in this species. The partial selfing behaviour of *P. media* may well be related to the tetraploidy in this species: Breakdown of self incompatibility in autotetraploid species is fairly common (Lewis 1949). *P. media* is the only tetraploid in the Netherlands (Table 2.2) and tetraploidy probably is relatively young (diploid populations are also known; see Rahn 1957; Van Dijk, Van Delden 1990).

The level of outcrossing in the self-compatible species *P. major* and *P. coronopus* have been estimated for ovules (Van Dijk 1987; Wolff 1988). Outcrossing rates are inferred from the spatial distribution of allozyme genotypes (see Sect. 7.2). Estimates of the outcrossing rate in *P. major* are available for two populations of subspecies *major* and one of subspecies *pleiosperma* (Table 2.2). The values vary between populations and they seem to be related to plant density (Van Dijk et al. 1988). Values for *P. coronopus* were close to 1 in all four populations studied, suggesting that, although self-compatible, this species hardly self-fertilizes its ovules.

Table 2.2 also shows the numbers of pollen grains and ovules per flower. The latter is one of the main criteria for distinguishing between subspecies in

Table. 2.2. Ploidy level and breeding system traits of *Plantago* species in The Netherlands

Species	Ploidy level	Gynodioecy	Ovule outcrossing rate	No. of pollen grains/ flower[a]	No. of ovules/ flower
Coronopus	2×	+	0.35 – 0.99[b]	20800	5
Lanceolata	2×	+	1	81600	2
Major	2×	–		25600	
sp. *major*			0 – 0.08[b]		5 – 13
sp. *pleiosperma*					14 – 35
Maritima	2×	+	1	42400	3 – 4
Media	4×	–	prob. 1	48800	4

[a] According to Hammer (1978).
[b] According to Wolff (1988; figures based on three and four populations for *P. coronopus* and *P. major*, respectively).

P. major (Van Dijk and Van Delden 1981). The pollen/ovule ratio is high in self-incompatible species and low for the predominantly selfing species *P. major*. This is consistent with observations on a large number of other species (Cruden 1977). Remarkably, the data for number of pollen grains per flower (Table 2.2) which have been determined with plants from Europe (Hammer 1978), differ widely from estimates made by Bassett and Crompton (1968) for North American populations of the same species (700 grains per anther in North America versus 11000 in Europe, averaged over species). The cause of this difference is unknown.

2.6.5 Gynodioecy

Not all gynodioecious *Plantago* species (Table 2.1) have gynodioecious populations in the Netherlands (Table 2.2). Apart from a few rare observations of (partially) male sterile individuals, *P. major* and *P. media* should be regarded as hermaphrodite in the population studies in this Volume.

Gynodioecy was originally defined as the joint occurrence of hermaphrodite and female (male sterile) plants in natural populations (Darwin 1877). The actual situation is more complex. One aspect that has received considerable attention is the gradual transition of gynodioecy via subdioecy (see e.g. Lloyd 1976). In *Plantago* this transition of a gradual decline of female fertility in hermaphrodites does not occur (Van Damme 1985). Another complication is the occurrence of partially male steriles (intermediates). Intermediates have been reported for virtually all gynodioecious species. Judged by the literature, they usually occur in marginal frequencies. However, in *Plantago* species intermediates may even outnumber completely male steriles and hermaphrodites.

Intermediate plants usually carry a mixture of fully sterile and fertile anthers; partially sterile anthers may also occur. This is the case in *P. coronopus*, *P. maritima* and in one of the intermediate types, IN1, of *P. lanceolata*. Anoth-

Table 2.3. Distribution (%) of the sex types male sterile (MS), partially male sterile or intermediate (IN) and hermaphrodites (H) of two plantain species in The Netherlands

Population	MS	IN	H	Sample size
Plantago coronopus				
Uddel	0	15.1	84.9	311
Kwade Hoek	4.7	24.3	71.0	427
Petten	4.6	41.7	53.7	307
Terschelling	4.9	67.6	27.5	117
Westduinen	20	32	48	100
Schiermonnikoog 1	24.1	30.2	45.7	176
Schiermonnikoog 2	18.5	6.2	75.3	178
Plantago maritima				
Kwade Hoek	3.5	78.5	18.0	200
Schiermonnikoog 3	3.0	94.0	3.0	200

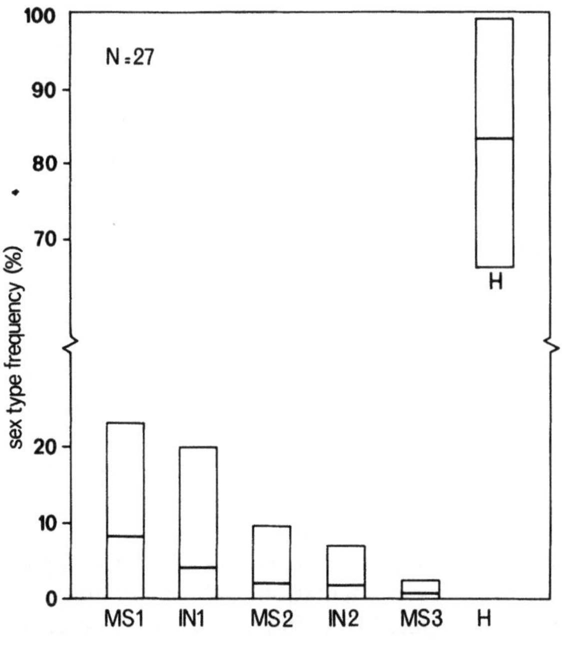

Fig. 2.5. Frequencies of three male sterility types (*MS*), two partial male sterility types (*IN*) and hermaphrodites (*H*) in natural populations of *Plantago lanceolata.* Of each sex type range and average (*thick line*) are shown. (Van Damme and Van Delden 1982; Van Damme 1983a and unpubl.)

er intermediate type, IN2, carries partially sterile anthers. Frequencies of male steriles and intermediates are shown in Table 2.3 and Fig. 2.5. In *P. coronopus* male sterile frequencies range from 0–24%. Rushton (1983) found in 56 populations a somewhat smaller range, up to 15%. The intermediate sex type is more frequent than male steriles. Particularly high frequencies of intermediates occur in the two *P. maritima* populations studied and male steriles and hermaphrodites seem to be rare sex types.

In *P. lanceolata* the situation is still more complex: three male sterility types occur, two of which are accompanied by intermediate sex types. The male sterility types can be distinguished by anther morphology, and they also differ in the moment of breakdown of spore development, the mode of inheritance and in life history traits such as survival rate and seed production (Van Damme 1983 a, b). Furthermore, the sex types differ in the frequencies observed in the field (Fig. 2.5). MS1 is the common type, followed by MS2 and finally by MS3. The intermediate sex types occur in somewhat lower frequencies than the corresponding male sterility types.

2.7 Questions Arising from Variation in General Biology

2.7.1 The Importance of Breeding Systems

The quality of the breeding system affects the population and the evolutionary biology of species. Some of the questions that arise from the variation in breeding systems in *Plantago* will be addressed in this Volume.

- Is the genetic variation in populations organized within or between individuals? What is the spatial distribution of genotypes? How large is the neighbourhood size? The breeding system has a direct effect on the level of consanguinity of the gametes fusing into zygotes, and thereby on the genetic structure of a population (Sects. 6.1 and 6.5).
- Are there effects, other than on outcrossing rate, of the breeding system? Classically, the effects of self-(in)compatibility and sexual dimorphism are thought to be exerted on gene transmission between generations and in particular on the degree of selfing. In *Plantago* species self-incompatibility and gynodioecy may occur simultaneously, so that the existence of two systems seems superfluous in prevention of self-fertilization. It appears that the breeding system may have a direct effect on the maintenance of genetic variation in important life history traits (Sect. 6.3).
- How does the breeding system of a species influence its ability for population differentiation? What strategies in adaptation to the environment are allowed under various breeding regimes? The importance of the breeding system in this respect is clear. Perpetuation of successful genotypes in *Plantago* is only possible in selfing species. In the generalist-specialist concept the breeding system variation has to be taken into account (Chap. 8).

2.7.2 The Importance of Phenotypic Plasticity

Next to genetic variation (see Sect. 2.7.1) phenotypic plasticity is important in establishment and maintenance of plant populations. The following questions may be raised on the subject of phenotypic plasticity.

- Do differences in phenotypic plasticity exist between species, ecotypes, genotypes, populations and within populations?
- Are differences in plasticity related to environmental conditions at the habitat? Are differences in plasticity related to regular or unpredictable changes of the habitat?
- Is phenotypic plasticity a compensating factor operating independently of each other?
- What is the physiological basis for phenotypic plasticity? Involvement of energy metabolism, hormones and phytochrome in regulation of plasticity seem likely.

3 Distribution of *Plantago* Species: Characterization of Sites

3.1 Phytosociology of *Plantago* Habitats in The Netherlands and the Relation with Habitat Characteristics

3.1.1 Introduction

A phytosociological description of the habitat of a species is a prerequisite for studies aiming to elucidate the mechanisms determining its habitat preference. At the beginning of the present multidisciplinary study, our understanding of the habitats of the *Plantago* species was of a floristic nature, as recorded e.g. in the *Flora of The Netherlands* (Heukels et al. 1983). Slightly more specific knowledge was available from a survey of Dutch plant communities (Westhoff and den Held 1969). To provide a detailed description of the phytosociology of the habitat, it was necessary to consult original literature and to carry out field observations.

Table 3.1 shows the habitat of the five *Plantago* species which occur in The Netherlands, according to the descriptions in the *Flora* (for *Plantago major*: Heukels and Van Ooststroom 1975; for its subspecies and the other species: Heukels et al. 1983). Such a qualitative description, however, does not allow conclusions about the optima and ecological amplitudes of the species.

More quantitative information is available for three *Plantago* species in various types of grassland. From 1934 to 1958, D.M. de Vries and associates (see Kruijne et al. 1967) combined a botanical analysis of 1577 parcels with a description of seven habitat factors: type of use, soil moisture conditions, type of soil, clay-humus factor, alkalinity, P-status and K-status. The type of management and degree of soil moisture were classified according to statements of the farmers, the other values were based on soil analysis. The vegetation of each parcel was analyzed by the combined frequency and order method (De Vries 1937): per field the frequency percentage (F%), i.e. the number of times a species was found in 100 samples per field, was established for each species. The occurrence of a species in the total number of fields was indicated by the percentage of fields in which the species was found (the presence percentage, P%). Also the average frequency (F%) of a species was calculated over the 1577 parcels (gaF). The 1577 fields were grouped according to four to five classes of habitat factors, and gaF (%) per species and per class was determined.

Table 3.1. Habitats of *Plantago* species (and subspecies of *P. major*) as described in Heukels et al. (1983)

P. coronopus	*P. maritima*	*P. lanceolata*	*P. media*	*P. m. major*	*P. m. pleiosperma*
On "green beaches", high salt marshes, sea dikes, in dry grassland and trodden places. Rather common in the coastal area, also along Lake IJssel; in inland rare and ephemerous	On salt marshes, beach plains behind the outer dune ridge, along sea dikes and in salty grass- and reedland on the landside of the dike. Common in Wadden and Delta area; occasionally along Lake IJssel	Along roads and dikes, in grassland and grassy dune valleys. Very common	In moderately dry grasslands on dikes and road verges, on limestone, sandy clay and calciferous sand. Rather common in river district and in southern Limburg, elsewhere very rare	Trampled sites, along and on roads and paths, in often-trodden grasslands and in ruderal places. Very common	On open, moist sand, loess and sandy clay, on moist arable land, on open places in grassland, in moist dune valleys, in river forelands, also ruderal. Rather common

Table 3.2 summarizes the data for the *Plantago* species found in these grasslands. *Plantago lanceolata* had its greatest frequency in pure hayfields, in both wet and dry habitats, at neutral or alkaline conditions and at a low nutritional status of P and K. *Plantago lanceolata* was rather indifferent to soil type and granular composition of the soil. *Plantago major* was most frequent in pastures of moderate soil moisture, which were not too acid and which were characterized by high levels of P and K. *Plantago media* had a strong preference for grazed grasslands on dry sandy clay of high alkalinity; the P-status of the soil was low and the K-status was higher.

The above material provided valuable data on *Plantago* in agricultural grassland. For the present purpose, data from other habitats and from all *Plantago* species were needed. Therefore a limited phytosociological, pedological and partly also micrometeorological survey of habitats in which one or more *Plantago* species were found was made (see Sect. 3.2 and 3.3). In addition a large sample of (mostly) published phytosociological tables covering the whole range of plant communities described for The Netherlands was used; this material is treated in the present section.

3.1.2 Material and Methods

In The Netherlands the plant communities are usually described according to the Braun-Blanquet method (Westhoff and van der Maarel 1973). For the majority of associations similar phytosociological tables are available. Our collec-

Table 3.2. General average frequency (ga F) and presence percentage (P%) of *Plantago* species in 1577 grassland fields and relative average frequencies for seven habitat factors grouped into four or five classes. (Kruijne et al. 1967)

	ga F	P%	Type of use				Degree of soil moisture			
			Hay field	Hay pasture	Alternate pasture	Pasture	Dry	Normal	Humid	Wet
No. of fields			130	396	422	534	158	768	452	178
P. lanceolata	4.8	47.0	207	93	49	51	123	69	86	122
P. major	1.0	31.9	20	50	103	227	81	144	123	52
P. media	0.1	2.9	46	9	75	270	276	88	+	16

	Type of soil				
	Sand	Sandy clay	Clay	Peaty soil	Peat
% humus	<21	<21	<21	21–30	>30
fraction <20 µ	0–10%	11–40%	>40%		
No. of fields	385	432	290	168	273
P. lanceolata	93	127	111	83	86
P. major	156	122	57	102	63
P. media	28	380	83	9	–

	Clay humus factor (%) = % humus + $\frac{1}{4}$ of fraction <20 µ				
	0–10	11–20	21–40	41–60	60–100
No. of fields	222	469	570	211	65
P. lanceolata	117	101	95	62	125
P. major	153	149	82	69	47
P. media	104	302	94	–	–

	Alkalinity (pH water)					P status (P citric acid) (mmol kg^{-1})				
	<5.0	5.0–	5.5–	6.0–	>7.0	0–20	21–32	33–50	51–80	>80
No. of fields	75	338	529	487	113	213	273	358	344	346
P. lanceolata	37	99	111	119	134	240	102	77	50	31
P. major	34	74	134	148	110	40	78	124	106	152
P. media	–	–	+	56	444	220	200	63	6	11

	K status (K value) (mmol kg^{-1})				
	0–13	14–19	20–27	28–37	>37
No. of fields	81	276	475	350	322
P. lanceolata	216	139	81	34	30
P. major	68	58	70	136	168
P. media	17	152	263	59	9

tion of tables, most of which has been published, comprises 361 tables, describing 181 out of the 278 associations listed by Westhoff and den Held (1969). The associations include the range of vegetation types from saline to non-saline environments and from open water to woodland. Swamp and water plant communities are under-represented in the collection since *Plantago* species do not occur in such habitats. A total of 1184 species, including 220 mosses and lichens, occur in the collected material. The mean number of relevés per table is 26, with a minimum of 10 and a maximum of 157.

For each species the frequency of occurrence in each table was calculated. Also, low frequencies can be traced because the tables contain complete species lists. The only problem is to find the most exploratory arrangement of the data. The sequence of associations (Westhoff and Den Held 1969), which follows a progression from simple to highly structured communities, is a first possibility (Fig. 3.1). A multi-dimensional arrangement of the data is more appropriate in order to find the response of the species to various ecological factors. This was achieved by ordination, based on the similarity of the species composition in the tables. Reciprocal averaging was used as ordination technique (Hill 1973). The program Decorana delivered four axes, which by the position of the communities and the species in the ordination diagrams (Haeck et al. 1982; Haeck et al. 1985) were interpreted as follows:

1. A moisture gradient ranging from communities of open water, via reed swamps and sedge communities, willow and alder scrubs, to communities of dry habitats.
2. A salinity gradient from low tidal saline mud flats via low and high tidal salt marshes to more and less saline dune slacks and all non-saline communities.
3. A soil fertility gradient from peat bogs and communities of drift-sands on one hand to arable land and woods on fertile soil on the other hand.
4. A gradient of increasing vegetation structure and ranging from open communities like arable land and dry grassland, via moist grasslands, ruderal vegetations and scrub communities, to various types of woodland.

The performance of *Plantago* species with regard to the above four factors could be visualized either by two-dimensional diagrams, showing the position of the *Plantago* sites (Fig. 3.2) or by construction of one-dimensional response curves per axis. For the latter purpose the ordination axes were divide into sections of equal length and the average frequency of a species was computed per section of the axis. In this way the occurrence of *Plantago* as presented in the above-mentioned tables could be converted to curves with an optimum along the four gradients (Fig. 3.3) which can be interpreted as environmental factors.

3.1.3 Results and Discussion

The frequency of the *Plantago* species is shown in Fig. 3.1, averaged per alliance, and arranged according to Westhoff and Den Held (1969). The description of the communities is very limited but the alliances in which *Plantago* spe-

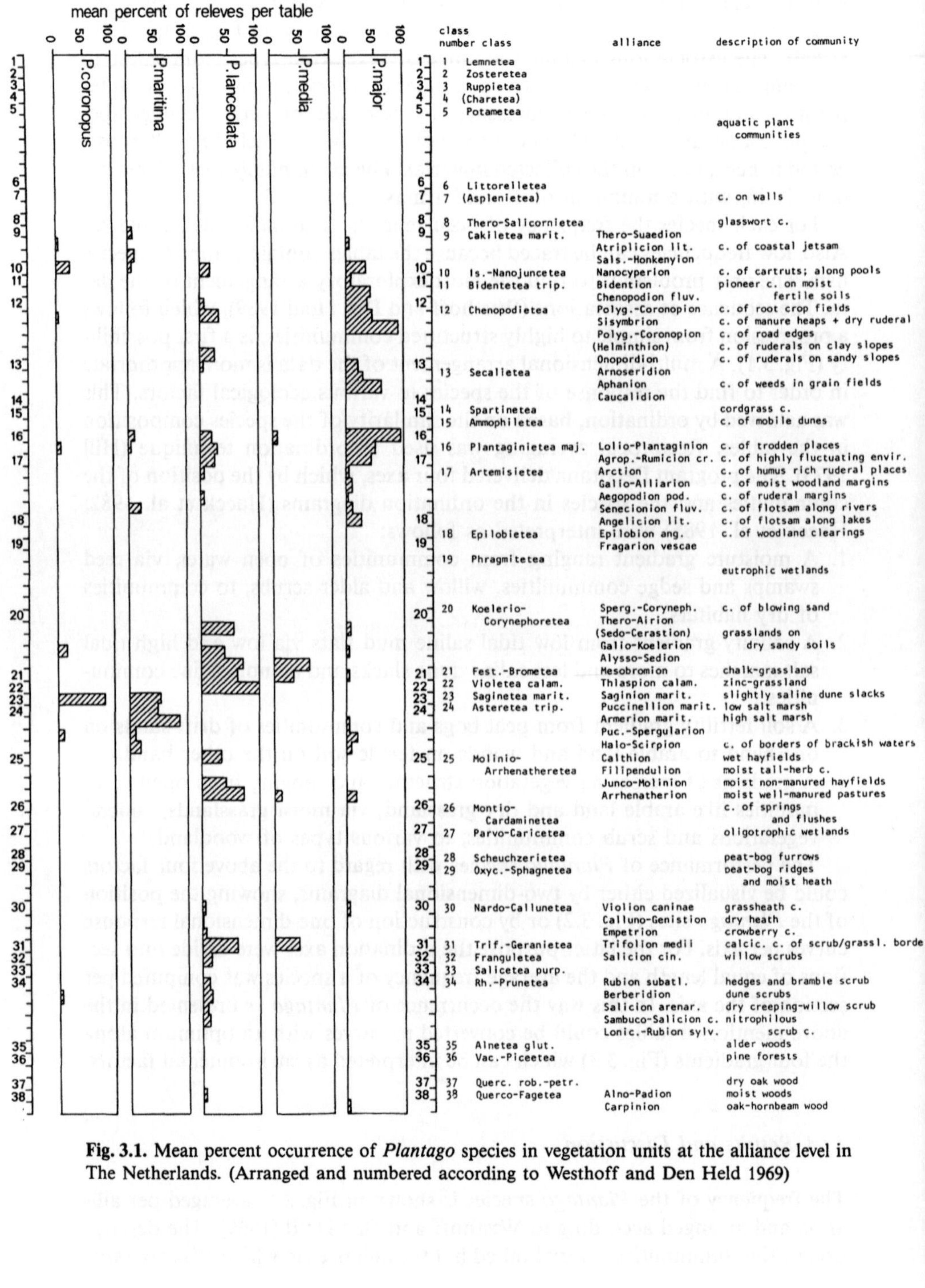

Fig. 3.1. Mean percent occurrence of *Plantago* species in vegetation units at the alliance level in The Netherlands. (Arranged and numbered according to Westhoff and Den Held 1969)

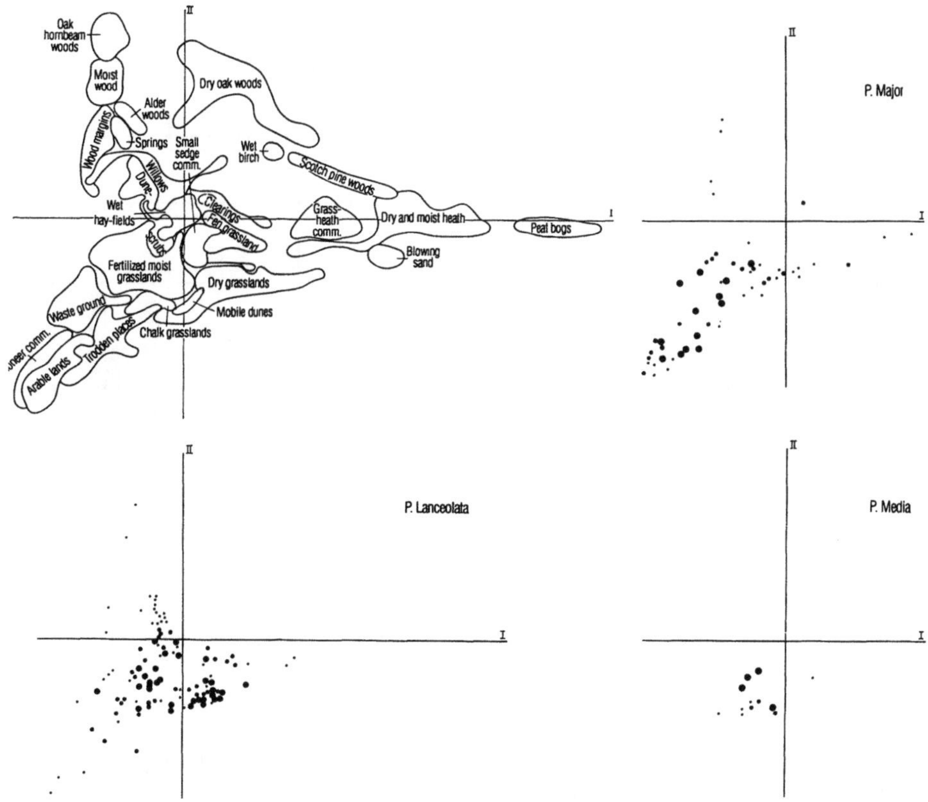

Fig. 3.2. Ordination diagram for axis 1 and 2 of the ordination with the extreme wet and saline communities excluded, and distribution of three *Plantago* species in this plane

cies occur may be distinguished and the results may be compared with the description of Table 3.1.

It can be seen that *P. coronopus* has been most frequently found in the vegetation of "green beaches" (Saginion maritimae and Nanocyperion) and much less, with about equal frequency, in communities of high tidal salt marshes (Puccinellio-Spergularion salinae and Armerion maritimae), of dry grassland (Galio-Koelerion) and of trodden places (Nanocyperion and Sisymbrion); *P. coronopus* occasionally occurred in communities of coastal jetsam (Thero-Suaedion).

P. maritima was most frequently observed in high tidal salt marshes (Armerion) and somewhat less frequently in low salt marshes and beach plain communities (Puccinellion maritimae, Saginion maritimae and Nanocyperion). *P. maritima* was found less frequently in saline grass- and reedlands (Lolio-Plantaginion, Agropyro-Rumicion crispi (= Lolio-Potentillion anserinae and Angelicion litoralis); it was also found on coastal jetsam (Atriplicion littoralis).

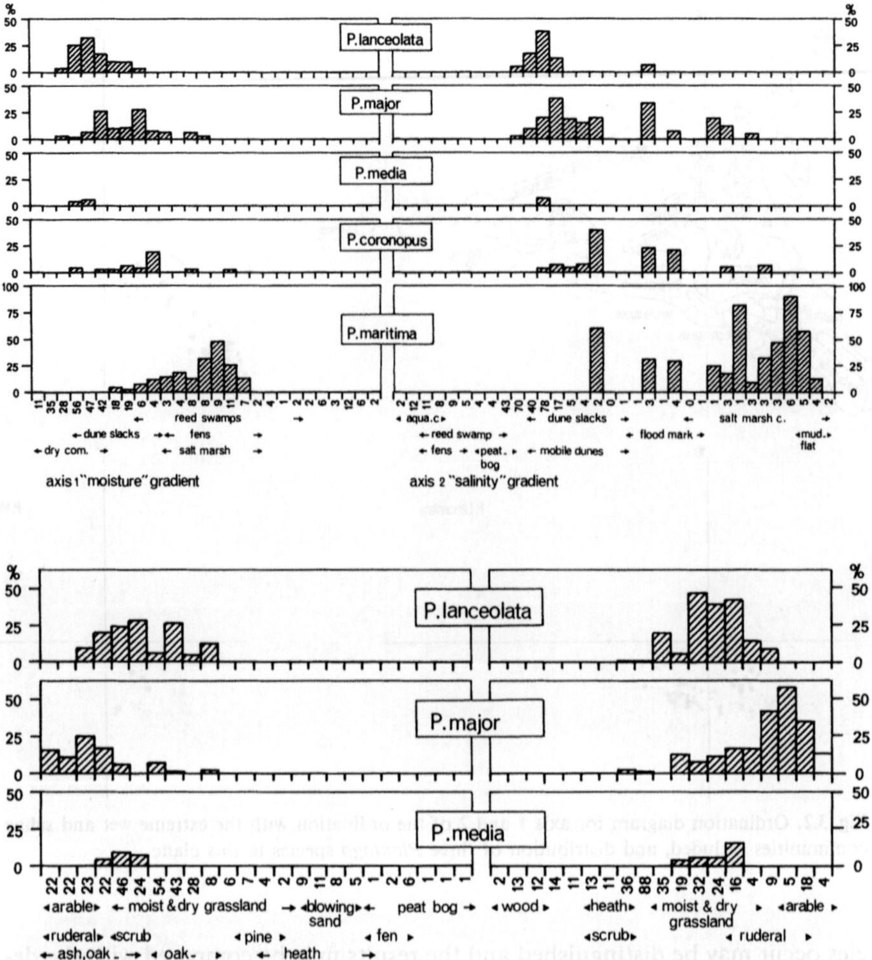

Fig. 3.3. Average frequency of occurrence of *Plantago* species in vegetation types along four ordination axes. At the bottom are indicated: *1* the number of tables per section of the axis; *2* the approximate position of various communities; 3 the overall gradient

P. lanceolata is known as a character species of the Molinio-Arrhenatheretea (moist grasslands) and in The Netherlands as a differentiating species of communities of dry grassland (Galio-Koelerion and Thlaspion calaminaris) and Juniper scrubs (*Prunetalia spinosae*). *P. lanceolata* was most frequent in more or less dry grassland (Koelerion-Corynephoretea, Mesobromion and Thlaspion), somewhat less frequent in moist grassland (wet hayfields of the Calthion and Junco-Molinion and moist pastures of the Arrhenatherion) and less frequent in different ruderal communities, belonging to the Sisymbrion,

Onopordion, Plantaginetea majoris and Artemisietea vulgaris, woodland margins (Trifolion medii) and scrubs (dune scrub: Berberidion and Salicion arenariae; willow scrub: Salicion cinereae).

P. media occurred most frequently in chalk grassland (Alysso-Sedion and Mesobromion) and its border with woods (Trifolion medii) and less frequently in dry subassociations of Arrhenatherion communities, as found on river dikes.

In most of the literature, the subspecies of *P. major* are not discriminated. Therefore, it is not possible to make different surveys for the subspecies. It may be inferred that *P. major* ssp. *major* was most frequent in "trampled sites along roads" (Polygono-Coronopion and Lolio-Plantaginion) and less frequent in "ruderal places" (Sisymbrion, Artemisietea) and "often trodden grassland" [the Poo-Lolietum, see Westhoff and Den Held (1969), and classified among the Agropyro-Rumicion crispi]. *P. major* ssp. *major* also occurred in low frequencies in tables of various other types of grassland (Arrhenatherion, Thero-Airion and Galio-Koelerion). Other high frequencies of *P. major* probably could be ascribed to subspecies *pleiosperma*: "moist arable land" (Secalietea, Polygono-Chenopodion), "moist dune valleys" (Nanocyperion) and "river forelands" [Agropyro-Rumicion crispi (Potentillion anserinae) and Bidention]. The occurrence of *P. major* in tables of Asteretea communities (Beeftink 1975) probably may also be attributed to ssp. *pleiosperma*.

The ordination technique allowed transformation of original phytosociological data into environmental response curves and Fig. 3.3 shows the results for the *Plantago* species. For *P. coronopus* and *P. maritima* the histograms for axes 3 and 4 are not shown, as the projection of the halophytic communities on these axes cannot be interpreted. *P. coronopus* had its optimum in slightly saline and relatively dry habitats compared with the habitats of *P. maritima*: saline dune-slack and high tidal salt marsh and dune grassland versus low and high tidal salt marshes. *P. media* had a very narrow amplitude for all factors; it was almost completely confined to grass communities on neutral and alkaline soils. *P. lanceolata* and *P. major* both had wide amplitudes, the former with an optimum in moderately fertile grasslands and ranging from ruderal communities to dune and willow scrubs, the latter having its optimum in fertile and open vegetation and being most frequent in trodden places; it was less frequent in various other open communities.

The present information does not lead directly to the conclusion that in *Plantago* ecotypic differences with respect to specific habitats exist. However, wide amplitude of *P. major* and *P. lanceolata* and the existence of multi-topped curves as in *P. major* do point to a certain degree of ecotypic differentiation in relation to the habitat.

As an example (Fig. 3.2), the distribution of three *Plantago* species is shown in a diagram as the first two axes of the ordination of the tables without the extreme wet and saline communities. This ordination reveals as most important gradients "soil fertility" along axis 1 and "vegetation structure" along axis 2. The differences between the three species in performance with regard to these (complexes of) environmental factors are clear.

The description of *Plantago* habitats in The Netherlands in general is in good agreement with those for other countries. Some differences are worth mentioning.

In South Sweden, *P. coronopus* has a strictly coastal distribution towards the northern boundary of its range. Inland occurrence is rare and more or less connected with human interference in North and Eastern France, The Netherlands, North Germany (Rauh and Senghas 1976) and Switzerland (Hess et al. 1972). It is more frequent and permanent in Great Britain and Western and Central France (Fournier 1946). From the *Plantago* species occurring in mediterranean France, *P. coronopus* has been found in the greatest number of communities (Braun-Blanquet et al. 1951). In contradiction, Clapham et al. (1962) point to a strictly maritime occurrence of *P. coronopus* in Ireland.

P. maritima is occasionally found in inland stations in pioneer communities on gypsum and marl soils (Oberdorfer 1979). The subspecies *serpentina* probably presents an ecotype (Tutin et al. 1964–1980) which occurs on mountains of Central and Southern Europe, above 2000 m. It is also found at lower altitudes in open grassland on stony alkaline soils (Sedo-Scleranthion; Oberdorfer 1979).

In the southern part of England, *P. media* seems to be confined to base-rich areas of old relatively undisturbed grassland where cultivation is at a minimum (Sagar and Harper 1964). In southern Germany (Oberdorfer 1979) and in Switzerland (Hess et al. 1972), *P. media* is as common as *P. lanceolata*. It occurs in lime and loam areas in permanent pasture on rather poor soil (Cynosurion and Arrhenatherion), in chalk grassland (Mesobromion) and in more acidophilous grassland (Violion). In northern Germany, as in The Netherlands, *P. media* is much less common and confined to base-rich soils.

In Germany, *P. lanceolata* is common in cow pastures and hayfields, in park lawns, on road verges and arable land, on fertile soils in communities of the Molinio-Arrhenatheretea (class character species), and in Lolio-Potentillion, Mesobromion and other contact communities; its subspecies *sphaerostachya* occurs in grassland on more dry sandy places (Sedo-Scleranthetea; Oberdorfer 1979).

P. major ssp. *pleiosperma* is rather common in pioneer communities on river banks, in moist arable land and temporarily flooded places along roads, and in too heavily stocked meadows (Lolio-Potentillion, often in contact with Nanocyperion and Bidention). It is a character species of the Plantaginetalia and is also found in meadows (Cynusurion) and ruderal communities (Chenopodietea) (Mølgaard 1976; Blom 1979; Oberdorfer 1979).

3.1.4 General Conclusions

Most *Plantago* species, although with the exception of *P. major* not strictly ruderal, seem to profit from a not too heavy disturbance in the vegetation in which they occur: *P. coronopus* grows most luxuriously near rabbit scratchings in dune grassland, *P. maritima* reaches its optimum along daily-flooded creeks in the salt marsh, *P. lanceolata* stands can reach very high densities and relatively large individuals are observed in newly made and recently disturbed road verges on different soil types. This species is also found in greatest numbers in communities of nutrient poor and moist or dry grassland near mole heaps, ant nests or places disturbed by man, e. g. by hay storage. The same holds more or less for *P. media*, which has even been found in loam pits in heath (Westhoff et al. 1970). Both subspecies of *P. major* are rapidly overgrown by grasses and other species if their habitat is not disturbed again in due time. On the beach plain *P. major* ssp. *pleiosperma* seems to profit from the activity of the yellow meadow ant, *Lasius flavus*.

All Dutch *Plantago* species occur optimally in rather contrasting habitats: *P. coronopus* in dry and open grassland and on moist beach plains; *P. maritima* along tidal creeks on the salt marsh and near the border of the dune ridge; *P. lanceolata* in wet hayfields as well as in dry road verges; *P. media* on grazed river dikes and in ungrazed chalk grassland; *P. major* ssp. *major* in meadows, lawns and road edges; and finally *P. major* ssp. *pleiosperma* on river banks and beach plains. Such a group of species offers excellent opportunities for further studies on the genetics, physiology and ecology of plant populations.

3.2 Chemical and Physical Characteristics of the Soil of *Plantago* Sites

3.2.1 Introduction

Detailed data on the nutrient status of soils of species with contrasting nutrient ecology are scarce. Results are often difficult to interpret due to differences in sampling and soil analyses. Statements about the preference of a species for a specific soil chemical factor are largely based on common knowledge: e. g. the occurrence of *P. media* in calcareous and *P. maritima* in saline habitats.

The present section summarizes results of soil chemical analyses of sites of the five *Plantago* species occurring in The Netherlands: *P. major, P. media, P. coronopus, P. lanceolata* and *P. maritima*. Analyses were carried out simultaneously and for all samples identically. The following subjects will be discussed.

a) A survey of a total of approximately 100 natural *Plantago* habitats in The Netherlands, providing information of these locations (Sect. 3.2.3).
b) A seasonal study of six selected *Plantago* sites; two sites of *P. major* ssp. *major*, of *P. major* ssp. *pleiosperma*, and of *P. lanceolata* (Sect. 3.2.4).

3.2.2 Methods

In May and June 1979, a soil-sampling program was carried out in various parts of The Netherlands (Fig. 3.4), covering 100 natural habitats of *Plantago* species. In general, three successive soil layers of 10 cm thickness were sampled and determinations were carried out for pH, carbonates, organic matter (loss

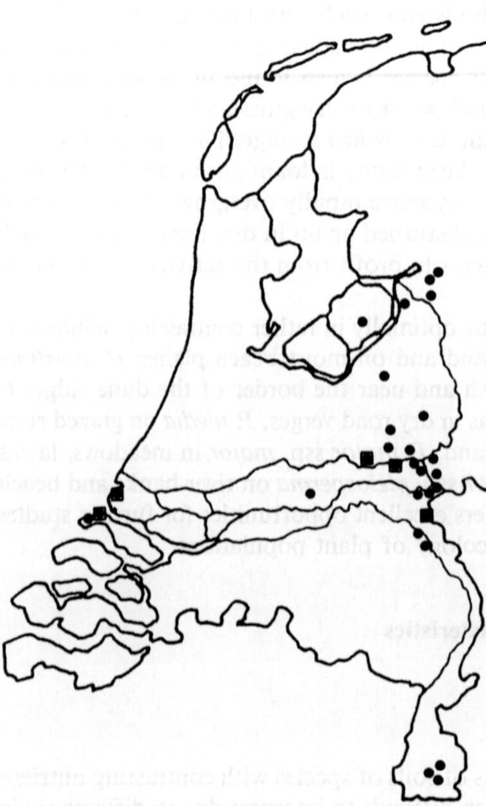

Fig. 3.4. Main locations of the investigated *Plantago* sites in The Netherlands. *Most points* comprise more than one site. *Square symbols* refer to the detailed seasonal study

on ignition; Davies 1974), total N, total P, total organic P, Olsen-P (Olsen et al. 1954), labile organic P (Bowman and Cole 1978), exchangeable cations, chloride, electrical conductivity, cation exchange capacity (CEC; determined with NH_4OAc 1 N at pH 7), granular composition (soil texture), and nitrogen mineralization potential. Pretreatment of the samples and analytical procedures have been previously described (Troelstra et al. 1981). Values obtained on a weight basis were transformed to a volume basis using bulk density estimates for the pretreated soil material. This was necessary since values on a weight basis may give a distorted picture on peaty locations, and from an ecological point of view a soil volume basis is often preferable.

A limited number of *Plantago* sites were investigated in greater detail in 1983 and throughout the season soil chemical and soil physical properties in successive layers of the soil profile were determined. These studies were confined to *P. major* and *P. lanceolata*; the aim was to gather quantitative data on fluctuations in soil factors of the natural habitats. Two hayfield situations of *P. lanceolata* and two path situations of *P. major* ssp. *major* were chosen, and extreme situations such as peaty locations were avoided. In addition, two coastal sites of *P. major* ssp. *pleiosperma* were investigated in a similar way and the results were compared with data on an inland location of the same subspecies.

In general, four successive soil layers were sampled, viz. 0–5, 5–10, 10–15 and 15–25 cm. Sampling frequency was either every week for soil moisture or every fortnight for mineral N, Olsen-P, labile organic P, exchangeable cations, pH, chloride and electrical conductivity. The results refer to an area of 3 to 4 m^2 at the most for each location. The areas were marked at the start of the season and on occasion samples were taken within a 15×15 cm square. Bulk samples were collected once at the start of the growing season in order to determine the general soil chemical characteristics of the site. Samples were collected in pre-weighed moisture content tins and shaking bottles, transported at 4°C to the laboratory, and immediately analyzed without further pretreatment. Where relevant, allowance was made for the moisture content of each sampling date and at each location, and the results are expressed on an oven-dry weight basis. Details of the analytical procedures are earlier described (Troelstra et al. 1984b).

3.2.3 A Survey of Natural Plantago Sites

Figure 3.5 summarizes the observed ranges and mean values for soil properties of natural habitats of *Plantago* species. There was a considerable degree of overlap of soil factors of habitats of the five species, except for: low pH, *P. lanceolata*; high $CaCO_3$ content, *P. media* and *P. lanceolata*; high organic matter and total N content, *P. lanceolata*; and high chloride, sodium, and magnesium levels, and high electrical conductivity, *P. coronopus* and *P. maritima*. Some sites of *P. major* revealed relatively high values for total P and Olsen-P, and, among habitats of the three glycophytic species, relatively high amounts of exchangeable K. Due to the occurrence of *P. lanceolata* in peat areas and in dry dune grasslands, this species showed a wide range in soil factors with respect to organic matter, total N, total organic P, labile organic P, and exchangeable Ca.

Mean values for the above-mentioned soil factors in the 0- to 10-cm soil layer (on a volume basis) confirm the above results: the above-mentioned differences between the species are easily recognized (Table 3.3). Textural analysis of the surface layer (of 0–10 cm) show a high degree of overlap, except for relatively fine textures in the habitats of *P. maritima* (Table 3.4).

The inorganic and organic phosphorus pool in the 0- to 10-cm layer of 65 natural habitats comprising all five *Plantago* species was completely fractionated. At low pH, Fe-P and/or Al-P were the predominant inorganic P forms, except for peaty locations or relatively young stages of soil development, where Ca-P prevailed. With increasing pH, the proportion of Ca-P increased to values above 90%. Among the organic P fractions (Bowman and Cole 1978),

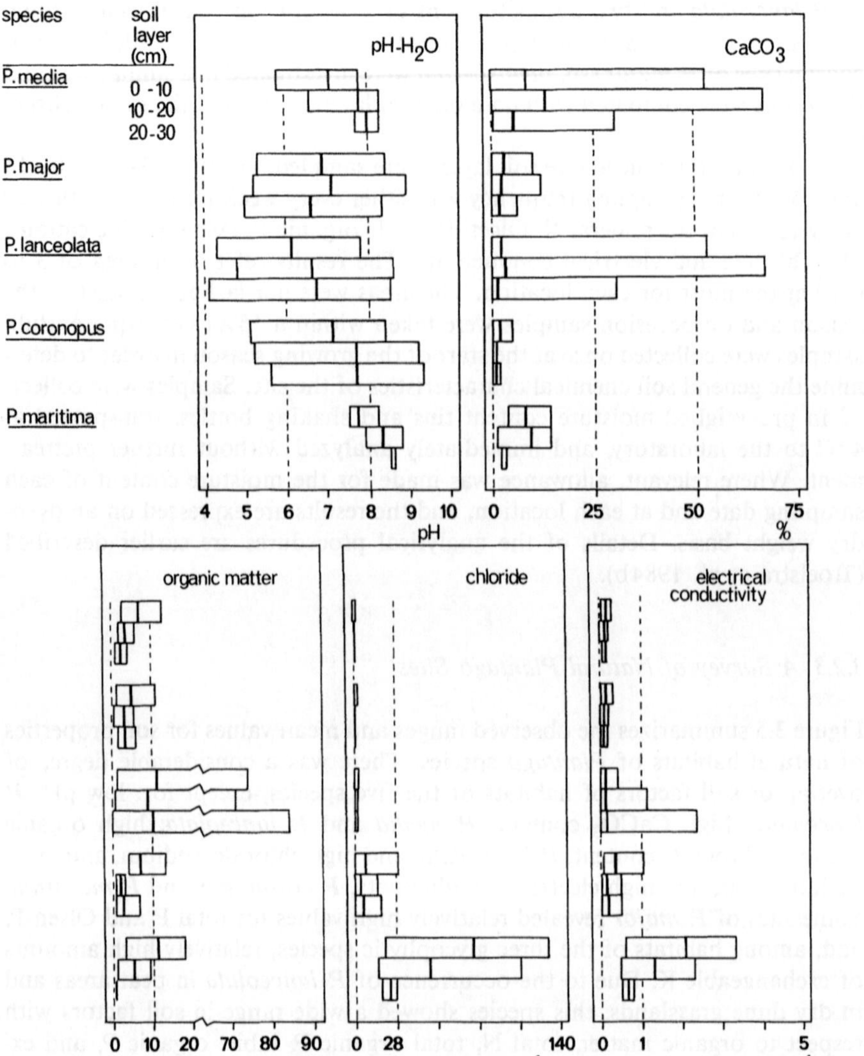

Fig. 3.5. **a** Ranges and mean values (*solid vertical bars*) of pH-H_2O, percentages $CaCO_3$ and organic matter, chloride contents, and electrical conductivities for three successive 10-cm soil layers collected from various natural habitats of *Plantago* species in The Netherlands. Weight basis values are expressed per kg dry soil. Number of sites investigated: 15 (*P. media*), 17 (*P. major*), 72 (*P. lanceolata*), 4 (*P. coronopus*), and 13 (*P. maritima*)

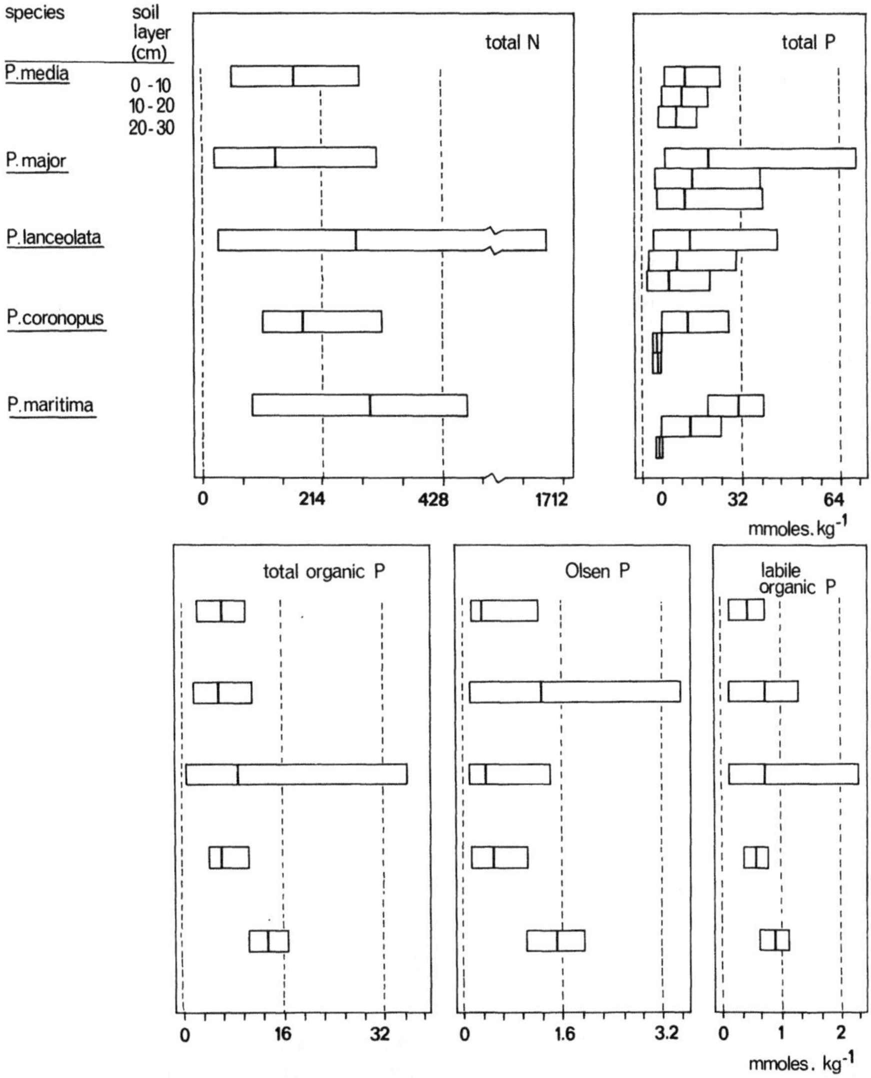

Fig. 3.5. b Ranges and mean values (*solid vertical bars*) of total N, total P, total and labile organic P, and Olsen-P for the 0–10 cm soil layer or three successive 10-cm soil layers (total P) collected from various natural habitats of *Plantago* species in The Netherlands. Values are expressed per kg dry soil. Number of sites investigated: 15 (*P. media*), 17 (*P. major*), 72 (*P. lanceolata*), 4 (*P. coronopus*), and 13 or 2 (*P. maritima*; P analyses refer to two sites only)

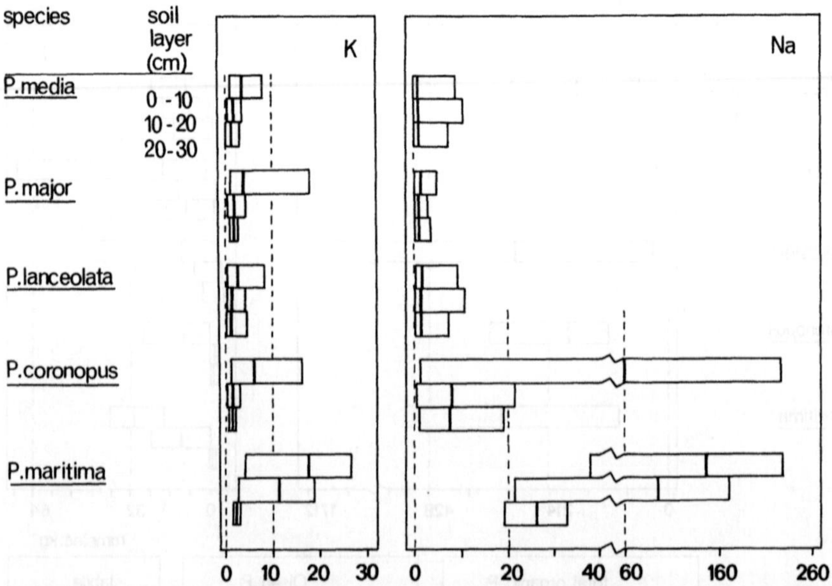

Fig. 3.5. c Ranges and mean values (*solid vertical bars*) of exchangeable cations for three successive 10-cm soil layers collected from various natural habitats of *Plantago* species in The Netherlands. Values are expressed per kg dry soil. Number of sites investigated as in Fig. 3.5a, but calcareous sites (% $CaCO_3 > 0.1$) have not been included in the Ca data

the moderately labile and moderately resistant fractions were predominant, whereas no clear trend could be observed over the total pH range of the sites from 4.5 to 8.4. Expressed as percentage of the sum of the organic P fractions, the following ranges were found for the four categories of organic P: labile, 5–37%, average 14%; moderately labile, 13–70%, average 41%; moderately resistant, 13–45%, average 32%; highly resistant, 4–22%, average 14%. In general, relatively high percentages of labile organic P were found at relatively low pH values.

The most important differences between habitats of the various *Plantago* species concerned the pH and salinity levels. In the field *P. lanceolata* occurred over a wide range of pH values, but more often at relatively low pH (pH-$H_2O < 6.0$) than at high pH values. This wide pH range (Fig. 3.5), together with the wide variety of textural classes indicates a wide ecological amplitude of this species. From pot experiments with *P. lanceolata*, using non-sterilized and sterilized soils from a particular area, De Nooij (Fig. 3.6) concluded that the observed low survival of seedlings probably was due to an abiotic factor, a low pH: pH-$H_2O < 4.6$; sites with *P. lanceolata* had a higher soil pH value than sites without *P. lanceolata*, 5.5(±0.06) and 5.0(±0.10), respectively.

Habitats of *P. media* and *P. major* generally had relatively high pH values; nitrate, rather than ammonium, will be the predominant source of N for these species (Schmidt 1982). Indeed, after incubation at 30 °C and at 14 to 45%

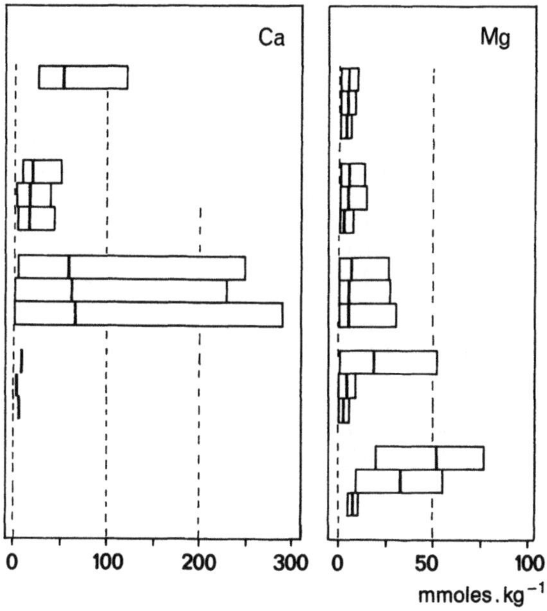

Fig. 3.5c (continued)

moisture content the accumulated mineral N in soils from habitats of *P. media* and *P. major* predominantly was in the form of nitrate (Table 3.5). *P. coronopus* and *P. maritima* also occurred frequently at relatively high pH values, coinciding with high levels of salinity. The chloride contents of Table 3.3 correspond with NaCl concentrations in the soil solution (at 25% moisture) of approximately 300 mM and 25 mM for habitats of *P. maritima* and *P. coronopus*, respectively. Since at high salinity levels (*P. maritima*) the process of nitrification was rather retarded (Table 3.5; Laura 1977), ammonium may be the predominant source of N for *P. maritima*.

The present results are in general line with those of Kruijne et al. (1967), who carried out a soil analysis of 1577 grasslands, in combination with botanical surveys of habitat classes. Ecological optima for each grassland species were indicated. Optimum values of some characteristics for *P. media, P. major*, and *P. lanceolata* are shown in Table 3.6. This table refers to the top 0 to 5 cm soil layer, a common practice of soil sampling for fertility research of grasslands. The results of Kruijne et al. (1967) indicated a preference of *P. major* for a moderately high P and K status. The percentage of fields (per habitat class) in which *P. lanceolata* was found increased with decreasing P and K status of the soil. Water culture experiments carried out in our laboratory with the species *P. major* ssp. *major* and *P. lanceolata* and with varying levels of P and K availability, supported to some extent the above field data. The yield of *P. major* ssp. *major* decreased more strongly with decreasing K availability than the yield of *P. lanceolata. P. major* also showed a tendency to react more strongly to a decrease of the P concentration than *P. lanceolata* (Troelstra et al. 1983a).

Table 3.3. Some selected soil properties (mean values) of the 0–10 cm soil layer of natural habitats of *Plantago* species in The Netherlands. Volume basis values are expressed per litre soil

Soil factor	Species				
	P. media	*P. major*	*P. lanceolata*	*P. coronopus*	*P. maritima*
pH-H_2O	7.1	6.9	6.1	7.1	7.7
% $CaCO_3$	8.8	2.9	2.3	1.9	7.9
% organic matter	6.7	5.0	11.3	7.0	11.2
Electrical conductivity 1:5 extract (mS cm^{-1})	0.16	0.13	0.13	0.36	2.07
Chloride (mmol l^{-1})	0.39	0.25	0.25	7.0	80
Total N (mmol l^{-1})	195	160	250	195	300
Total P (mmol l^{-1})	17	27	16	16	29
Total organic P (mmol l^{-1})	7.4	7.4	8.9	6.8	12.6
Olsen-P (mmol l^{-1})	0.32	1.58	0.42	0.52	1.42
Labile organic P (mmol l^{-1})	0.48	0.90	0.77	0.65	0.81
K (mmol l^{-1})	4.1	4.9	2.8	5.8	17.4
Na (mmol l^{-1})	1.3	1.6	1.7	58	146
Ca[a] (mmol l^{-1})	61	24	52	9.3	–
Mg (mmol l^{-1})	6.2	7.1	6.5	18	52
CEC[a] (mEq l^{-1})	156	74	131	60	–

[a] Calcareous soils (% $CaCO_3 > 0.1$) not included.

Table 3.4. $CaCO_3$, organic matter, and grain size distribution for soils (0–10 cm layer) collected from various natural habitats of *Plantago* species in The Netherlands; data for *P. maritima* refer to two sites only. Values are expressed as % of the total soil weight, after drying at 105 °C

Soil component (%)	Species				
	P. media	*P. major*	*P. lanceolata*	*P. coronopus*	*P. maritima*
$CaCO_3$	9.0	2.3	2.3	1.9	7.5
Organic matter	6.7	5.0	11.3	7.0	13.5
Particle size fraction					
>53 μ	53 (5–91)[a]	53 (13–93)	58 (0–95)	72 (38–90)	25 (12–38)
16–53 μ	13 (1–37)	18 (1–56)	11 (1–58)	8 (1–17)	19 (17–22)
2–16 μ	8 (1–16)	10 (1–27)	8 (0–31)	5 (1–12)	15 (12–18)
<2 μ	10 (1–18)	10 (1–34)	10 (0–54)	6 (1–15)	19 (15–23)
Clay-humus factor[b]	11 (5–20)	12 (2–37)	16 (2–77)	10 (6–20)	22 (20–25)

[a] Mean (range).
[b] Clay-humus factor = % organic matter + $\frac{1}{4}$ of fraction < 16 μ.

Ionic contents of plant material of *Plantago major* collected in the field also revealed relatively high contents of P and K. The question can be raised whether there is a direct relationship pertaining to higher P and K requirements for *P. major* than for *P. lanceolata.* The alternative explanation is the more solitary occurrence of *P. major* and the adaptation to more compacted

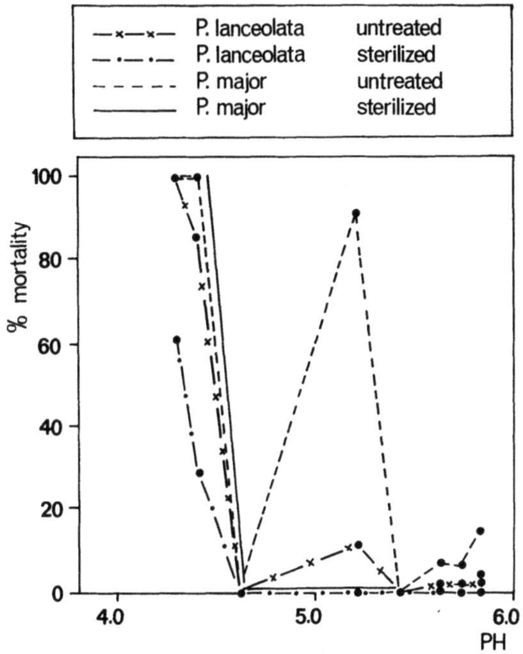

Fig. 3.6. The relationship between pH-H_2O and mortality of seedlings of *Plantago lanceolata* and *P. major* ssp. *major* in soils originating from + and − *P. lanceolata* sites within a wet hayfield area (Merrevliet). The percentage organic matter in the 0–15 cm layer (>85%) fell outside the range shown in Fig. 3.5a, due to the fact that this site had not been included in the general survey. (M.P. De Nooij unpubl. results)

Table 3.5. Mean nitrification degree (% NO_3 in total N mineralization) of net production of mineral N in soils (0–10 cm) collected from various natural habitats of *Plantago* species in The Netherlands. Soils were aerobically incubated in the laboratory at 30 °C and 10%–45% moisture (soil dry weight), the latter depending on soil type; peat(y) samples were mixed with quartz sand

Period of incubation (weeks)	Species				
	P. media	*P. major*	*P. lancelata*	*P. coronopus*	*P. maritima*
2	74	72	26	63	9
4	89	83	32	82	30
6	96	88	38	92	57

sites, resulting in less competition and better soil transport of nutrients to the roots. *P. major* sometimes is a poor competitor for nutrients and/or light; a relatively low availability of nutrients may be caused by the presence of a dense vegetation rather than a low availability of nutrients in the soil.

Königshofer (1981) analyzed 22 *Plantago* sites in Austria (Table 3.7). The figures refer to water extractions and the agreement with our study is obvious:

Table 3.6. Optimal values for some soil properties (0–5 cm depth) of grassland habitats of three *Plantago* species. (After data of Kruijne et al. 1967)

Soil factor	Species		
	P. media	*P. major*	*P. lanceolata*
$pH\text{-}H_2O$	>7.00	6.05–7.00	>5.05[a]
P citric acid ($mmol\ kg^{-1}$)	0–4.5	>4.8	<2.6 (4.5)
K value ($mmol\ kg^{-1}$)	3–6	>6	<3 (4)
Clay humus factor[b]	11–20	<20	<40 and >60

[a] No distinct preference.
[b] Clay-humus factor = % organic matter + $\frac{1}{4}$ of fraction <16 μ.

Table 3.7. Some selected soil properties of natural habitats of four *Plantago* species in Austria. (After data of Königshofer 1981)

Soil factor	Species			
	P. media	*P. major*	*P. lanceolata*	*P. maritima*
K ($mmol\ kg^{-1}$)	0.6–2.8	0.5–2.8	0.4–3.2	1.3–20
Na ($mmol\ kg^{-1}$)	0.1–0.3	0.1–1.9	0.1–0.3	0.2–215
Ca ($mmol\ kg^{-1}$)	4.0–16	0.6–9.0	0.6–18	0.5–11
Mg ($mmol\ kg^{-1}$)	0.3–4.3	0.2–2.9	0.2–2.7	0.5–4.0
SO_4 ($mmol\ kg^{-1}$)	0.05–0.30	0.05	0.05–0.10	1.8–83
Cl ($mmol\ kg^{-1}$)	0.1–1.0	0.1–0.4	0.1–0.9	0.9–36

a high degree of overlap of soil factors among the glycophytic species and relatively high levels of K, Na, and Cl at sites of *P. maritima*. Values for the pH were not reported, but most of the sites probably were calcareous. According to Königshofer (1981), *P. media* and *P. major* prefer calcareous sites. In agreement with the present results, however, their occurrence was not restricted to calcareous habitats (Königshofer 1981).

Ordination of phytosociological data revealed soil fertility as one of the major factors (see Sect. 3.1 and Haeck et al. 1982). This interpretation is partly supported by correlation of the factor scores of the floristic data with those on the soil chemical factors of our set of about 100 *Plantago* sites. Table 3.8 shows very significant but low correlations for Olsen-P, total P, pH and K, but not for total N and other factors. However, individual data sets for these significant factors showed a large scatter (Haeck 1984), which emphasizes the fact that species composition is in part the result of complex and interacting influences, such as soil structure, vegetation management and history, and biotic factors. The outcome of such correlative studies should therefore be viewed with caution; the more so as widely different vegetation and soil types were compared and the chemical data, expressed on a soil weight basis, referred to dried and sieved soil samples.

Table 3.8. Values for Kendall rank correlation (τ with probabilities P) of factor scores of 105 relevés on third ordination axis with soil chemical values of these sample sites

	τ	P
pH-H_2O	0.427	≪0.0001
pH-KCl	0.407	≪0.0001
Total N	0.012	0.857
Total P	0.354	≪0.0001
Total organic P	0.116	0.083
Olsen-P	0.374	≪0.0001
Labile organic P	0.015	0.818
% organic matter	0.045	0.497
K	0.245	0.0003
Na	0.007	0.912
Mg	0.208	0.0018
Conductivity	0.172	0.0098

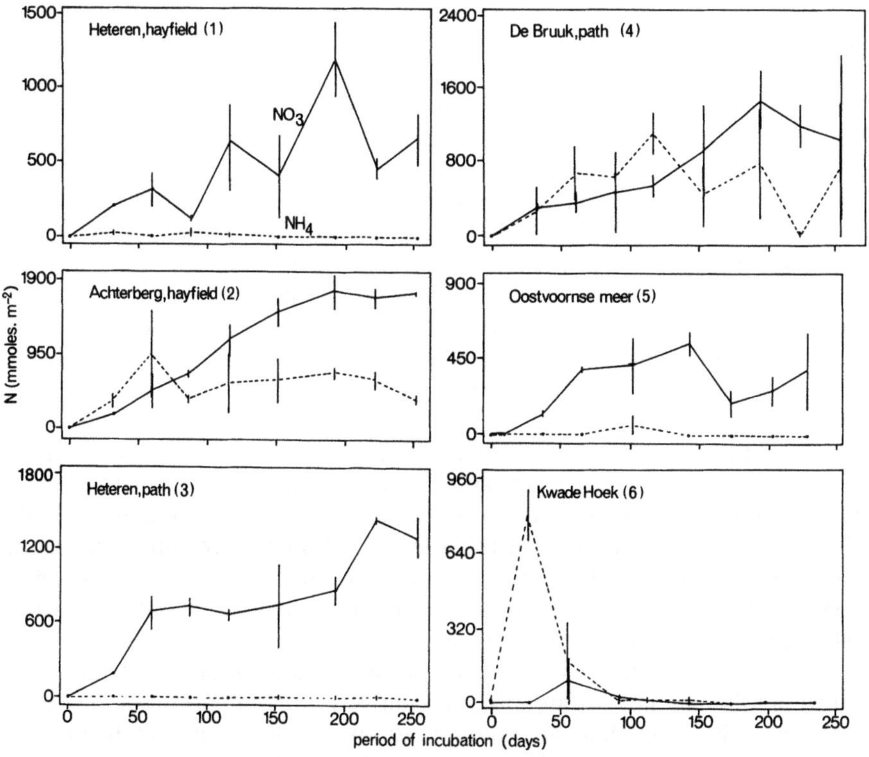

Fig. 3.7. Time course of mineral N in intact 0–10 cm soil cores from six *Plantago* sites during incubation in the laboratory at 30 °C and field moisture content (present at sampling). Samples were collected 3/31/83 (sites 1 and 2), 3/30/83 (sites 3 and 4), 4/15/83 (site 5), and 5/30/83 (site 6). Both *points* and *vertical bars* refer to the results of duplicate determinations (*vertical bars* connect the duplicate values)

The additional soil analysis of 21 non-*Plantago* sites, close to the vicinity of *Plantago* sites, revealed no clear differences between the sites with or without *Plantago*. Differences in microclimate are not likely to play a major role in the absence or presence of *Plantago* at these sites. On the other hand, it should be noted that part of the non-*Plantago* sites showed either extremely low or rather high nutrient levels. The influence of disturbing factors like trampling by cattle and flooding cannot be excluded at some of the studied non-*Plantago* sites. Moreover, even under constant management the species composition of a vegetation may reveal large fluctuations in time (van den Bergh 1979); in one hayfield site of the present survey, *P. lanceolata* individuals were absent at the time of sampling, while they were abundant the years before.

3.2.4 Seasonal Analysis of Plantago Sites

A comparative study was undertaken to obtain insight into the nutrient supply throughout the season at six different *Plantago* sites. Table 3.9 summarizes the soil chemical characteristics of the sites. Total N and P and chloride were expressed on an area basis. The organic matter content was very high in the salt marsh (site 6), relatively low on the beach plain (site 5), and relatively high at one of the inland locations (site 4). Site 3 (*P. major* ssp. *major*) and site 1 (*P. lanceolata*) had relatively high pH values, whereas the soil reaction of sites 2 and 4 was more acidic. Free $CaCO_3$ was present at the four sites with the high pH values, especially at the two coastal locations. The amount of total N per square meter was highest at De Bruuk, whereas total P was relatively high at Heteren (hayfield) and relatively low at Achterberg and Oostvoornse Meer. Chloride levels and electrical conductivities were high at both coastal locations, particularly in the salt marsh. Measured ranges of chloride concentrations in the soil solution during the season were about 5 to 100 mM (seasonal average of about 30) and 50 to 150 mM (seasonal average of about 85) for sites 5 and 6, respectively.

In addition to the two coastal locations of *P. major* spp. *pleiosperma*, three successive 5-cm soil layers were sampled at an inland study site (Angeren), a sandy riverbank trampled and grazed by cattle. Prolonged inundation by the river made seasonal sampling at this site in 1983 practically impossible. Compared to the coastal locations, the amounts of N and P at the Angeren site were relatively high. Plant analyses indicated that this site was the only one with a relatively high N availability throughout the season.

Table 3.10 shows the ranges or mean values of the seasonal determinations on fresh soil material. Nitrate was quantitatively more important in the deeper (>5 cm) soil layers than ammonium, except for locations De Bruuk and Kwade Hoek. Both forms of N showed a relatively high concentration at the De Bruuk. NO_3 concentration also was relatively high at Heteren (hayfield). Of course, data represent values at a particular instant and the N availability for the plant will depend on net mineralization rate, transport conditions as dependent on soil moisture conditions and ion mobility: NH_4 is less mobile

Table 3.9. Some properties of four successive soil layers at six *Plantago* sites in The Netherlands. Sites 1 and 2 (*P. lanceolata*); sites 3 and 4 (*P. major* ssp. *major*); sites 5 and 6 (*P. major* ssp. *pleiosperma*)

Site	Soil layer (cm)	pH		Organic matter (%)	$CaCO_3$ (%)	Total N (mol m^{-2})	Total P (mol m^{-2})	Chloride[a] (mol m^{-2})	Electrical conductivity[a] 1:5 extract (mS cm^{-1})
		H_2O	KCl						
Heteren (1)	0–5	7.0	6.4	7.5	0.7	14.5	2.16	0.14	0.15
Hayfield	5–10	7.2	6.5	4.8	0.7	12.0	2.19	0.07	0.10
Clay	10–15	7.3	6.5	4.1	0.9	11.1	2.29	0.09	0.10
	15–25	7.6	6.7	3.6	1.1	18.9	4.06	0.15	0.10
Achterberg (2)	0–5	4.9	4.1	8.0	–	9.57	0.68	0.05	0.04
Hayfield	5–10	5.2	4.3	3.6	–	5.50	0.45	0.03	0.02
Sand	10–15	5.3	4.3	3.1	–	5.71	0.39	0.02	0.02
	15–25	5.3	4.4	2.9	–	8.43	0.71	0.04	0.02
Heteren (3)	0–5	6.8	6.1	6.0	0.1	13.6	1.48	0.09	0.13
Path	5–10	7.5	6.8	2.5	0.6	6.86	1.29	0.05	0.12
Sandy clay	10–15	7.8	7.1	1.5	0.4	4.79	1.03	0.05	0.11
	15–25	N.D.	N.D.	N.D.	N.D.	N.D.	N.D.	0.08	0.11
De Bruuk (4)	0–5	5.0	4.3	19.8	–	24.6	1.16	0.10	0.06
Path	5–10	5.2	4.4	8.6	–	15.4	0.94	0.07	0.05
Peaty loam	10–15	5.4	4.6	5.5	–	12.3	0.97	0.08	0.04
	15–25	5.8	4.9	4.1	–	12.3	1.10	0.14	0.04
Oostvoornse Meer (5)	0–5	7.9	7.7	2.9	6.5	5.86	0.45	0.57	0.32
Embanked	5–10	8.5	8.3	0.6	7.1	1.64	0.45	0.31	0.17
Beach plain	10–15	8.6	8.6	0.4	6.3	0.79	0.45	0.35	0.16
Sand	15–25	8.8	8.8	0.2	6.1	0.86	0.87	0.93	0.18
Kwade Hoek (6)	0–5	7.1	6.9	37.6	5.2	14.9	0.77	3.65	1.47
Salt marsh	5–10	7.1	6.9	25.9	2.4	19.9	0.87	3.22	1.22
	10–15	7.3	7.0	15.8	2.7	15.8	0.71	2.65	1.02
	15–25	8.1	7.9	1.2	2.8	4.79	0.71	3.37	0.57

[a] Mean values of seasonal determinations on fresh soil material (n = 9–13).
N.D. = not determined.

Table 3.10. Moisture content range throughout the season and mean values of seasonal determinations of pH, mineral N, 'available' P, and exchangeable cations in four successive soil layers at six *Plantago* sites in The Netherlands. Sites 1 and 2 (*P. lanceolata*); sites 3 and 4 (*P. major* ssp. *major*); sites 5 and 6 (*P. major* ssp. *pleiosperma*). Determinations were carried out on fresh soil samples

Site	Soil layer (cm)	Moisture content range (% oven-dry soil)	Mean values of seasonal determinations (n = 9 – 13)								
			pH-H_2O	NH_4	NO_3	Olsen-P	Labile organic P	K	Na	Ca	Mg
				mmol m^{-2}				mol m^{-2}			
Heteren (1)	0 – 5	16 – 78	7.1	17.9	16.4	48.4	21.6	0.54	0.10	(7.0)[a]	1.40
Hayfield	5 – 10	15 – 36	7.5	5.7	13.6	37.7	12.3	0.28	0.07	(6.9)	1.34
Clay	10 – 15	15 – 32	7.6	5.0	16.4	33.5	N.D.	0.19	0.06	(7.8)	1.26
	15 – 25	15 – 30	7.8	7.9	22.9	39.4	N.D.	0.31	0.13	(17.4)	2.09
Achterberg (2)	0 – 5	8 – 53	5.1	10.0	9.3	21.3	73.9	0.11	0.08	0.58	0.09
Hayfield	5 – 10	6 – 26	5.4	5.0	6.4	24.5	54.5	0.03	0.03	0.51	0.03
Sand	10 – 15	7 – 30	5.4	6.4	10.0	30.3	N.D.	0.01	0.03	0.44	0.02
	15 – 25	8 – 28	5.6	9.3	12.9	45.8	N.D.	0.02	0.05	0.89	0.02
Heteren (3)	0 – 5	8 – 42	7.3	8.6	10.0	28.7	12.6	0.28	0.04	(5.8)	0.50
Path	5 – 10	2 – 27	7.7	5.7	7.9	23.2	6.77	0.13	0.03	(5.1)	0.39
Sandy clay	10 – 15	5 – 21	8.0	5.7	9.3	27.4	N.D.	0.14	0.03	(3.9)	0.27
	15 – 25	7 – 22	8.0	22.9	15.7	51.6	N.D.	0.34	0.03	(6.6)	0.45
De Bruuk (4)	0 – 5	25 – 156	5.6	33.6	14.3	5.16	49.7	0.18	0.11	3.1	0.45
Path	5 – 10	19 – 92	5.8	27.1	13.6	2.90	44.2	0.08	0.08	3.8	0.47
Peaty loam	10 – 15	16 – 57	5.9	23.6	12.9	2.58	N.D.	0.06	0.11	5.0	0.53
	15 – 25	19 – 83	6.1	23.6	19.3	3.23	N.D.	0.06	0.20	9.3	0.79
Oostvoornse Meer (5)	0 – 5	23 – 57	7.6	5.7	5.7	1.61	4.19	0.21	0.44	(10.3)	0.29
Embanked	5 – 10	13 – 31	8.3	2.9	4.3	1.61	1.61	0.12	0.19	(11.0)	0.20
Beach plain	10 – 15	12 – 29	8.5	2.9	5.7	1.61	N.D.	0.11	0.21	(11.9)	0.19
Sand	15 – 25	10 – 32	8.6	5.7	9.3	3.87	N.D.	0.20	0.53	(22.8)	0.39
Kwade Hoek (6)	0 – 5	206 – 389	7.4	16.4	7.1	10.0	9.35	0.44	4.57	(8.2)	1.55
Salt marsh	5 – 10	103 – 207	7.5	10.7	6.4	3.87	4.52	0.33	3.46	(6.9)	1.26
	10 – 15	48 – 131	7.6	10.0	6.4	2.58	N.D.	0.29	2.44	(7.1)	0.96
	15 – 25	27 – 43	7.9	16.4	16.4	3.23	N.D.	0.27	2.92	(18.7)	1.06

[a] $CaCO_3$ present.
N.D. = not determined

than NO_3, due to adsorption by clay and organic matter. The amounts of available inorganic P (Olsen-P) were relatively low at De Bruuk and Oostvoornse Meer, whereas labile organic P was high at De Bruuk and at Achterberg. The amount of labile organic P was remarkably low at the Kwade Hoek site. The amount of exchangeable potassium was relatively high at the Heteren (hayfield) and Kwade Hoek sites, and extremely low at Achterberg. At the latter site, Ca and Mg values were also low. The measured seasonal ranges of the amounts of N, P, and K reflect again the relatively low amounts of mineral N and available P (Table 3.10) at Oostvoornse Meer; in combination with the relatively high pH and $CaCO_3$ content (Table 3.9) they indicate the young stage of soil development at this site.

Intact 0- to 10-cm soil cores were incubated in the laboratory. Nitrate predominantly accumulated at the two Heteren sites and the Oostvoornse Meer site (Fig. 3.7). At the other two inland locations, Achterberg and De Bruuk,

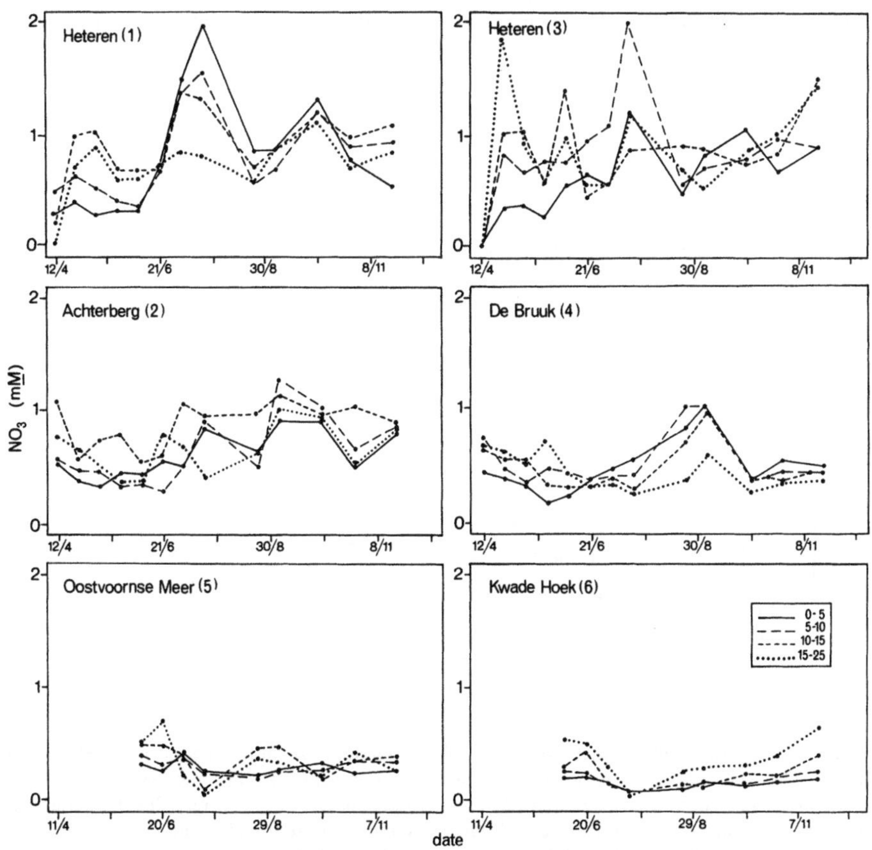

Fig. 3.8. Seasonal pattern of nitrate concentration of the soil solution in four successive soil layers at six *Plantago* sites in The Netherlands during 1983. Concentrations were calculated assuming all nitrate to be dissolved and evenly distributed in the soil water

NH_4 accumulated as well; during the first 100 days, NH_4 was quantitatively at least as important as NO_3. At the Kwade Hoek site the predominant N form was NH_4, and, after 100 days of incubation, mineral N was completely immobilized. The accumulation of mineral N under more or less optimal conditions, expressed as mmol m^{-2}, with 71 mmol N m^{-2} = 10 kg N ha^{-1}, were considerable for the inland locations when compared to average agricultural fertilizer inputs.

The production of above-ground biomass (total vegetation) at Heteren (hayfield) was approximately twice the amount produced at Achterberg (Troelstra et al. 1984a). Comparison of chemical analyses of successive clippings at these sites with availability of nutrients as indicated by soil analyses, suggested that in fact only K at Achterberg appeared to be a possible production-limiting factor (Troelstra et al. 1984a).

Figure 3.9 shows the N-mineralization potential of a site in an old inner dune grassland, Westduinen. Intact 0- to 10-cm soil cores were aerobically incubated, either in the laboratory at 30 °C, or in the field. Saturation levels of NH_4-N and NO_3-N were attained at an earlier stage in the laboratory test.

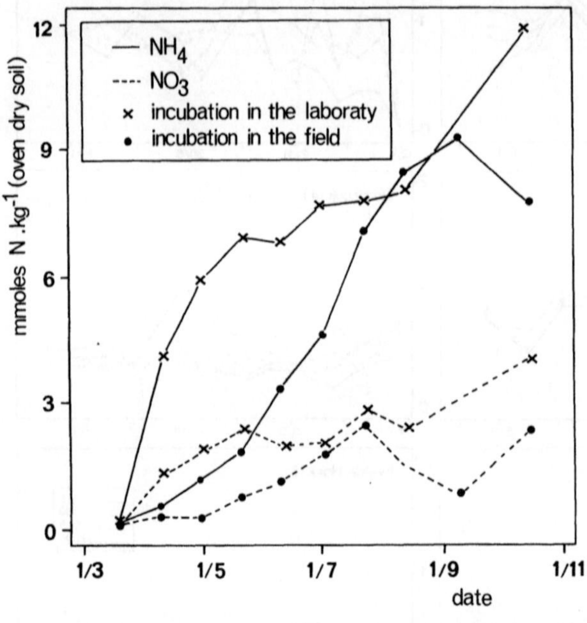

Fig. 3.9. Comparison of two methods of aerobic incubation, viz. in the laboratory at 30 °C and in situ. The accumulation of mineral N was measured in intact 0–10 cm soil cores collected from a site in an inner dune grassland area (Westduinen) on the Island of Goeree (The Netherlands). Samples were put in polypropylene centrifuge tubes and either returned to the site (field incubation) or transported to the laboratory (laboratory incubation). Data of sampling and incubation 3/17/80; soil moisture approximately 22% (dry weight). Soil parameters for this site: pH-H_2O 4.2; % organic matter 6.3; total N 140 mmol kg^{-1}. (S. R. Troelstra and R. Wagenaar unpubl. results)

Approximately the same levels were attained in the field at a later date. It seems justified to consider the results of laboratory incubation procedures as indicative for nitrogen mineralization potentials which may be reached in the field. However, moisture conditions of the actual situation in the field with pronounced short- and long-term moisture fluctuations may differ from that indicated by the above cumulative field incubation procedure. The low levels of mineral N initially found with field incubation are mainly due to suboptimal field temperature.

Adsorption of nitrate by positively charged soil constituents as clay mineral edges and oxide surfaces is often negligible (Bolt and Bruggenwert 1976). Nitrate will mainly be present in the soil solution or as solid salt deposit in dry soil zones. When nitrate is dissolved in the soil water, concentrations of nitrate can be calculated as 100a/b mM, where a = mmol NO_3 kg^{-1} and b is the % soil moisture. The results of such calculations are shown for the six sites (Fig. 3.8). Because of the heterogeneous nature of the soil profile at site 3 (a disturbed site with alternation of clay and sand layers), spatial variability of soil chemical parameters can be considerable as illustrated by the irregular pattern of nitrate concentrations. At the Heteren sites, peak levels of about 2 mM were calculated. At Achterberg and De Bruuk concentrations were generally not higher than 1 mM, and relatively low values, <0.5 mM, were calculated for Oostvoornse Meer and Kwade Hoek. The calculations were made using average data for the various soil layers. Even higher values may be expected on a small-scale level within the soil layers, but high concentrations of nitrate will frequently coincide with unfavourable conditions for transport of nitrate. The nitrate concentrations found were relatively low compared with the normal range in fertile soils, 2 to 20 mM (Mengel and Kirkby 1978).

Variation in the availability of nitrate to the plant root in soils with continous changes in nitrate transport parameters merits consideration. In the field soil moisture conditions will be very important in regulation of the supply of nitrate to the roots. In the field high concentrations of nitrate were recorded, when compared with the K_m value for nitrate of *Plantago* roots. The effective supply of nitrate to the plant will be less luxurious and continuous than in water culture experiments as shown by the low levels of NO_3 in the plant throughout the season at these locations. Model calculations suggest that at nitrate levels above 0.07 mmol kg^{-1}, a maximum potential inflow of nitrate would be achieved, whatever the morphology and dry weight of the roots (Robinson and Rorison 1983). However, these calculations referred to single plants growing in isolation and in a soil water content range of 0.2 to 0.3, conditions seldom met in non-cultivated soils.

The seasonal variability comprises both temporal and spatial elements. To obtain an idea of the spatial variability in mineral N, on a single occasion 16 0–10-cm samples were taken from a 0.36 m^2 area at four sites (Table 3.11). NH_4 was quantitatively more important than NO_3. Moreover, the coefficient of variation was about twice as large for NH_4 as for NO_3. The relatively high CV values for site 3 may be due to a higher spatial variability at this site. Among the four sites the calculated concentrations of NO_3 in the "soil solu-

Table 3.11. Small scale variability of mineral N in the 0–10 cm soil layer at four *Plantago* sites in The Netherlands (n = 16). Samples were taken from 15×15 cm subunits within a total area of either 60×60 cm (sites 1, 2, and 4) or 120×30 cm (site 3). Date of sampling: May 1, 1984 (sites 1 and 3) or May 16, 1984 (sites 2 and 4)

Site	NH_4	NO_3	Moisture content % (oven-dry soil)	NO_3 (mM)	NH_4/NO_3
	mmol kg^{-1} (oven-dry soil)				
Heteren (1) Hayfield Clay	0.21 (43)[a]	0.14 (18)	20.8 (13)	0.70 (22)	1.50
Achterberg (2) Hayfield Sand	0.08 (43)	0.06 (24)	18.3 (9)	0.35 (26)	1.33
Heteren (3) Path Sandy clay	0.15 (62)	0.11 (27)	19.3 (33)	0.65 (38)	1.36
De Bruuk (4) Path Peaty loam	0.67 (47)	0.20 (26)	74.2 (15)	0.27 (17)	3.35

[a] Coefficient of variation.

tion" were higher by a factor 2 at the two Heteren sites, in agreement with other results (Fig. 3.8).

3.2.5 Conclusions

The main site characteristics for the five *Plantago* species are summarized in Table 3.12. It is hard to arrive at any new soil characteristics from the present general survey of *Plantago* sites and the seasonal data of the selected *Plantago* sites except for the well-known differences between the habitats of these species as pH, carbonate, salinity and soil texture. *P. lanceolata* has the most widespread occurrence in non-saline habitats, and part of the *P. lanceolata* sites coincides with the total pH range of the *P. media* sites, so that both species often occur together. Among the glycophytic species, the occurrence of *P. media* is restricted to sites that are not too low in exchangeable Ca.

There are indications that *P. major* is preferably found at relatively nutrient-rich sites (Kruijne et al. 1967; Haeck et al. 1982). The present study indicates a fairly wide ecological amplitude for this species with respect to nutrient levels. The restriction of *P. major* ssp. *major* to relatively open and compacted sites may in an indirect way, via favourable moisture conditions and absence of competitive effects, result in a high percentage of sites with higher levels of N, P and K. It is not necessary to assume actual adaptation to higher nutrient levels. It has been suggested that moisture conditions could be more variable with *P. lanceolata* than with *P. major* (Woldendorp 1983), leading to a less

Table 3.12. Some general soil characteristics (0 – 10-cm layer) of habitats of five *Plantago* species in The Netherlands. The indicated outlines refer to salient features

Species	Preferred pH range	Approximate % of sites investigated containing >0.1% $CaCO_3$	Salinity (chloride, sodium, electrical conductivity)	Soil texture	Organic matter	P	K
P. media	Almost neutral to alkaline	70	(Very) low	Variable, but not too fine	Variable, but in the low range	Low	Low
P. major	Moderately acidic to alkaline	60	(Very) low	Variable	Variable	Relatively high, but wide potential range	Low to moderate
P. lanceolata	Strongly acidic to alkaline	30	(Very) low	Very variable	Very variable	Low	Low
P. cor-onopus	(Moderately acid to) alkaline	75	Moderate	Variable, but not too fine	Variable, but in the low range	Low	Low to moderate
P. maritima	Alkaline	100	(Very) high	Relatively fine	Variable	Relatively high	Moderate

steady nutrient supply in the habitat of the former species. It is difficult to arrive at such a conclusion from the present data on soil moisture content fluctuations and soil texture at the four inland sites. Also, the chemical composition of the plants at the different sites does not lead to such a conclusion. In some aspects there were more similarities than differences between the two species: differences among the four sites indicated soil chemical differences rather than species differences. A single site of *P. major* ssp. *major* (De Bruuk) was relatively low in "available" P, which contrasts with the observations by Kruijne et al. (1967). Although the K status of site 2 (Achterberg, hayfield) could be too low for *P. major* ssp. *major* , it cannot be suggested that K is decisive for the absence of this species at this particular site.

Although *P. maritima* seems to be restricted to sites of high salinity, the species has been reported to occur in non-saline habitats such as in Wales and Austria (Bos 1980; Königshofer 1981). Königshofer (1981) mentions two instances of the combined occurrence of three *Plantago* species, viz. *P. media, P. lanceolata* and *P. maritima.* The occurrence of *P. maritima* together with *P. major* and/or *P. lanceolata* has also been reported for Austrian sites (Königshofer 1981; J.W. Woldendorp pers. commun.). *P. maritima* tolerates high salinity but salinity is not required for optimal growth (Weissenböck 1969; Königshofer 1981; Blom 1983). Königshofer (1981) distinguishes two levels of salinity, high saline and low saline sites, each corresponding with *P. maritima* "forms". At the high saline sites, *P. maritima* plants mainly accumulated sulphate, while at the low saline sites chloride was predominant in the plant.

As to the locations of *P. major* ssp. *major* and *P. major* ssp. *pleiosperma,* competition by neighbouring plants and physical disturbance by flooding and trampling probably are of greater importance than conditions of nutrient supply during the growing season. It may be of interest to gather more data on the actual rooting pattern, the condition of the roots and the relevant rooting depth in such situations. The present study shows that it is difficult to relate differences between the sites of these two species to a single well-defined abiotic soil chemical factor.

3.3 Micrometeorological Characterization of *Plantago* Sites

3.3.1 Introduction

The microclimate which a plant experiences depends not only on the overall climate of the area where the species occurs, but also largely on the surrounding vegetation and the moisture conditions of the soil. For a description of the sites of the *Plantago* species, it was necessary to measure temperature, humidity and solar radiation in the vegetation. This section 3.3. provides general information on the microclimate in grasslands and more specific information on the grassland sites where *Plantago* species occur.

3.3.2 Instruments and Methods

Air temperature and humidity were measured with a miniature psychrometer (Stoutjesdijk 1980). The percentage of transmitted solar radiation was measured using silicon photocells with transparent white plastic discs. To give an average reading, five of these cells, each whith a diameter of 1 cm, were mounted side by side on an oblong plastic strip. The silicon cells were sensitive between 400 and 1 100 nm wavelenght with a maximum sensitivity at 700 nm. Because of their smaller size they were preferred above a caloric instrument of uniform sensitivity.

3.3.3 Results and Discussion

Temperature and saturation deficit, S.D., inside the vegetation at 1 cm height above the soil or litter surface were compared with those of the air, 1 m above the vegetation. This procedure allowed comparison of measurements taken under different conditions of air temperature and humidity. In Fig. 3.10 the temperature difference, Δt, between the air inside the vegetation and at 1 m height above, is plotted against the difference of the vapour pressure deficits, ΔS.D. The measurements, combined in Fig. 3.10, were all made under strong global

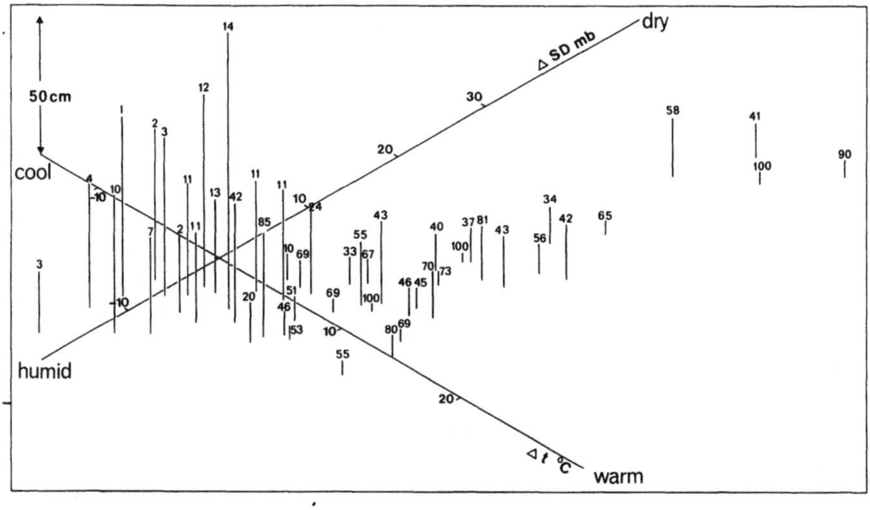

Fig. 3.10. Three-dimensional diagram comparing the differences between temperature (Δt) and saturation deficit (ΔS.D.) in the vegetation with those 1 m above the vegetation for various heights of the vegetation. The *length of the bars* indicates the height of the vegetation and the *numbers at the top of each bar* the percentage of radiation transmitted. Each *number* at the top of a bar represents the mean value of three to five sets of measurements taken in stands with similar values of Δt and ΔS.D. and of similar height and density. Each set of measurements consists of 10 to 20 readings taken on one spot within a few minutes

radiation (600–900 W m^{-2}), at least 24 h after the last rainfall and never after a severe drought. The diagram furthermore shows how Δt and ΔS.D. are related to the height and density of the vegetation.

The diagram can be divided into four quadrants: cool and dry, cool and humid, warm and dry, and warm and humid. Cool means that Δt was negative, i.e. that the air temperature in the vegetation was lower than at 1 m height. Humid means that ΔS.D. was negative, and the saturation deficit in the vegetation was lower than in the air above the vegetation. For warm and dry, respectively, the values of Δt and ΔS.D. were positive.

The combination cool and dry cannot occur in conditions with a strongly positive net radiation. The vapour pressure in the vegetation was always higher than that in the free air above it, and thus the S.D. could only be increased when the temperature was raised, overriding the effect of the high vapour pressure.

The actual data on Δt versus ΔS.D. were arranged in a rather narrow oblong area; there was a good correlation between these two parameters. There was also a good correlation between height and density of the vegetation and the two parameters, Δt and ΔS.D.

In the cool humid quadrant heavily manured production grasslands are present. *Lolium perenne* and *Phleum pratense* are the dominant species. The density of the vegetation was high; usually less than 3% of the sunlight was transmitted by the vegetation. The value of Δt was negative to slightly positive in these grasslands and the ΔS.D. was always strongly negative.

The traditional hayfields were manured less intensively and they had a higher transmission of sunlight (7 to 14%). The temperature was always above ambient air temperature, up to 4 °C; the S.D. was always reduced, often to a considerable extent.

The wet hayfields with an open vegetation structure and low productivity were found in the warm humid sector; Δt values of 5 to 8 °C were observed. The saturation deficit was reduced, that is ΔS.D. became negative, but not as strongly as in the denser grasslands. The measurements were made in May and June. The height of the vegetation was then about 40 cm and the transmission of this inhomogeneous vegetation was between 40 and 80%. It is remarkable the as soon as the moss layer (*Hypnum cupressiforme* or *Sphagnum*) became superficially dry, dry and warm conditions prevailed in this type of grassland. Temperature and humidity close to the surface were determined much more by the structure of the vegetation than by the water content of the soil. Later in summer the vegetation became denser and higher. The temperature and humidity conditons were then similar to those in high and dense grassland.

As expected, the dry poor grasslands were represented in the warm and dry sector. The vapour pressure in the vegetation was always higher than in the free air, but the effect of the higher temperature was overriding, leading to a higher saturation deficit.

The lower part of the dry and warm sector referred mainly to vegetations with species like *Festuca ovina*, *F. rubra*, *Anthoxantum odoratum* and *Luzula campestris.* More important than the botanical composition were the height and density of the vegetation: rather low, open vegetations with a dry soil surface, even though the upper centimeters of the soil might have a high water content.

The highest values of ΔS.D. and Δt were obtained in vegetations on nutrient-poor and sandy soils with little humus and a low water-holding capacity. The dominating species were *Corynephorus canescens, Festuca ovina, F. rubra* and *Agrostis canina.*

Shortly after a rain period the dry grasslands were warm and humid but it did not take more than 24 h of warm sunny weather to make them warm and dry again. For dry grassland on peaty soil the transfer from warm/humid to warm/dry lasted somewhat longer. Under warm and dry conditions Δt reached 13 °C and still ΔS.D. was -4.7 mbar. The diagram with the data on Δt and ΔS.D. within and above the vegetation (Fig. 3.10) was redrawn in Fig. 3.11 in a two-dimensional form; only the line enclosing all data was drawn. A set of 70 series of measurements for which no reliable transmission data were available was also used, since it did not change the general pattern.

Plantago lanceolata (heavy broken line) occurred over almost the whole field of combinations of Δt and ΔS.D., except for the driest and hottest part of the range; it was also absent in those grasslands that reach a high density and cool humid conditions in May and June. Open and wet grassland in which *P. lanceolata* was found may reach a high density late in summer with cool and humid conditions. The wide amplitude of Δt and ΔS.D. for *P. lanceolata* was expected. *P. lanceolata* has a wide distribution; all over Europe except for the extreme north and eastward it is a regular element of the Pontic steppe flora (Walter 1968).

Similar curves encircling the data for sites with *P. major, P. media* and *P. coronopus* were made as well. Compared with *P. lanceolata,* the range of conditions of *P. major* (thin broken line) sites seemed to be clearly reduced. *P. major* was found on trampled sites; even when the soil surface is bare and dry, overheating may not be expected to such a large extent as with *P. lanceolata* sites.

Plantage media (thin drawn line) is a less frequent species in the area of investigation. The properties of the grasslands in which it was found were rather different, viz. hayfield and grazed meadow. Although its geographical range is nearly as wide as that of *P. lanceolata* (but probably not extending as far north), it has a narrow distributional range in the Netherlands. This was not reflected in the climatic characteristics of the sites concerned here. *P. media* had a wide distribution before intensive agricultural methods were used (Goethart and Jongmans 1902, 1906).

Plantage coronopus (dotted line) is a typical coastal dune species. It was found on only one single inland site in the area studied. The data for this site are shown since they may represent rather typical climatic conditions of sites of *P. coronopus* on the seacoast as well (Noë and Blom 1982). On the slightly trampled sites with open vegetation on a sandy soil, moderately warm and dry conditions are expected. The distribution of *P. coronopus* is more limited to southern regions that that of the other species; it only reaches to southern Scandinavia and it is widely distributed in the Mediterranean area. *P. coronopus* var. *crassipes* is found in the Negev dessert.

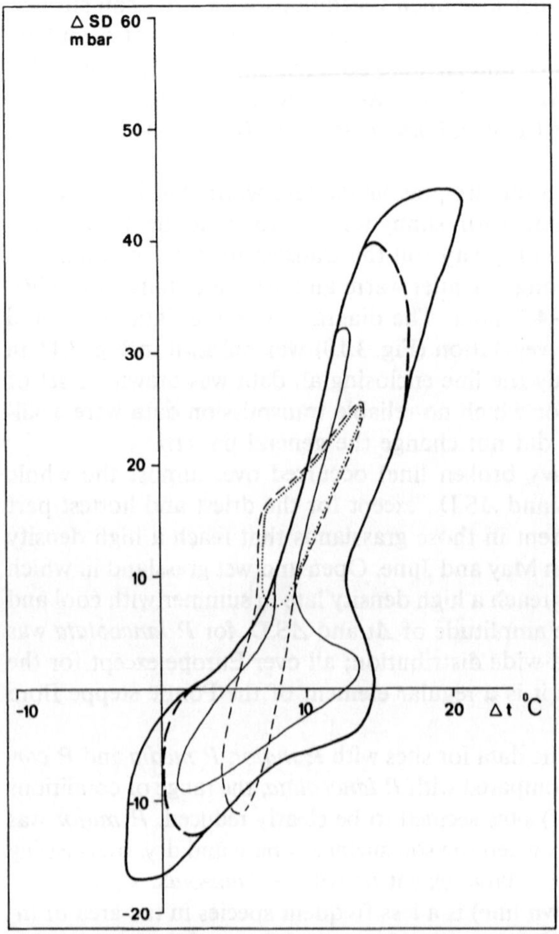

Fig. 3.11. Enveloping line for the measurements of Fig. 3.10 (*heavily drawn*). Additional measurements outside this range were included. The range of *P. lanceolata* is shown by a *heavy broken line*. The range of *P. major* is shown by a *thin broken line*, of *P. media* by a *thin drawn line*, and of *P. coronopus* by a *dotted line*

3.4. Interactions with Organisms Other than Plants

3.4.1 Introduction

The interactions with organisms other than plants are very diverse. They will be arranged according to taxonomical groups as microorganisms (including nematodes), and animals (invertebrateas and vertebrates, including man). General lists of the organisms interacting with *Plantago* species have been given by Dodds (1953) for *P. coronopus* and by Sagar and Harper (1964) for *P. major, P. media,* and *P. lanceolata* and by De Nooij (1987a) for *P. lanceolata.*

Detailed studies of the interaction of specific organisms with populations of *Plantago* plants have been published: on the influence of trampling by mammals on the occurrence of some species (Blom 1979); on the interaction between *P. lanceolata* and the fungus *Fusarium moniliforme* var. *subglutinans* (Alexander 1982); and on the plant *P. lanceolata*–fungus *Phomopsis subordinaria*–weevil *Ceutorrhynchidius troglodytes* interaction (De Nooij 1987a).

3.4.2 Interactions with Microorganisms

Several authors have formulated hypotheses about the impact of microorganisms, especially plant pathogens, on natural populations (Harper 1977; Newman 1978; Burdon and Shattock 1980; Burdon 1982, 1987). In this section microorganisms of *Plantago* species are listed and discussed according to their relevance for the population biology of the host species.

Viruses and Mycoplasma-Like Organisms. Hammond (1982) reviewed the occurence of viruses in *Plantago* species. Plantains are infected by a wide range of viruses (Table 3.13) from at least 19 viral groups. The potential importance of *Plantago* species as reservoirs of economically important viruses is greater than that of many other weeds, because many plantains are perennials, and thus can act as sources of viruses between annual crop plantings (Linn 1940). In many cases the infections are systemic and symptomless, or produce very mild and indistinct symptoms. Many viruses are readily mechanically transmissible, but in natural situations, transmission may be most frequently via insects (Kloeper and Garrot 1981), nematodes (Rush 1970) or fungal vectors (Tomlinson and Garrett 1964).

Viruses like Tobacco Mosaic Virus and Cucumber Mosaic Virus are found worldwide, largely because of host and host-habitat diversity. Conversely, some viruses that may affect plantains are limited in their host ranges, and appear to be closely adapted to their natural hosts. Such specialized relationships tend to be geographically restricted and are more commensal than parasitic in nature (Hammond 1982).

Table 3.13. Numbers of viruses isolated from *Plantago* species. (After Hammond 1982)

Species	No. of viruses	Species	No. of viruses
P. arenaria	1	*P. major*	28
P. asiatica	1	*P. media*	3
P. coronopus	2	*P. psyllium*	5
P. depressa	1	*P. racemosa*	1
P. erecta	2	*P. rugelii*	3
P. indica	2	*P. sempervirens*	5
P. insularis	1	*P. serpentina*	1
P. lanceolata	35	*P. virginica*	2

The frequency of infection of plantains and the frequent lack of obvious detrimental effects of the plants suggest that plantains have developed a high tolerance to most endemic viruses. This suggests a long association with those viruses and their vectors (Holmes 1954; Matthews 1970).

Bacteria. Apart from studies on free-living bacteria in the soil environment of plantains (Sect. 3.2), few records on bacteria influencing *Plantago* species are available. Christie et al. (1978) found that the abundance of root-surface bacteria on *P. lanceolata* increased in mixed culture with other grassland species, compared to monoculture. Bacterial abundance was positively correlated with nitrogen content in the plant. No evidence was given about the types of bacteria in these observations.

USDA (1960) reports a leaf spot on *P. major* caused by *Xanthomonas plantaginis* Burth. Thornberry and Anderson (1937) isolated *Phytomonas plantaginis* from brown leaf spots on *P. lanceolata.*

Fungi. Mycologists have isolated many fungi from wild plants from a mycosystematical point of view. Fungi isolated from plantains are listed in Table 3.14.

Mycorrhizal Fungi. Fries (1944) reported the occurrence of mycorrhizas on the root systems of *P. lanceolata, P. coronopus* and *P. maritima,* without reference to any ecological relevance of his observations. Bloss (1982) found that *P. ovata,* inoculated with *Glomus fasciculatus, G. macrocarpus* and *G. mossae,* produced higher shoots and larger seeds than uninfested plants.

Newman et al. (1981) investigated the relationship between mycorrhizal abundance on roots of *P. lanceolata* and the properties of soil and vegetation at the sampling sites. Mycorrhizal abundance was negatively correlated to the extent of the vascular plant cover at the sampling site and it was positively correlated to leaf length of the host plant. These factors probably arise from indirect, rather than direct relationships: leaf lenght was negatively correlated with the quantum flux density at the plant base, while mycorrhizal infection tended to be greater in shaded plants.

Christie et al. (1978) studied the impact of neighbouring plants on each others' endomycorrhizas. Mycorrhizal infection of *P. lanceolata* was lower in mixed culture than in monoculture. Newman et al. (1981) concluded that this negative effect of other species on the reduction of mycorrhizal abundance on *P. lanceolata* was general and not specific for grasses. Transport of nutrients between plants was stimulated by mycorrhizas: not only are there hyphal connections between roots of *P. lanceolata* and *Lolium perenne,* but the mycorrhizas also increased phosphorus (^{32}P) transport between these species (Heap and Newman 1979, 1980). Whittingham and Read (1982) demonstrated that differences in ^{32}P transport between mycorrhizal and non-mycorrhizal *P. lanceolata* and *Festuca ovina* plants were significant 48 h after application of ^{32}P to the donor plant. Phosphorus can leak from the donor root into the soil solution, be taken up by the hyphe growing from the sink plant and then be translocated

Table 3.14. Fungi isolated from *Plantago* species[a]

Fungus species	*Plantago* species					Reference
	cor	lan	maj	mar	med	
Aecidium molluginis Wurth.	–	l, ru	–	–	–	Gäumann (1959)
Ascochyta plantaginicola Melnik	–	l	l	–	–	Melnik (1977)
Cerospora plantaginis Sacc.	–	l, n	l, n	–	l, n	Chupp (1953)
Cladosporium cladosporioides (Fries.) De Vries	–	sd	–	–	–	Cavers et al. (1980)
Coniothyrium scapidesum Sacc. et Speg.	–	st	–	–	–	Bestagno et al. (1958)
Erysiphe sordida Junell	–	l, m	l, m	l, m	l, m	Braun (1987)
Fusarium moniliforme var. *subglutinans*	–	e	–	–	–	Alexander et al. (1984)
Leveillula taurica (Lév.) Arnaud	–	l, n	–	–	–	Braun (1987)
Mycosphaerella plantaginis (Sollm.) Vest.	–	st	st	–	st	Tomilin (1979)
Nodulosphaeria modesta (Desm.) Munk	–	st	–	–	–	Holm (1957)
Olpidium brassicae	–	–	r	–	–	Keyworth (1962)
Peronospora alta Fuckel	–	l	l	–	l	Francis (1981)
Phomopsis subordinaria (Desm.) Trav.	–	st	st	–	st	De Nooij and Van der Aa (1987); Grove (1935)
Phyllosticta plantaginis Sacc.	–	–	l	–	–	Grove (1935)
Physarum cinereum (Batsch.) Pers.	–	l, st	l, st	–	–	Nannenga-Bremenkamp (1974)
Pleospora herbarum (Fr.) Rab.	–	–	st, e	st, e	–	Wehmeyer (1961)
Puccinia cynodontis Desm.	l, ru	l, ru	l, ru	–	l, ru	Gäumann (1959)
Puccinia sesleriae Reichardt	–	l, ru	–	–	–	Gäumann (1959)
Pyrenopeziza plantaginis Fuck.	–	l	–	–	–	Hütter (1958)
Ramularia plantaginea Sacc. et Berl.	–	l, n	l, n	–	–	Rabenhorst (1907)
Rhizoctonia sp.	–	r	–	–	–	Campbell et al. (1982)
Sclerotinia trifoliorum Eriks.	–	r	r	–	r	Frandsen (1946)
Septoria plantaginea Passer	–	l	l	–	–	Grove (1935)
Sphaceloma plantaginis Jenk. et Bit.	–	l	l	–	–	Jehle et al. (1947)
Sphaeroteca plantaginis (Cast.) Junell	l, m	l, m	l, m	–	l, m	Braun (1987)
Synchitrium aureum Schröt.	–	l, g	l, g	–	–	Oudemans (1923)
Synchitrium plantagineum Sacc. et Speg.	–	l, g	–	–	–	Karling (1961)
Thanathephorus langlei-regis Reid	–	l, n	–	–	–	Reid (1970)
Tilletia glomerulata Coc. et Mor.	–	l, sm	–	–	–	Duran and Fisher (1961)
Verticillium albo-atrum Reinke et Berth.	–	r	r	–	–	Heale and Isaac (1963)

[a] e = ear, g = gall, l = leaf, m = mildew, n = necrosis, r = root, ru = rust, sd = seed, sm = smut, st = stem

into the plant. Another possibility is that phosphorus passes into the fungus colonizing the source root and is then translocated through interconnecting hyphae to the sink plant, without entering the soil solution. Whittingham and Read (1982) suggest that the second mechanism is of greater importance than the first. It is concluded that mycorrhizal infection can stimulate plant growth (*P. ovata)* and that it plays a role in the interaction between neighbouring plants. Infection of *P. major* with VAM fungi lead to increased P concentrations in roots and shoots, and increased fungal respiration. Cytokinins were not involved in VAM stimulated growth (Baas 1989, Baas and Kuiper 1989, Baas and Lambers 1988, Baas et al. 1989).

Parasitic Fungi. Many of the fungi listed in Table 3.14 do not cause severe symptoms and the population biological relevance of these "diseases" is not clear. Alexander (1984) and Alexander et al. (1984) studied in populations of *P. lanceolata* the impact of an inflorescence disease caused by *Fusarium moniliforme* var. *subglutinans.* With increasing host density the proportion of diseased plants remained constant or decreased. Diseased plants were more likely to be male sterile. Half-sib families, planted in the field and derived from genotypes with a history of disease, were slightly more likely to become diseased than plants from healthy genotypes. As the fungus reduced seed production, variation in incidence of disease among different genotypes may be of evolutionary importance to the host population.

De Nooij (1987a) studied an ear stalk disease on *P. lanceolata,* caused by *Phomopsis subordinaria* (Desm.) Trav. Weevils transport the fungus into the stalk in or just below the ear (De Nooij 1988). After injection, the necrosis devélops downward. Depending on the flowering stage at the moment of infection, seed set is partly or totally reduced (De Nooij and van der Aa 1987).

Both studies on diseases in *P. lanceolata* consider the generative phase of the plant. The impact of soil-borne disease on seeds, seedlings and vegetative growth are of minor importance. Two fungal species, *Fusarium oxysporum* and *Gliocladium roseum,* could be regularly isolated from mouldy *P. lanceolata* seeds in grassland soil (De Nooij and Van Gulik 1984). No variation in responce to inoculation with both fungi apart or in a mixed inoculum could be shown among and between populations, between seeds of different sizes and between seeds of different ages. Therefore these fungi should be regarded as saprophytic colonizers of dead seeds. No impact of soil pathogens on survival of seedlings could be demonstrated (De Nooij 1987b). Pasteurized and unpasteurized soil samples, sampled at different standtypes in several populations of *P. lanceolata,* showed no differences in seedling mortality between *P. lanceolata* seed populations. Seedling survival appeared to be associated with abiotic soil parameters (see also Sect. 3.2.3).

Soils sampled in a *P. lanceolata* population at sites with different densities of the species yielded differences in vegetative growth of *P. lanceolata* plants grown in the greenhouse (De Nooij et al. 1986). These differences were partly eliminated when the soils were sterilized. Soil treatment with fungicides showed no elimination of differences in plant growth. However, the impact of

Table 3.15. Nematodes on *Plantago* species

Species	Nematode	Symptoms[a]
P. lanceolata	*Anguilulina dipsaci*	Leaf galls
	Ditylenchus dipsaci (Kuehn) Filip	Leaf/stem
	Meloidogyne sp.	Root (knots)
	Meloidogyne incognita Kof et White	–
P. major	*Anguilulina dipsaci*	–
	Aphelenchoides ritzemabosi	–
	Ditylenchus destructor Thorne	–
	Hexatylus viviparus Goodey	–
	Meloidogyne sp.	Root (knots)
	Pratylenchus penetrans Cobb.	–
	Pratylenchus pratensis Filipjev	–
P. maritima	*Ditylenchus dipsaci* (Kuehn) Filip	–
P. media	*Anguilulina dipsaci*	–
	Meloidogyne sp.	–
P. rugelii	*Ditylenchus dipsaci* (Kuehn) Filip	–

[a] – means: no symptom mentioned in literature.

(non-fungal) harmful root microorganisms in the rhizosphere was by far overruled by abiotic factors such as the nutritional status of the untreated soils. As minor pathogens (Salt 1979) can be condidered as generalists, not specializing on a single host genotype or species, no strong evolutionary consequences can be expected.

Nematodes. Symptoms on *Plantago* species caused by nematodes are listed in Table 3.15. The nematodes are not specialized to a single host species. All records concern very minor symptoms; *Plantago* species seem to possess a high tolerance against nematodes.

3.4.3 Interactions with Animals

Influences of animals on the *Plantago* plants were often encountered in the field studies. It was not always possible to decide which specific animal was responsible for a certain effect. This was especially the case for herbivory: damage to the plants by herbivores was often seen as holes in leaves or as missing parts of leaves, ears or ear-stalks.

In the pasture habitats such symptoms can be attributed to grazing by e.g. cattle, geese or rabbits. In the hayfields, herbivory may be mainly due to invertebrates such as slugs eating from the leaves; in some cases vertebrates, especially birds, were found to predate on the ears (Van Groenendaal 1985a).

Apart from herbivory, animals may influence the plant populations in other ways, e.g. by causing soil compaction, by trampling, by pollination, by seed dispersal, or by creating habitat microsites favourable for germination.

Molluscs (Slugs and Snails). Molluscs are polyphagous herbivores which do not indiscriminately feed on all species of plants. Mølgaard (Sect. 6.3) determined the relative preference of two mollusc species for a range of food plants: *Plantago* species rank low, which is probably due to the presence of phenolic compounds like caffeic acid in the leaves. The relative preference for a plant is often measured by offering leaf disks to the herbivore under controlled circumstances. After a certain time the consumed leaf area (or dry weight) of the test and reference plants are determined and an acceptability index (A.I.) is calculated as the ratio between these two values (reference plant, A.I. = 1; A.I. ratios for species of *Plantago* are shown in Table 3.16). It is evident that the results differ with the chosen reference species (Richardson and Whittaker 1962) and with the herbivore species. These data support the finding of Mølgaard that the acceptability of *Plantago* species is low in comparison with members of the Compositae (Asteraceae); differences also exist between *Plantago* species. In general, *P. major* is more acceptable to molluscs than *P. lanceolata*: the position of *P. media* is not clear. In experiments by C.G. Bruggemann it was found that differences in acceptability exist between different populations of *P. media, P. major* and *P. lanceolata* (Table 3.16), and that also considerable differences can occur between plants within the same population. Differences between populations of *P. major* ssp. *pleiosperma* in damage to leaves were also found in the field by Mølgaard (Sect. 6.3).

It has been suggested that the influence of molluscs and especially of slugs on *Plantago* populations in the field is great. This suggestion is difficult to prove, because the slugs are seldom caught "red-handed". In a few cases mass mortality of seedlings of *P. lanceolata* was found in the presence of slugs. Seedlings were found to have been bitten through at the base of the stem. This could be the work of slugs, but it cannot be ruled out that this damage is caused by "cutworms" – larvae of noctuid moths – or by other species of nuctuid moths with larvae living on or in the soil. Graveland (pers. commun.) found the same type of damage to seedlings in a fen habitat, where the snail *Succinea putris* (L.) was the most important herbivore early in the season.

In the field, slugs (most often *Agriolimax reticulatus*) were regularly found on rosettes of adult *P. lanceolata*. It is not probable that they have an influence on the survival of these plants, but via defoliation they may influence seed production. An increase of the number of side rosettes often results when the apex of a plant is damaged.

Insects. A list of insect species associated with *Plantago* is given by Sagar and Harper (1964). In our experience the most important insect herbivores are found among the weevils (Coleoptera, Curculionidae) and the Gallmidges (Diptera, Cecidomyidae) to which flea beetles (Coleoptera, Chrysomelidae) can be added (Van Gijzen and Voorhoeve unpubl. data). The important beetles are mentioned in Table 3.17. Most observations were done on *P. lanceolata*. The adults of many species of weevils eat from the leaves, making holes or nibbling along the edges. Some adults puncture the flower stalks (*Ceuthorrhynchidius troglodytes, Mecinus pyraster*) and may thus cause entrance holes for the fungus *Phomopsis subordinaria* (De Nooij 1988). The larvae of *M. pyraster* tunnel the flower stalks of *P. lanceolata* while those of *M. collaris* form a gall

Table 3.16. Relative preference of slugs and snails for *Plantago* species. Mean acceptability index (A.I.), calculated over leaf area, as found in experiments by different authors. When two values are given these refer to different populations tested against the same population of the reference plant

Test species				Reference species[a]	Herbivore	Reference
	P. major	*P. lanceolata*	*P. media*			
	0	0	0	*Lactuca sativa* L.	*Agriolimax caruanae* (Pollonera)	Dirzo (1980)
	0.35	0.09		*Taraxacum officinale* (Weber)	*Agriolimax reticulatus* (Müller)	Richardson and Whittaker (1982)
		0.93		*Plantago major*	*Agriolimax reticulatus*	Richardson and Whittaker (1982)
	19.1			*Plantago lanceolata*	*Agriolimax reticulatus*	Richardson and Whittaker (1982)
	0.12	0.04	0.38	*Hieracium pilosella* L.	*Cepea nemoralis* (L.)	Grime et al. (1968)
ssp. *major*	ssp. *pleiosperma*					
0.27**		0.28**		*Taraxacum officinale*	*Agriolimax reticulatus*	C.G. Bruggemann (unpubl.)
0.59**				*Taraxacum officinale*	*Arion rufus* (L.)	C.G. Bruggemann (unpubl.)
		0.35**		*Crepis capillaris* (L.) (Wallr.)	*Agriolimax reticulatus*	C.G. Bruggemann (unpubl.)
	2.17*	0.52, 1.91	0.54	*Plantago major m.* (Site B)	*Agriolimax reticulatus*	C.G. Bruggemann (unpubl.)
	2.25**, 1.22	0.06**	0.89	*Plantago major m.* (Site B)	*Arion rufus*	C.G. Bruggemann (unpubl.)
1.97**		2.36**, 1.50		*Plantago lanceol.* (Site A)	*Arion rufus*	C.G. Bruggemann (unpubl.)

[a] All reference species other than *Plantago* belong to the family Compositae.
* $P < 0.05$.
** $P < 0.01$.

Table 3.17. Insects commonly found as herbivores on *Plantago* species. (T. van Gijzen, W. Voorhoeve, unpubl.)

	Species of *Plantago*					Plant part attacked		
	major	*lanc.*	*media*	*coron.*	*marit.*	Ear	Ear stalk	Leaves
Coleoptera Curculionidae								
Ceuthorhynchidius troglodytes[a] (F.)	·	+	·	·	×	l	a	a
C. rufulus[a] Dufour	·	+	·	·	×	l?	·	a
C. thalhammeri[a] Schultz	·	·	·	·	×	l?	·	a
Mecinus pyraster[a] (Herbst.)	·	+	×	·	·	·	l	a
M. collaris[b] Germ.	×	×	×	×	+	·	l	a
Gymnaetron labile[a] (Herbst.)	·	+	×	·	·	l	·	·
G. pascuorum[a] (Gyll.)	·	+	×	·	·	l	·	·
Philopedon plagiatus (Schall.)	·	+	+	·	×	·	·	al
Hypera plantaginis (De G.)	·	+	+	·	·	·	·	al
Coleoptera Chrysomelidae								
Longitarsus pratensis (Panz.)	+	+	×	·	×	·	·	al
L. melanocephalus (Deg.)	+	+	·	·	·	·	·	a
Diptera Agromyzidae								
Phytomyza plantaginicaulis[a] (Hg.)	·	+	·	·	+	·	l	·
Diptera Cecidomyidae								
Dasyneura schmidtii[a] (Rübs.)	×	+	×	·	·	l	·	·
Hemiptera Heteroptera								
Adelphocoris lineolatus (Goeze.)	·	+	·	·	·	al	·	·
Plagiognathus chrysanthemi (Wolff.)	·	+	·	·	·	al	·	·

[a] Monophagous on the genus, a: adult; l: larva; x: data from the literature; +: data corroborated with own observations

in the flower stalks of *P. maritima* (Doctors van Leeuwen 1984). The larvae of some species of weevils live in the ears, predominantly eating the young seeds. The most abundant is *C. troglodytes*, the larvae being occasionally found in 20% of the ears; thus seed output of *P. lanceolata* populations is affected.

Another important predator of seeds of *P. lanceolata* is the gallmidge *Dasyneura schmidtii* (Rübs.), the larvae being present in large numbers in nearly all ears of populations of *P. lanceolata*. It is plausible that a proportion of "bad" seeds found in demographic studies had been partly eaten by larvae of this species; the seeds afterwards shrivelled.

Several species of insects can be found on flowering ears of *Plantago* plants. Most of these eat pollen, such as, e.g., the beetles *Oedemera lurida* (Oedemeridae) and *Leptura livida* (Cerambicidae). Abundant among the pollen-eating visitors are Syrphid flies (*Melanostoma* and *Platycheirus*). These flies visit *P. lanceolata, P. media* and *P. major*, especially in the period July to October. Stelleman (1982) has shown that these species affect pollination in *P. lanceolata*; they are probably important in habitats sheltered from the wind. In windy places pollination by wind is far more important.

Vertebrates. Vertebrates can influence *Plantago* plants in various ways. Herbivory can be very important. However, few observations exist on populations of *Plantago*. Mammals are the most important herbivorous vertebrates in this respect, but also grazing by geese can be important (e.g. Prins et al. 1980), although Van Groenendael (1985a, b) mentions important predation of ears of *P. lanceolata* in a wet hayfield by birds as wood pigeons, ducks and pheasants. In a dune grassland grazed by cattle, horse and rabbits, he found damage to the rosettes of *P. lanceolata,* especially the larger ones. The damage to prostrate ears (dune grassland) was less than to upright ears (hayfield). Further information on ecotypic variation of *P. lanceolata* in relation to grazing and mowing is given in Sections 4.1 and 7.3.

P. lanceolata seems to be palatable to sheep; they eat whole rosettes, biting them off with their teeth near the ground (Milton 1933, cited in Sagar and Harper 1964). Cows, which grasp their food with the tongue, cannot do this. Both cattle and sheep remove spikes of *P. lanceolata* and of *P. media,* but they seem to ignore spikes of *P. major* (Sagar and Harper 1964). About half of the ingested seeds of *P. lanceolata* can still germinate after passage through a cow's digestive tract (Petersen 1910, cited in Sagar and Harper 1964). This may be important for gene flow.

Haymaking by man is also a kind of herbivory. The plants themselves are able to survive the removal of most of their leaves, but part of the seeds may be removed, as in *P. media* which flowers late. In *P. lanceolata* most seeds are ripe at the moment of haymaking and they fall out of the ears during haymaking; this process assists in dispersal of seeds over the field. In *P. lanceolata* genotypic differences exist between populations in the onset of flowering in relation to mowing time (Van der Toorn, Sect. 7.3).

An important side effect of grazing by domestic mammals is trampling of the vegetation and soil compaction. Trampling by man on paths has similar effects. This can have profound influences on the local distribution of *Plantago* species (Blom 1979 and Sect. 4.2). Effects contrary to trampling are caused by burrowing activities of animals as moles and voles, through which gaps with a loose soil can be created in dense vegetations, which are favourable for germination (see Sect. 4.2).

3.4.4 Population Dynamics of Biotic Interaction

In the previous sections several examples of interactions between species of the genus *Plantago* and other organisms have been described. The interaction between *P. lanceolata*, the fungus *Phomopsis subordinaria* and the weevil *Ceutorrhynchidius troglodytes* has been studied in more detail (De Nooij 1987a). The intensity of *Phomopsis* disease in the field is influenced by four different types of interaction: plant-fungus, plant-insect, fungus-insect, and plant-fungus-insect. The population dynamics of the interaction between the three partners of the *P. lanceolata-Ph. subordinaria-C. troglodytes* system has been investigated (De Nooij and Van Damme 1988a, b; De Nooij 1988).

Susceptibility within and among *P. lanceolata* populations together with pathogenicity within and among *Ph. subordinaria* populations has been determined. Three populations of *P. lanceolata* that differed in intensity of disease were studied in the field. The sites were characterized as follows.

- Road side: ruderal vegetation type, intensity of disease amounted to 80% (plants with at least one diseased stalk); location Baarn.
- Verges of playing fields: ruderal vegetation type, intensity of disease amounted to 60%; location Papendal.
- Dune grassland: extensively grazed by cattle, intensity of disease amounted to 1 to 2%; location Westduinen.

At each of the three sites, ten *P. lanceolata* plants were sampled along a transect at distances of 10 m. Each plant was cloned into three daughter plants. Five stalks of each plant were inoculated with a suspension of conidia of the same isolate of the pathogen. To determine susceptibility, the size of the lesions (which was correlated with a decreased seed production; De Nooij and Van der Aa 1987) was used as the main characteristic. The size of the lesions varied from genotype to genotype (Table 3.18). There was no difference in mean size of the lesions among the three populations; significant variation was restricted within the populations. The Vg/Vp ratio, or "clonal repeatability" (Falconer 1981) was 0.68; a maximum of 68% of the total variation could be attributed to genetic differences. Fungal growth into rosettes and death of side rosettes, another characteristic for susceptibility, was not correlated with lesion size (De Nooij and Van Damme 1988a). This suggests that the genetics of the different components of susceptibility may be determined independently. The fact that susceptibility to *Phomopsis* disease is expressed in several non-correlating characteristics implies that estimation of the fitness of a genotype under pathogen pressure can only be achieved by field trials, in which the plants are exposed to the complete set of disease-inducing factors.

Table 3.18. Variation in size of lesions caused by the fungus *Phomopsis subordinaria* (isolate 3B), among populations and among genotypes within populations of *Plantago lanceolata*, expressed as mm necrosis 4 weeks after inoculation. Each entry is the mean of 15 replications. (De Nooij and Van Damme 1988a)

Population Baarn		Population Papendal		Population Westduinen	
Genotype	Necrosis	Genotype	Necrosis	Genotype	Necrosis
B11	193	P6	194	W15	203
B15	197	P1	198	W2	206
B14	207	P12	208	W4	206
B17	211	P9	210	W1	211
B4	221	P62	211	W7	215
B20	228	P63	212	W9	219
B19	238	P66	214	W8	226
B2	242	P39	216	W11	230
B16	245	P35	223	W12	230
B13	246	P64	227	W3	237
Population mean	222		211		218
C.V.	0.11		0.11		0.10

	MS	df	F	P
Among populations	0.00824	2/27	0.75	>0.25 NS
Among genotypes within populations	0.01102	27/55	7.25	<0.001 ***
Among slips within genotypes	0.00152	55/266	0.62	>0.75 NS
Error	0.00245			

C.V. = coefficient of variation. For statistical analysis data were transformed with log (x+1).
*** $P<0.001$

Besides the variation in susceptibility within the host, the intensity of disease in the field was also expected to be influenced by variation in pathogenicity within the fungus (De Nooij and Van Damme 1988b). In each of the three populations strains of the pathogen were sampled by isolating conidia from 12 diseases stalks, collected at distances of at least 5 m. Development of the 36 pathogen isolates was followed in three different genotypes of *P. lanceolata*, which were selected according to degree of susceptibility, in the above experiment (Table 3.18).

The isolates of the pathogen varied within and among populations (Table 3.19). At the population level, the isolates from Baarn caused larger lesions than those from the other two locations. This cannot explain the differences in intensity of disease in the field between the three populations: the isolates from Westduinen showed a relatively high pathogenicity, while the intensity of disease in the field only was 1 to 2%, compared with 60 to 80% at the other two locations.

Occurrence and intensity of disease was also expected to be determined by attractiveness of the host genotypes to the weevils, which transmit disease.

Table 3.19. Variation in lesion size among isolates of *Phomopsis subordinaria* and genotypes of *Plantago lanceolata*, expressed as the length (mm) of the necrosis in the scape 4 weeks after inoculation, F values from analysis of variance. Each entry is the mean of ten replications. (De Nooij and Van Damme 1988b)

Fungus population	Baarn				Papendal				Westduinen			
	Isolate	Host genotype			Isolate	Host genotype			Isolate	Host genotype		
		B11	B16	P35		B11	B16	P35		B11	B16	P35
	2A	205	267	237	3B	182	224	214	7D	205	247	242
	2E	225	258	219	3C	200	252	234	7E	206	265	240
	2F	176	293	213	3D	214	232	229	7F	196	230	198
	2G	229	282	283	3E	191	236	201	7G	227	252	216
	2H	207	249	242	3F	198	207	141	7H	198	229	198
	2I	210	266	247	3G	216	247	242	7I	218	263	227
	2K	234	268	233	3H	180	248	206	7J	223	253	221
	2L	220	246	272	3I	229	263	197	7K	223	263	265
	2M	212	286	189	3J	229	240	237	7L	222	267	227
	2N	230	314	202	3K	220	210	198	7M	216	237	235
	2O	205	267	213	3L	206	255	239	7N	215	253	226
	2P	224	259	201	3M	214	252	243	7O	187	226	200
Average per												
– Host genotype:		215	271	229		207	238	215		211	249	225
– Fungus population:			238				220				228	

	F	df
Plant genotype	155.81***	2/972
Fungus isolate		
Within populations	7.70***	35/972
Between populations	6.35**	2/99
Interaction		
Genotype × isolate within populations	3.47***	70/972
Genotype × isolate between populations	0.93	4/99

** $P<0.01$; *** $P<0.001$.

These interrelationships between plant, fungus, and insect were studied in a garden experiment; two host genotypes, differing in susceptibility to the fungus, were grown in cages together with weevils and with two isolates of the pathogen which differed in pathogenicity (De Nooij 1988). The development of the disease in the highly susceptible host genotype was not faster than in the moderately susceptible genotype (Fig. 3.12; Table 3.20). However, there was more disease in the cages with the highly pathogenic *Phomopsis* isolate than in cages with the moderately pathogenic isolate. When exposed to both the pathogen and the weevil, the reduction in seed production did not differ for the 2 genotypes (Table 3.21): the susceptible host suffered as much (or less) seed reduction as the less susceptible genotype.

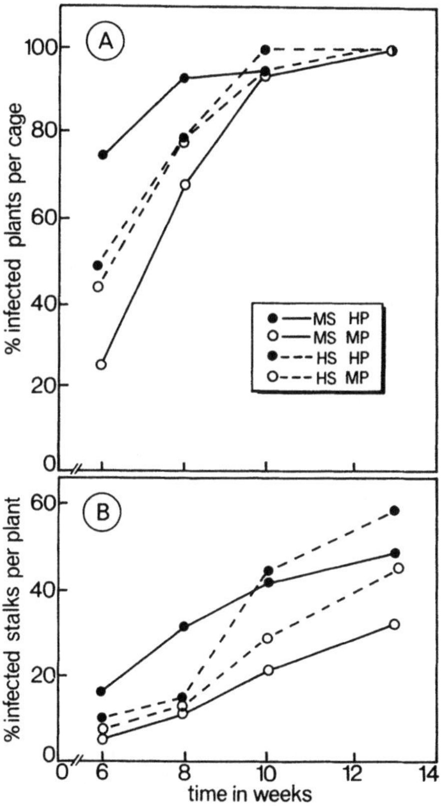

Fig. 3.12. *Phomopsis* infection in a relatively highly (*HS*) and moderately (*MS*) susceptible genotype of *Plantago lanceolata* under pathogen pressure of a highly (*HP*) or moderately (*MP*) pathogenic isolate of *Phomopsis subordinaria* in cages where weevils were responsible for pathogen transmission. **A** Percentage infected plants per cage. **B** Percentage infected stalks per plant. (De Nooij 1988)

The effect of the plant on the weevil and vice versa was primarily expressed in the feeding and oviposition behaviour of the weevils (Tables 3.22 and 3.23). The more susceptible host (to the fungus) was more suitable for the weevil. Susceptibility to the pathogen ran parallel to susceptibility to the weevil and the effect of the pathogen on the susceptible host was intensified.

The impact of the fungus on the weevil and vice versa was diverse: in host plants infected by *Ph. subordinaria* the weevils could die in the larval stage, because the ears died and did not contain enough food. There was less food available for weevils in the cages with the highly pathogenic isolate (Table 3.22), and more adults of the new generation were caught in cages with the moderately than in those with the highly pathogenic isolate. This suggests less impact on weevil behaviour from the moderately pathogenic *Phomopsis* isolate. A direct effect of isolate on weevil survival could not be demonstrated in vivo. The weevil transmits "its own enemy". The pathogen benefits by its association with the weevil; without this vector it would not be able to utilize *P. lanceolata.*

Co-adaptation between the three partners in this system may be expected when changes in the genetic variation in a characteristic of one of the partners

Table 3.20. Analysis of data from Fig. 3.12 (De Nooij 1988)

Plant	Fungus	% Infected plants per cage		% Infected stalks per plant	
		Y-intercept	Slope	Y-intercept	Slope
MS	HP	60.9	3.2	−19.0	6.1
MS	MP	−25.5	10.5	−19.3	4.0
HS	HP	16.3	7.1	−48.6	8.9
HS	MP	8.1	7.7	−30.9	5.8
$F_{y\text{-intercept}}$		1.35 NS		75.90***[a]	
F_{slope}		1.49 NS		6.96*	
F plant genotype		0.04 NS		0.01 NS	
F fungus isolate		2.61 NS		51.80***	
F time		17.32***		94.17***	
F plant × fungus		0.50 NS		6.71**	
F plant × time		0.18 NS		3.46*	
F fungus × time		1.12 NS		1.14 NS	
P plant × fungus × time		0.85 NS		0.79 NS	

[a] The F value is based on the comparison of the line HS, MP and MS, HP which do not differ significantly in slope.
* $P<0.05$, ** $P<0.01$, *** $P<0.001$.

Table 3.21. Ear parameters (per plant) of a highly (HS) and moderately (MS) susceptible genotype of *Plantago lanceolata* grown in cages under pathogen pressure of a highly (HP) or moderately (MP) pathogenic isolate of *Phomopsis subordinaria*, transmitted into the host by weevils, and of control plants (C) grown in cages without pathogen and weevils. Data have been transformed with log (x + 1) before analysis. Number of plants observed = 108. (De Nooij 1988)

Parameter	MS			HS		
	HP	MP	C	HP	MP	C
Mean ear weight (mg)	65	83	92	132	152	182
Mean ear length (mm)	16	18	20	34	37	44
Number of viable seeds per plant	69	141	444	288	286	957
Seed weight viable seeds (mg)	1.04	1.05	1.10	1.36	1.34	1.26

	F_{plant}	F_{fungus}		$F_{interaction}$	
		HP + MP vs. C	HP vs. MP	HP + MP vs. C	HP vs. MP
Mean ear weight	158.41***	17.91***	8.81**	0.06 NS	0.63 NS
Mean ear length	329.74***	19.17***	4.06*	0.43 NS	0.09 NS
Number of viable seeds per plant	34.02***	38.65***	1.95 NS	0.46 NS	2.04 NS
Seed weigth viable seeds	60.87***	0.15 NS	0.01 NS	4.17*	0.12 NS

* $P<0.05$, ** $P<0.01$, *** $P<0.001$.

Table 3.22. The mean number of feeding wounds per stalk caused by *Ceutorhynchidius troglodytes* in the 2 cm just below the ear in a highly (HS) or moderately (MS) fungus-susceptible genotype of *Plantago lanceolata* in cages under pathogen pressure of a highly (HP) or moderately (MP) pathogenic isolate of *Phomopsis subordinaria*. (De Nooij 1988)

Plant genotype	Fungus isolate	*Phomopsis*-infected stalks	Healthy stalks
MS	HP	0.65	0.60
MS	MP	0.80	1.00
HS	HP	0.65	1.62
HS	MP	0.84	2.22
F plant genotyp			11.93***
F fungus isolate			4.21*
F infection			14.19***
F plant × fungus			0.14 NS
F plant × infection			11.16***
F fungus × infection			1.00 NS
F plant × fungus × infection			0.07 NS
Error MS			0.98

*$P<0.05$, ***$P<0.001$.
n = 18 (the mean of the infected and the healthy stalks of each plant in the six cages with the pathogen isolates).

Table 3.23. Correlation (r) between the number of weevil feeding wounds in healthy stalks of *Plantago lanceolata* in the 2 cm just below the ear and some ear parameters in the host genotypes MS (moderately susceptible of *Phomopsis subordinaria*) and HS (highly susceptible)

Parameter	MS		HS	
	r	n	r	n
Ear weight	−0.09	194	−0.12	244
Ear length	−0.07	194	−0.19**	244
% Aborted seeds	0.02	162	0.15*	234
Number of viable seeds	−0.12	194	−0.14*	244
Seed weight viable seeds	−0.15*	194	−0.11	244

*$P<0.05$, **$P<0.01$.
n = Number of ears observed.

influence the fitness of the other partners. Which characteristics are important in this tripartite system? The contribution of descendants to subsequent generations of the plant is determined by the size of lesions caused by the pathogen, the ability of the pathogen to spread from one rosette into other rosettes of the same individual (De Nooij and Van Damme 1988a), the number of ears (De Nooij 1987a), and the attractiveness of the plant to the weevil for feeding and oviposition (De Nooij 1988). For the pathogen it is determined by its rate of growth and reproduction in the host (De Nooij and Van Damme 1988b), and by its attractiveness and/or its capacity to get affixed to the weevils that transmit the pathogen and inoculate new stalks (De Nooij 1988). For the weevil it is determined by the number of eggs laid in the host, the mortality of eggs and/or larvae in diseased scapes of the host, and the capacity to avoid transmission of the pathogen to healthy scapes where the weevil may lay its eggs (De Nooij 1988). In each of the organisms, the characteristics involved are rather diverse and it is likely that the genetics of these features are determined independently of each other.

4 Ecology of *Plantago* Populations

4.1 The Demographic Structure of Populations

4.1.1 Introduction

The performance of plants in the field differs due to the environmental conditions and genetic variation between plants. Demographic studies provide basic material for investigations of the ecophysiological aspects of the adaptation to the environment and of the selection processes which have given rise to adaptation. The demographic work in the project includes a comparison on different levels: between species, between populations of a single species and, in some cases, comparison between spatial subdivisions within a single population. The last mentioned studies have been carried out to determine the scale on which genetic differentiation occurs and will be treated in Chapter 7. They are essential to define accurately a population. Theoretically a population can be defined as a group of individuals which (potentially) interbreed, and which is more or less separated from other groups (Bakker 1964). The results of the detailed studies showed that inhomogeneity in the habitat or variation in the management of the vegetation can cause genetic variation between the local plant groups. When demographic plots are chosen in a relatively homogeneous vegetation, with greatest distances in the order of metres, no genetically determined variation between the plots can be expected. Thus plots with a distance of in most cases less than 10 m (in the two Achterberg sites maximally 25 m) in a visually homogeneous vegetation as used in our demographic studies were regarded as representing a single population.

4.1.2 Comparative Demography Between and Within Species

Comparative demographic studies can be made in different ways. When the intention is to describe differences between species, populations of these species can be studied in the same location. With large differences between species the method can give clear results, but in other cases the influence of the local environment may be dominating, as will be shown for *P. media* and *P. lanceolata*. When the environmental influence is the main concern of the study, a single species can be analyzed in different habitats. These approaches may be combined by studying more than a single species in a range of habitats.

The combination of both methods could not be entirely realized for the *Plantago* species: most of the species occur in habitats that differ considerably from each other. However, *P. lanceolata* often occurs together with one of the other species at the same site; the comparisons in the same habitat were therefore restricted to this species and one of the other species. For *P. media* two sites and for *P. coronopus* and *P. major* ssp. *major* single sites were chosen in which the species occurred together with *P. lanceolata*. Also, a population of *P. major* was studied which occurred adjacently to a population of *P. lanceolata*. Finally, data are available from a second site where *P. coronopus* and *P. lanceolata* were studied together (van der Toorn et al. 1980; van Groenendael 1985a).

The description of sites and habitats is given in Table 4.1. The plants were followed individually in permanent plots of 1/8 m^2. The number of plots varied between the sites because of differences in plant density. The presence of adult plants at the start of the observations in the spring of 1978 and the appearance of seedlings and side rosettes were recorded four times during the seasons of 1978 and 1979. Also, the number of damaged and undamaged leaves and the length and width of the longest leaf, the number of spikes and their length and stage of development and seed production were recorded. The survival of all individual rosettes that were present in the autumn of 1979 was followed during the spring and autumn of later years, in most populations until the spring of 1982. A further comparison between *P. lanceolata* and *P. media* was carried out in three different habitats from 1980 to 1986.

Table 4.1. Description of the habitats, with selected environmental factors

No. and site	Species	No. of demographic plots	Soil type	Management	Vegetation height (cm)
Uddel	*coronopus, lanceolata*	3	Sand	None, lightly trodden	8
Bruuk path	*major* ssp. *major, lanceolata*	6	Peaty loam	Mown once (August), trodden	14
Heteren hayfield	*lanceolata*	8	Clay	Mown once (early July)	67
Heteren path	*major* ssp. *major*	3	Clay	Mown once (July), trodden	30
Westervoort	*media, lanceolata*	8	Sand/clay	None, grazed by rabbits	39
Pannerden	*media, lanceolata*	8	Sandy loam	Lightly grazed by ponies	28
Bruuk hayfield	*lanceolata*	6	Peaty loam	Mown once (August)	60
Achterberg pasture	*lanceolata*	10	Sand	Grazed by cattle	9
Achterberg hayfield	*lanceolata*	15	Sand	Mown once (end June)	26

The demography of *P. maritima* and *P. major* spp. *pleiosperma* was investigated separately.

4.1.3 Comparison of Four Species

The analysis of the relationships between the demographic properties of the populations on the one hand and the environmental factors on the other proved to be complicated because of the strong correlation between the variables within each of the two groups of data. Therefore, to search for the characteristics that give the largest contribution to the differences between the populations, a multivariate statistical method, the principal component analysis, was used. In this analysis the original (correlated) variables were grouped together, forming a new set of independent variables, called principal components. The contribution of each of the original variables to the new variables is known, as well as the relative contribution of each of the principal components to the total variance. The correlation of the principal components with the environmental factors can be calculated (Williamson 1972) as was the case with the original characteristics.

In principle, it should be possible to obtain an estimate for each of the demographic characteristics in each of the plots. This could not be realized in all cases. The number of plants in a certain category of plants (e.g. seedlings of a certain cohort) was often small or nil, and no reliable estimate could be

Table 4.1 (continued)

Light transmission (%)	Soil moisture (% of volume)		% Organic matter	Total nitrogen (ppm)	Total phosphorus (ppm)	Potassium (mEq/100 g)
	Summer	Winter				
64	16	16	5.0	1470	193	0.13
14	40	58	10.1	4100	444	0.25
7	32	39	5.5	2310	907	0.60
18	47	45	4.5	1460	672	0.47
12	25	38	6.6	2660	612	0.42
10	21	33	10.3	4190	878	0.52
5	61	68	12.7	4250	578	0.25
54	28	33	5.9	1660	278	0.10
31	30	34	5.0	1440	235	0.13

obtained for, e.g., survival. The problem could be also met by use of principal component analysis; a single mean value per population is used for each characteristic. The calculations were based on the correlation matrix between the values of the characteristics, which means that the contribution of each variable was independent of the range of its values.

The principal component analysis of species differences has been carried out for the data of the localities where two species occurred together, viz. the five upper sites, mentioned in Table 4.2. For clarity of the argument, data for turnover have been included, but they have not been used in the calculations; they were calculated using data on density and survival that are already represented in the analysis. The principal component axes 1 and 2 explained 32 and 26% of the total variance of the data, respectively, while for axis 3 this was 16% and for the other axes 10% or less. Data on the first two axes are shown in Fig. 4.1, yielding the relative contribution of each of the original characteris-

Table 4.2. Demography of four *Plantago* species. Sites were chosen where *P. lanceolata* alone was present or in combination with another *Plantago* species. Data of *P. lanceolata* are given in brackets

Density adults (no. of adult plants (genets) per m^2 in spring)

	1978		1979	
P. coronopus				
Uddel	803	(284)	72	(239)
P. major				
Bruuk path	129	(58)	146	(79)
Heteren	128	(59)	103	(64)
P. media				
Westervoort	139	(87)	150	(88)
Pannerden	175	(155)	101	(126)
P. lanceolata				
Bruuk hayfield		(311)		(171)
Achterberg pasture		(105)		(167)
Achterberg hayfield		(111)		(146)

Density juveniles (no. of seedlings and juveniles per m^2)

	1978		1979	
P. coronopus				
Uddel	1467	(342)	603	(628)
P. major				
Bruuk path	1656	(360)	986	(21)
Heteren	261	(99)	1275	(54)
P. media				
Westervoort	305	(269)	286	(177)
Pannerden	22	(41)	12	(39)
P. lanceolata				
Bruuk hayfield		(290)		(96)
Achterberg pasture		(174)		(118)
Achterberg hayfield		(71)		(67)

Table 4.2 (continued)

Side rosettes (fraction of plants with side rosettes)

	1978		1979	
P. coronopus				
Uddel	0.04	(0.15)	0.15	(0.47)
P. m. major				
Bruuk path	0.01	(0.12)	0.04	(0.15)
Heteren	0.00	(0.07)	0.00	(0.16)
P. media				
Westervoort	0.01	(0.04)	0.04	(0.05)
Pannerden	0.00	(0.08)	0.00	(0.09)
P. lanceolata				
Bruuk hayfield		(0.01)		(0.35)
Achterberg pasture		(0.08)		(0.10)
Achterberg hayfield		(0.05)		(0.11)

Natality (fraction of adult plants flowering and number of seeds per flowering plant)

	Fraction flowering				Seeds/fl. plant			
	1978		1979		1978		1979	
P. coronopus								
Uddel	0.52	(0.57)	0.57	(0.14)	17	(39)	44	(14)
P. m. major								
Bruuk path	0.48	(0.27)	0.56	(0.10)	408	(39)	523	(50)
Heteren	0.41	(0.17)	0.28	(0.25)	2309	(229)	1908	(166)
P. media								
Westervoort	0.32	(0.22)	0.39	(0.14)	40	(62)	300	(129)
Pannerden	0.08	(0.15)	0.07	(0.11)	36	(32)	75	(49)
P. lanceolata								
Bruuk hayfield		(0.01)		(0.24)		(7)		(38)
Achterberg pasture		(0.19)		(0.21)		(53)		(54)
Achterberg hayfield		(0.30)		(0.23)		(104)		(81)

Survival of genets during summer (April – October)

	Juveniles				Adults			
	1978		1979		1978		1979	
P. coronopus								
Uddel	0.57	(0.71)	0.64	(0.70)	0.77	(0.89)	1.00	(0.92)
P. m. major								
Bruuk path	0.03	(0.16)	0.05	(0.22)	0.85	(0.67)	0.81	(0.54)
Heteren	0.14	(0.06)	0.04	(0.11)	0.85	(0.75)	0.73	(0.56)
P. media								
Westervoort	0.42	(0.24)	0.21	(0.15)	0.94	(0.77)	0.91	(0.80)
Pannerden	0.58	(0.05)	0.09	(0.17)	0.77	(0.90)	0.76	(0.81)
P. lanceolata								
Bruuk hayfield		(0.01)		(0.36)		(0.67)		(0.88)
Achterberg pasture		(0.66)		(0.36)		(0.94)		(0.83)
Achterberg hayfield		(0.45)		(0.61)		(0.96)		(0.93)

Table 4.2 (continued)

Survival of genets during winter (October – April)

	Juveniles				Adults			
	1978/1979		1979/1980		1978/1979		1979/1980	
P. coronopus								
Uddel	0.03	(0.29)	0.73	(0.76)	0.05	(0.64)	0.71	(0.91)
P. m. major								
Bruuk path	0.78	(0.72)	0.21	(0.33)	0.92	(0.75)	0.87	(0.55)
Heteren	0.13	(1.00)	0.00	(1.00)	0.85	(1.00)	0.88	(1.00)
P. media								
Westervoort	0.18	(0.39)	0.63	(0.57)	0.96	(0.78)	0.94	(0.79)
Pannerden	0.58	(0.50)	1.00	(0.57)	0.70	(0.84)	0.92	(0.90)
P. lanceolata								
Bruuk hayfield		(0.63)		(0.10)		(0.79)		(0.83)
Achterberg pasture		(0.64)		(0.63)		(0.91)		(0.93)
Achterberg hayfield		(0.93)		(0.73)		(0.96)		(0.90)

Turnover (fraction of genets that have survived the winter taken up by plants of less than 1 year old)

	1978/1979		1979/1980	
P. coronopus				
Uddel	0.81	(0.33)	0.88	(0.64)
P. major				
Bruuk path	0.30	(0.63)	0.09	(0.06)
Heteren	0.11	(0.31)	0.00	(0.14)
P. media				
Westervoort	0.16	(0.41)	0.32	(0.38)
Pannerden	0.07	(0.28)	0.03	(0.04)
P. lanceolata				
Bruuk hayfield		(0.06)		(0.03)
Achterberg pasture		(0.46)		(0.22)
Achterberg hayfield		(0.30)		(0.22)

tics to the axes (A), and the score of the populations on these axes (B). Axis 1 was positively correlated with the density and the survival of adults and juveniles in the two summers and the percentage of adults that flowered, and negatively with survival in the winter of 1978/1979 (the winter was severe, in contrast to the mild winter of 1979/1980). The other characteristics were more or less indifferent. Axis 2 shows positive correlations with seed production, with percentage flowering and with the density of juveniles. A positive, but weak, correlation was seen with adult survival; negative correlations were found for the survival of juveniles in both seasons and for the percentage of plants with side rosettes.

From Fig. 4.1B it is clear that *P. coronopus* scored highest on axis 1 and *P. m. major* on axis 2; the *P. media* populations scored close to the origin of both axes. The *P. lanceolata* populations showed a considerable variation on

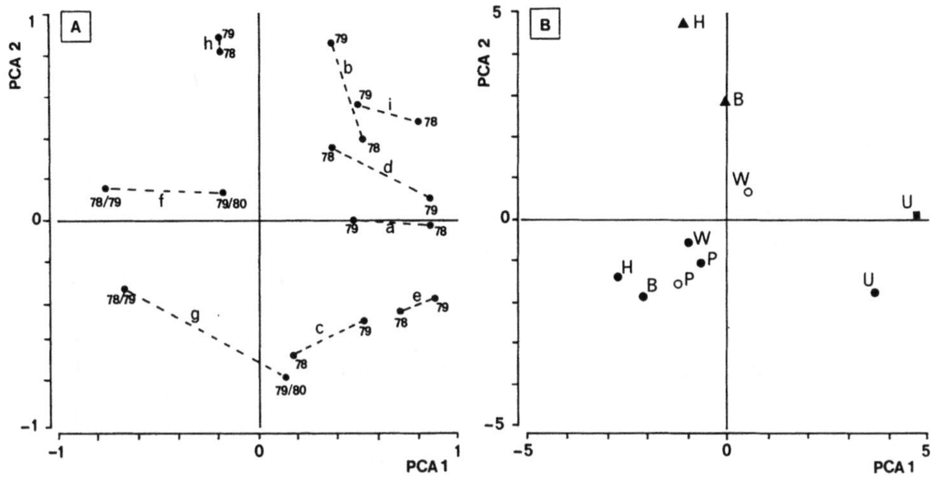

Fig. 4.1. Principal component analysis of demographic characteristics, determined in populations of four *Plantago* species. The first two principal components are shown. **A** Plot of 18 demographic characteristics. *a* Density of adults in spring; *b* density of juveniles in spring; *c* percentage of plants with side rosettes; *d* survival of adults during summer; *e* survival of juveniles during summer; *f* survival of adults during winter; *g* survival of juveniles during winter; *h* number of seeds per flowering plant; *i* percentage of flowering plants. **B** Scores of the populations on the principal component axes. Sites: *U* Uddel; *W* Westervoort; *P* Pannerden; *H* Heteren; *B* Bruuk path. ■ *P. coronopus*; ▲ *P. m. major*; ○ *P. media*; ● *P. lanceolata*

axis 1, and less on axis 2. The Uddel population was closely positioned to the co-occurring population of *P. coronopus*, while the same tendency was seen for the populations of *P. media* sites but not for those of the *P. m. major* sites.

From the combination of the data of Fig. 4.1 A and B and Table 4.2 the following short characterization of the demographic behaviour of the species can be derived.

***P. coronopus*.** A relatively high density of adult and juvenile plants; high percentage of flowering plants; medium to low seed production per flowering plant. Medium occurrence of side rosettes. High survival of juveniles and adults in summer, a very low survival of plants in the severe winter of 1978/1979 and a low to medium survival in the mild winter of 1979/1980. These characteristics together result in a high turnover of *P. coronopus* plants.

***P. major* ssp. *major*.** An average density of adult plants; a high percentage of flowering plants and a high seed production per flowering plants. As a consequence, a high density of juveniles, especially in 1979, was observed. Incidence of side rosettes was low. Adult survival was relatively high, in summer and in winter; survival of juveniles was low in both seasons, so that notwithstanding the high natality the turnover of *P. major* ssp. *major* was low.

P. media. The population showed no clearly distinctive features, relative to both axes; also with regard to the co-occurring *P. lanceolata* the differences were slight, especially in the Pannerden populations. It is evident that on the basis of the present data no clear distinction can be made in the demographic behaviour of the two species. The apparent similarity of the two species may be partly due to differences in behaviour in the two sites where they were studied. In the Westervoort situation the differences between the two species were somewhat greater: *P. media* had a slightly greater percentage of flowering, a better survival of adult plants in both seasons and a slightly lower occurrence of side rosettes. The turnover was lower in *P. media* than in *P. lanceolata*. The differences found in Westervoort were negligible or even reversed in the Pannerden site; the population density of both species was declining, probably because of the very dense structure of the vegetation and the fact that little rejuvenation occurred. Later observations in other sites also showed clear differences between the species in relatively open vegetation, and not in dense and rough vegetation.

P. lanceolata differed from the other *Plantago* species by a relatively high incidence of side rosettes. This was particularly clear in Uddel in 1979 after the severe winter when side rosettes developed after (frost) damage of the principal rosette. In a Heteren hayfield damage of the main rosettes by slugs has been observed and the populations had rather low to medium scores for percentages of flowering, seed production and density of juveniles. In comparison with *P. coronopus* and *P. major* the survival of juveniles was high. All characteristics, primarily correlated with the first axis, varied considerably between the different populations and will be analyzed further on.

Table 4.3. Half-life (in years) of adult plants (genets) of *Plantago* populations in various sites

Site, habitat and period of observation			*Plantago* species			
			coronopus	*major*	*media*	*lanceolata*
Uddel	Road verge	1978–82	0.48			2.07
Westduinen[a]	Dune pasture	1978–80	0.41			2.58
Bruuk	Path	1978–81		2.01		1.10
Heteren	Path	1978–81		1.90		
	Hayfield	1978–82				2.95
Westervoort	Rough herbage	1978–82			7.87	2.37
Pannerden	Pasture	1978–82			2.25	2.20
Brummen	Hayfield	1981–86			4.49	1.04
	Closely grazed pasture	1981–86			14.28	2.32
	Undergrazed pasture	1981–86			2.65	1.85
Bruuk	Hayfield	1978–82				2.15
Achterberg	Hayfield	1978–82				7.18
	Pasture	1978–82				3.14
Merrevliet[a]	Hayfield	1978–81				4.08

[a] Van Groenendael (1985a).

Data are available on the longevity of the plants over a period longer than 2 years. Of all plants present during the spring census of 1980 the presence or absence was recorded in the autumn and spring of following years until the spring of 1982. From these observations half lives of adult plants (Table 4.3) were calculated by regression. We conclude that adult plants of *P. coronopus* lived much shorter than those of the other *Plantago* species. In *P. major* ssp. *major* longevity was about the same as in *P. lanceolata*, while *P. media* lived longer in relatively open habitats and about as long as *P. lanceolata* in dense vegetation. Half life of juveniles was difficult to calculate because mortality was much greater during the first year than in later years. We have the impression that after the first year the mortality rate was independent of age.

In the following sections the separate species will be treated in more detail, especially with regard to differences between populations.

4.1.4 Further Analysis of Separate Species and the Variation Within a Species

4.1.4.1 *Plantago lanceolata*

P. lanceolata had the widest ecological amplitude and therefore was well suited to study the influence of variation due to habitat factors. The analysis has been given in the same way as in the species comparison, and three populations from sites where *P. lanceolata* occurred alone are included (Tables 4.1 and 4.2). The first principal component covered 41% of the total variance and the second 25%. Since the following principal components only explained 14% or less, the presentation of the results was restricted to the first two principal components (Fig. 4.2A). The values for characteristics that corresponded to each other often appeared to be close together: adult density in spring (a), the occurrence of side rosettes (c), and seed production per flowering plant (h) in the 2 years as well as the survival of adults and seedlings in the different seasons. The most important contribution to the first principal component was given by density of juveniles and adults, by survival in summer (positive), in the severe winter of 1978/79 and by seed production per flowering plant (negative). Survival in the mild winter of 1979/80 contributed highly to the second principal component; percentage of flowering, seed production and survival during the summer of 1978 and the winter of 1978/79 only gave small contributions.

The mean scores on the pc-axes of each population were also calculated (Fig. 4.2B), and could be related to the environmental factors of the habitat of the populations (Fig. 4.2C). Axis 1 was significantly correlated with the light transmission of the vegetation during its maximum development in summer and axis 2 with the, strongly intercorrelated, soil factors "content of organic material", "N-content" and "moisture content in winter".

Plant density and survival in summer (Fig. 4.2A, C) can be expected to have higher values in the open vegetations with a high light transmission. This could also be seen in direct correlations between survival and light transmission in

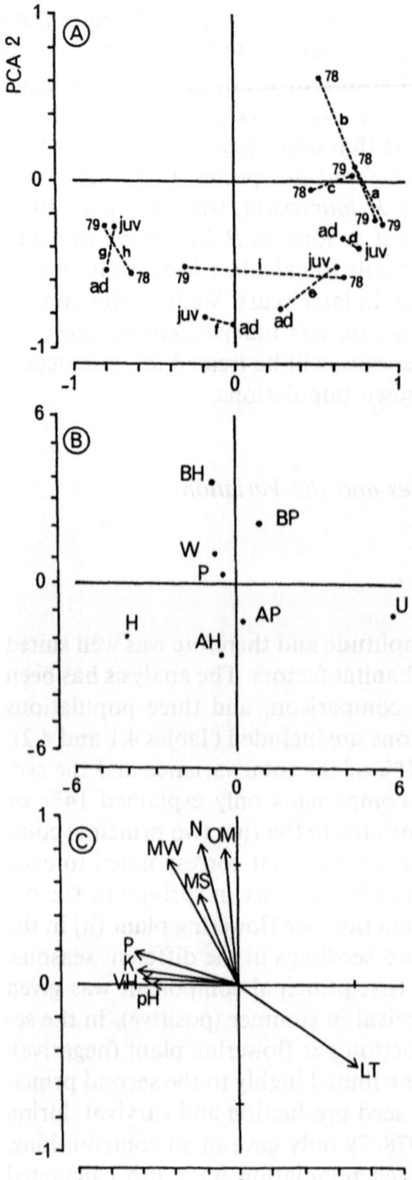

Fig. 4.2. Principal component analysis of demographic characteristics, determined in eight populations of *Plantago lanceolata*. The first two principal components are shown. **A** Plot of 18 demographic characteristics; *juv* juvenile plants; *ad* adult plants. *a* density of adults in spring; *b* density of juveniles in spring; *c* percentage of plants with side rosettes; *d* survival during the summer of 1979; *e* survival during the summer of 1978; *f* survival during the winter of 1979/1980; *g* survival during the winter of 1978/1979; *h* number of seeds per flowering plant; *i* percentage of flowering in adult plants. **B** Scores of the populations on the principal component axes. Sites: *U* Uddel; *AP* Achterberg, pasture; *AH* Achterberg, hayfield; *W* Westervoort; *P* Pannerden; *H* Heteren, hayfield; *BH* Bruuk, hayfield; *BP* Bruuk, path. **C** Plot of correlation coefficients of habitat factors with the population scores on the axes. *LT* light transmission to soil level (June–early July); *VH* vegetation height (June–early July). Soil factors: *OM* organic matter content; *N* total nitrogen concentration; *MS* moisture in summer (percentage of soil volume in June–early July); *MW* moisture in winter (percentage of soil volume in February–early April 1980); *P* phosphorus concentration; *K* potassium concentration; *pH* pH-KCl; all soil factors are given for the top 10 cm of the soil. The threshold for significance ($P = 0.05$) is given as a transverse line on the axes. (From Mook et al. 1989)

summer (Fig. 4.3) in which the group of seedlings corresponded with that of juveniles used in the foregoing analysis. The other plants had to be divided into two size groups according to the number of leaves during the spring census. The regression was becoming less clear with increasing size of the plants, and it was not significant for the plants with four leaves and more. This observa-

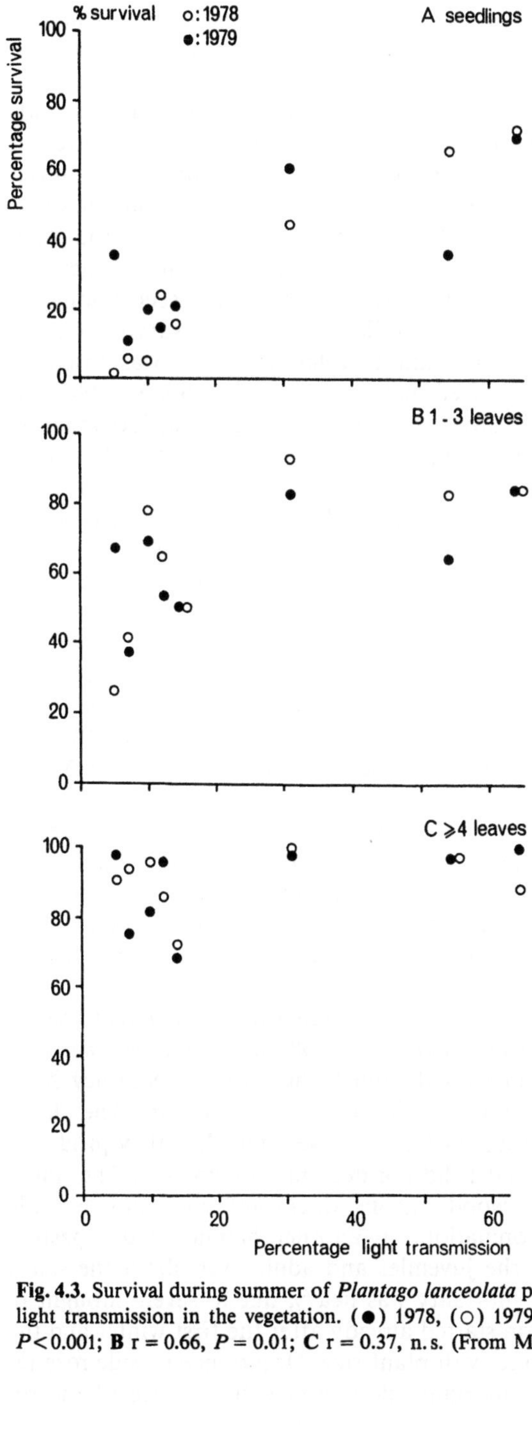

Fig. 4.3. Survival during summer of *Plantago lanceolata* plants of different size in relation to the light transmission in the vegetation. (●) 1978, (○) 1979. Correlation coefficients: **A** r = 0.87, $P < 0.001$; **B** r = 0.66, $P = 0.01$; **C** r = 0.37, n. s. (From Mook et al. 1989)

tion was related to the morphological variation of *P. lanceolata* (Van Tienderen 1989; see also Sect. 7.3): plants with more leaves also tend to have longer leaves, which is of advantage in dense vegetations because there leaves tend to grow vertically. The small plants will get insufficient light, which probably will affect survival negatively (Van der Toorn and Pons 1988). It should be stressed that the high survival in open vegetation in summer may not be generalized to situations like a very dry period or a soil that is incapable of retaining sufficient moisture; survival of juvenile and adult plants during a dry period (May 1980) in an open dune habitat was catastrophically low (Van Groenendael 1985 a). During the same period no unusually low survival was seen among the adult plants in the present studies. A similar case of low survival due to drought was seen on a river dune in summer 1986, and again was absent in other less dry sites only a few hundred metres away (Van Tanderen 1989). In conclusion, adverse influences of drought occur only in soils which retain little water and only in dry years, which occur relatively seldom in The Netherlands. Such a situation, in dune areas once in about 7 years in April and May, is critical for plants in a early stage of development (Ernst and Van Andel 1985).

The negative correlation between survival during the severe winter (1978/79) and light transmission may plausibly be explained by the cover given by dense vegetation and by the ability to retain snow as a protection against dry eastern winds.

Survival in the mild winter of 1979/80 (axis 2) was negatively correlated with soil organic matter content (and thus with total N-content) and soil moisture in winter. The water saturation of the soil in the peaty habitats (BH and BP in Fig. 4.2B) could probably have led to the relatively high mortality (Mook et al. 1989).

In conclusion, the demography of *P. lanceolata* was most affected by density of the vegetation in summer (and possibly also in severe winters) and water saturation of the soil in mild winters. There are no indications that competition for soil nutrients played an important role as suggested by Tilman (1985); in the present study survival was highest in the open vegetations, which were relatively poor in soil nutrients (Fig. 4.2C).

Population dynamics of *P. lanceolata* was studied in two different coastal habitats during 1979 to 1981 (Van Groenendael 1985 a, b). In a very wet hayfield on peat soil the juvenile ramets had a much shorter life expectancy, half-life of 6 months, than the adult ramets, with a half-life of 4 years. The plants needed at least 3 years before they started to flower, and then they produced a low number of spikes; the number did not depend on plant size. The spikes were vulnerable to predation by wood pigeons, ducks and pheasants. No side rosettes were found in this population. In another habitat, a dry pasture on dune sand, the half-life of the juveniles and adults was about the same, 1.5 and 1.7 years, respectively; mortality was heavy, due to severe drought in the second season. Reproduction started already after the first winter and the percentage of flowering increased with plant size. Occurrence of side rosettes also increased with plant size; mortality decreased with the size of the rosettes.

Similar correlations between flowering and plant age and plant size were available from two Achterberg populations (Tables 4.1 and 4.2) which were followed for more than 6 years; the seedlings had a high survival rate. There was no flowering in the year of germination of the plants. At least 5 years has passed before most of the plants had flowered at least once (Table 4.4A). A direct correlation between flowering and plant size was also found in these populations (Table 4.4B). The percentage of flowering increased with the number of leaves per rosette, as well as with the size of the rosette, the latter measured as the length of the longest leaf. The number of leaves probably was more important than leaf size (Table 4.4B). Size will also depend on age and the influence of the two factors cannot be separated.

4.1.4.2 *Plantago media*

The demography of *P. media* is of interest, because initially only small or negligible differences were found between *P. media* and *P. lanceolata* (see Sect. 4.1.3). Therefore a separate study was made along the IJssel (Brummen) in the period 1981 to 1986. The sites were situated on a river dune with sandy soil and comprised a hayfield and a pasture grazed by horses; lightly and closely grazed spots were chosen, the first with a very dense vegetation (Table 4.5). The maximum number of seeds per capsule was 7; mean seed weight varied between 0.4 and 0.8 mg. No flowering was observed in juvenile plants. In the relatively open hayfield and the closely grazed pasture, *P. media* had a higher percentage of flowering adult plants and a greater production of seeds per plant than *P. lanceolata*. In the rough and dense herbage of the lightly grazed pasture, the percentage of flowering plants was similar for both species. With regard to the survival of juveniles no clear differences between the species were found, but adult survival of *P. media* was much greater than that of *P. lanceolata*, especially in the closely grazed pasture and the hayfield (see half-life, calculated from extinction curves of adult plants in Table 4.3). In the hayfield and even more so in the closely grazed pasture, *P. media* had a much higher life expectancy than *P. lanceolata*. Just as in the dense vegetation of Pannerden, the longevity of *P. media* in the lightly grazed pasture was of the same order as that of *P. lanceolata*. The low survival of *P. media* in rough herbage was not exclusively due to height of the vegetation or to low light penetration. Competition with other herbs possibly was more detrimental than competition with grasses; especially the extensive mats of *Convolvulus arvensis* could have reduced survival.

Although no exceptional mortality due to drought was observed, wilting of both *Plantago* species was noted; it was often more severe in *P. lanceolata* than in *P. media*, which may be caused by the fact that *P. media* has a deeper-growing root system than *P. lanceolata*.

Table 4.4A. Age-dependent flowering in *Plantago lanceolata* (study area: Achterberg; 1978–1983)

	Hayfield Plant age (no. of winter seasons survived)						Pasture Plant age (no. of winter seasons survived)					
	0	1	2	3	4	5+	0	1	2	3	4	5+
Flowering plants: first time of flowering	0	8	46	34	19	5	0	6	10	17	8	0
Flowering plants having flowered before	0	0	7	23	21	13	0	0	3	9	10	10
Non-flowering plants having flowered before	0	0	1	2	15	9	0	0	1	2	7	4
Non-flowering plants not having flowered before	401	376	167	66	15	4	222	150	84	40	22	9
Total number in age class	401	384	221	125	70	31	222	156	98	68	47	23
Fraction of plants flowering first time	0.000	0.021	0.208	0.272	0.271	0.161	0.000	0.038	0.102	0.250	0.170	0.000
Fraction of flowering plants total	0.000	0.021	0.240	0.456	0.571	0.581	0.000	0.038	0.133	0.382	0.383	0.435

Table 4.4B. Size-dependent flowering of adult *P. lanceolata*. Fraction of flowering rosettes in relation to number of leaves and length of longest leaf. Data: Achterberg hayfield, spring census 1978–1983. No. of rosettes in brackets

No. of leaves	Length of longest leaf (mm)						Total
	–39	40–49	50–59	60–69	70–79	79–	
4	0.06 (50)	0.15 (47)	0.21 (57)	0.14 (36)	0.26 (19)	0.16 (32)	0.15 (241)
5	0.09 (11)	0.15 (27)	0.29 (41)	0.29 (35)	0.41 (22)	0.44 (27)	0.29 (163)
6	0.29 (7)	0.38 (24)	0.45 (29)	0.52 (25)	0.62 (26)	0.75 (16)	0.51 (127)
7+	0.70 (10)	0.70 (20)	0.74 (46)	0.65 (46)	0.79 (33)	0.78 (68)	0.74 (223)
Total	0.17 (78)	0.29 (118)	0.41 (173)	0.41 (142)	0.56 (100)	0.57 (143)	0.42 (754)

Table 4.5. Comparison of reproduction of *Plantago media* and *P. lanceolata* in three habitats near Brummen in 1982

Habitat	Management	Vegetation	Max. vegetation height (cm) (July)	% Light penetration (July)	Species	Fraction flowering plants	Seed production/flowering plant
Hayfield	Mown once a year (August)	Predominantly grass not productive	55	6	*P. media*	0.76	1242
					P. lanc.	0.28	95
Closely grazed pasture	Grazed by horse occasionally cattle	Low grass	18	23	*P. media*	0.71	49
					P. lanc.	0.18	11
Undergrazed pasture	Occasionally grazed by horse	Dense, high % forbs, increasing in later years	27	17	*P. media*	0.63	362
					P. lanc.	0.57	473

4.1.4.3 *Plantago major*

Both subspecies of *P. major* occurred in relatively open habitats and the above ground parts died off in winter. In *P. major* ssp. *major* the number of seeds per capsule was 6 to 14 and the mean seed weight varied between 0.20 and 0.32 mg. Populations are known in which some plants flowered within the year of germination. In the above-described Heteren and Bruuk path habitats, no flowering occurred in the zero age class: it started in 1-year-old plants, and seed production was relatively high. Juvenile mortality was high in summer, and most of the plants that did survive until autumn had hardly grown since the early seedling stage. Even when three or four small leaves were formed, growth was minimal. Only seedlings which had attained a leaf length of at least 5 cm survived the winter; survival in the juvenile stage was very low. Van der Toorn and Pons (1988) compared emergence and survival of juveniles of the subspecies *major* with those of *P. lanceolata* in a grass vegetation. *P. major* did not establish itself because of its low competitive ability, due to late germination and the small size of its leaves. Further details on emergence and survival are given in Section 4.2.4, on the effects of light in Section 5.6.

The supspecies *pleiosperma* in general had more than 14 (up to 30) seeds per capsule (Mølgaard 1976); exceptions to this rule were found in coastal populations (see below). Mean seed weight was between 0.11 and 0.21 mg. This subspecies was studied in two widely different habitats: in coastal plains in the SW part of The Netherlands and on the banks of the river Rhine (Lotz 1989; Sect. 7.4.). The two types differed in the mean number of seeds per capsule, being 22 in the river population and 8 in one of the coastal populations.

In another river population, 23% of the juvenile plants flowered, against 95% of the adult plants. Seed production depended to a large extent on the size, and thus on the age of the plants. In adult plants a mean of 14000 seeds per plant was found, but individuals with up to 60000 seeds were noted; in a site with almost exclusively juvenile plants, about 700 seeds per flowering plant were counted. The spikes of the plants in the latter population appeared over a rather long summer period or over two distinct periods in summer and autumn. The survival in the riverbank sites as a rule was high in summer, and erratic but most often low (a few percent) in winter, due to floods. A flood may wash away the soil around a root, or it may deposit a layer of sand, through which the plant could not penetrate. After floods a new population could be established from seed. In some years floods did not occur and survival was greater.

4.1.4.4 *Plantago maritima*

On the SW coast of The Netherlands two populations of *P. maritima* were studied (Blom 1983). The first population was situated at the edge of the inner dunes on sandy soil and was only flooded at extremely high tides; it was rarely grazed by cattle and, in winter, by rabbits. This population reproduced mainly by seeds: 80% of the adult plants flowered with a production of 1260 seeds per flowering plant. The seedlings germinated easily and had a relatively high survival rate (36%). The plants formed very few side rosettes and the survival rate of adult plants was relatively low (40 to 75%). The second population had a completely contrasting life history: it occurred on sandy clay in a low zone that was frequently flooded; it was heavily grazed and trampled by cattle. In this population the seed production was very low and seedlings had a low survival rate (4%); the plants formed complexes of many rosettes, which had a high survival rate (86 to 92%).

Experiments by Blom (1983) revealed that the differences were entirely phenotypic. The first population was typical for relatively undisturbed situations. Germination was faster on the sandy clay where the second population occurred, but it was not greater. In this site seed production was kept low by grazing, and seedling survival also was low, as the result of trampling. Trampling and inundation by salt water stimulated formation of side rosettes; the genets with their many ramets were less vulnerable.

Jerling (1985 a) and Jerling and Liljelund (1984) studied the population dynamics of *P. maritima* on the Baltic coast of Sweden along a gradient ranging from an open and frequently flooded zone to a high zone free from flooding with a dense grass vegetation. In the lower parts, flowering, seed production and seedling establishment were high and also mortality was high, due to flooding. In the high zone seed production and seedling establishment were low due to the interception of light by other plants, while grazing by cattle affected seed production negatively. Population density was highest in the intermediate zone.

The seed number per capsule was more often than not two, and a mean seed weight between 0.9 and 1.4 mg was found. Details on seedling emergence and establishment are given in Section 4.2.5, while phenotypic variation is treated in Section 7.1.

4.1.4.5 *Plantago coronopus*

Apart from the inland population discussed above, observations have also been carried out on a coastal population in the southwestern part of The Netherlands (Van der Toorn et al. 1980). Moreover, demographic data are available in the literature: Schat (1982) studied *P. coronopus* in five coastal sites along a gradient ranging from a low and often flooded coastal plain to a high and only rarely flooded sand dune. Waite (1984) and Hutchings and Waite (1985) describe the demography of two populations on the south coast of England.

The seeds of *P. coronopus* are the lightest of the five species, mean seed weights ranging from 0.05 to 0.07 mg. Up to five seeds can be formed in a capsule (Dodds 1953) and the spikes are often small; the number of spikes can be large. The plants in the studied dune habitat were much larger than in the inland site and they showed a seed production of up to 1000 per plant. Flowering may start early in the life history. In the inland site 19% of the spring cohort of seedlings of 1979 flowered the same year. Flowering in summer was observed in plants that had germinated in the foregoing winter (Hutchings and Waite 1986). In the coastal population flowering in juveniles was negligible. Schat (1982) found no flowering at zero age; the flowering percentage gradually increased up to the age of 4 years.

The survival of both juvenile and adult plants in the inland site of Uddel was very low in the severe winter of 1978/1979 (Table 4.2) and the low half lifetime given in Table 4.3 was due to winter mortality. In the same period such low survival values were not found in the coastal site of Westduinen, probably because of the milder conditions near the sea. Schat (1982) also found a lower survival in the winter of 1978/1979 than in 1979/1980, especially in three inundated habitats (0% survival in the first as against 2 to 20% in the second winter); he attributed this to the length of the period of inundation and the susceptibility of the plants to stagnant water. In the driest site, which did not become inundated, survival in general was higher, but it again differed between winters (22% in the first and 88% in the second winter). In an experiment, Schat (1982) found that plants of *P. coronopus* survived flooding better in the presence than in the absence of *Juncus maritimus*, with its highly aerenchymatous roots. As in *P. lanceolata*, this suggests that mortality due to inundation was caused by shortage of oxygen.

Schat (1982), who did part of his observations in the dry summer of 1980, concluded that *P. coronopus* was tolerant to drought. Vermuë (pers. commun.) found in the Westduinen sand dune site that in the very dry summer season of 1976 *P. coronopus* was more resistant to drought than *P. lanceolata*. Waite (1984), on the other hand, recorded a high mortality in one of his populations of *P. coronopus* during the same drought period of 1976, which led to the death of all plants. In this site the soil was shallow and contained a high proportion of shingle. Apart from the conclusion about differences in drought resistance between the two species, these data stress the importance of soil characteristics in the effect of dry periods on survival of the plants.

4.1.5 Conclusions

Although a large variation – due to environmental plasticity – exists in the demographic characteristics within the species, it is possible to draw general conclusions on the life history tactics of the separate species. Table 4.6 gives a summary of some characteristics per species. The species are arranged according to their seed weight. It appears that, except for the last two species (*P. maritima* and *P. lanceolata*), the differences are great, and over the whole range there is a more than twentyfold variation. The other properties are correlated with seed weight, in accordance with correlations observed in other plant species (Silvertown 1981, 1989). A negative correlation between seed weight and seed production (seeds per capsule and mean number of seeds per flowering plant) exists and positive correlations with survival of the seedlings and half life of the adult plants. Also, there is a correlation with age at first reproduc-

Table 4.6. Summary of life-history characteristics of *Plantago* species. Data are approximate and taken from own observations and from the literature cited in the text. Where possible the range has been given and in certain cases the mean ± SE (when at least 4 observations were available)

Species	Seed weight (mg)	Seeds per capsule	Mean seed number per flowering plant[a]	Survival of seedling to adult	Half-life adults (years)
P. coronopus	0.05 – 0.07	1 – 5	1000	0.02 – 0.47	0.41 – 0.48
P. m. pleiosperma (river habitat)	0.11 – 0.21	14 – >30	14000	0.00 – 0.10	<0.3
P. m. major	0.20 – 0.32	3 – 13	2500	0.00 – 0.02 (0.01 ± 0.005)	1.9 – 2.0
P. media	0.35 – 0.8	1 – 7	1250	0.00 – 0.34 (0.08 ± 0.03)	2.3 – 14.3 (6.3 ± 2.2)
P. maritima	0.9 – 1.4	2	1260	0.04 – 0.36	1.2 – 5.9
P. lanceolata	0.7 – 2.0 (2.9)	1 – 2	470	0.01 – 0.64 (0.20 ± 0.04)	1.0 – 7.1 (2.7 ± 0.43)

[a] Maximum found per year in most productive habitat.

tion; *P. coronopus* and *P. m. pleiosperma* (and in some situations *P. m. major)* may flower in the year of germination, while in *P. media* and *P. lanceolata* it only occurs in plants of 1 year old or older.

Secondly, the life history can be regarded in connection with the habitat in which the species occurs. The density of the vegetation in which the species are found increases with seed weight. The experimental study of Van der Toorn and Pons (1988) in which it was concluded that *P. lanceolata* was better adapted to dense vegetations than *P. m. major* with regard to both establishment and survival has already been mentioned.

These correlations are not always strict. For instance, *P. coronopus*, with the lightest seeds, did not have the greatest seed production. Also, the half-life of adults was higher in *P. media* than in *P. lanceolata,* which had a higher seed weight and lower seed production. At the same time the last species has a higher seedling survival than *P. media.* Strict correlations cannot be expected, however, because the requirements of the species for survival in their respective habitats are much more complex than those included in the demographic studies alone.

Similar tendencies as found in the comparison between species are also apparent within a single species. In *P. lanceolata,* which shows the greatest variation in the values for the demographic properties, seed weights are lower in open habitats and higher in the dense habitats and the values for seed production, and survival of seedlings and adult plants vary accordingly. Seed weight and seed production are not only environmentally but also, in part, genetically determined (Sect. 7.3.).

4.2 Germination and Establishment

4.2.1 Germination of Plantago Species as Affected by Seed Age, Temperature and Light

Although the percentage of germination in *Plantago* seeds is high under laboratory conditions and in the field, there are environmental factors which strongly influence the germination capacity. Especially age of the seeds, temperature, illumination, and moisture content of the substrate are important for radicle emergence and seedling establishment.

Radicle emergence of fresh seeds was significantly lower than that of 1- to 5-year-old seeds (Table 4.7). The germination capacity of seeds older than 4 years strongly decreased. The results of Table 4.7 were obtained with seeds stored at room temperature. Sagar and Harper (1960) showed that a prechilling

Table 4.7. Germination percentages[a] under different light conditions in relation to the age of dry-stored seeds of six *Plantago* species. All results were obtained at 25 °C 14 days after sowing. Most tests were performed with 10 × 50 seeds in each year

<table>
<tr><th colspan="2">Seed age in years</th><th>0[b]</th><th>1</th><th>2</th><th>3</th><th>4</th><th>5</th><th>6</th></tr>
<tr><td rowspan="4">P. lanceolata</td><td>a</td><td>10–55</td><td>52–76</td><td colspan="5">76–86</td></tr>
<tr><td>b</td><td>22–59</td><td colspan="6">86–94</td></tr>
<tr><td>c</td><td>–</td><td rowspan="2">57–75</td><td rowspan="2" colspan="5">76–100</td></tr>
<tr><td>d</td><td>–</td></tr>
<tr><td rowspan="4">P. coronopus</td><td>a</td><td>32–97</td><td colspan="4">84–91</td><td colspan="2">65–92</td></tr>
<tr><td>b</td><td colspan="2">31–67</td><td>66–98</td><td colspan="2">24–57</td><td colspan="2">14–27</td></tr>
<tr><td>c</td><td colspan="7">0</td></tr>
<tr><td>d</td><td colspan="7">0</td></tr>
<tr><td rowspan="4">P. major ssp. major</td><td>a</td><td>0–19</td><td colspan="4">89–99</td><td colspan="2">18–58</td></tr>
<tr><td>b</td><td>0</td><td colspan="4">69–87</td><td colspan="2">5–15</td></tr>
<tr><td>c</td><td colspan="3">–</td><td colspan="2">47–89</td><td colspan="2">0–11</td></tr>
<tr><td>d</td><td colspan="7">0</td></tr>
</table>

Table 4.7 (continued)

<table>
<tr><th colspan="2">Seed age in years</th><th>0[b]</th><th>1</th><th>2</th><th>3</th><th>4</th><th>5</th><th>6</th></tr>
<tr><td rowspan="4">P. major ssp. pleiosperma</td><td>a</td><td>0–12</td><td colspan="2">93–100</td><td colspan="3">56–71</td><td>–</td></tr>
<tr><td>b</td><td>0–7</td><td colspan="2">64–100</td><td colspan="3">0–36</td><td>–</td></tr>
<tr><td>c</td><td colspan="3">–</td><td colspan="3">0–34</td><td>–</td></tr>
<tr><td>d</td><td colspan="7">0</td></tr>
<tr><td rowspan="4">P. media</td><td>a</td><td>44–83</td><td colspan="5">81–89</td><td>38–73</td></tr>
<tr><td>b</td><td>5–36</td><td colspan="5">75–86</td><td>22–69</td></tr>
<tr><td>c</td><td>–</td><td colspan="5" rowspan="2">46–73</td><td rowspan="2">11–51</td></tr>
<tr><td>d</td><td>–</td></tr>
<tr><td rowspan="4">P. maritima</td><td>a</td><td colspan="7">86–93</td></tr>
<tr><td>b</td><td colspan="7">69–82</td></tr>
<tr><td>c</td><td colspan="5">–</td><td colspan="2" rowspan="2">0–12</td></tr>
<tr><td>d</td><td colspan="5">–</td></tr>
<tr><td colspan="2">Seed age in years</td><td>0</td><td>1</td><td>2</td><td>3</td><td>4</td><td>5</td><td>6</td></tr>
</table>

[a] The 95% confidence intervals are given (one interval for more years taken together means that no differences in germination were found).
[b] Freshly collected seeds.
a: Daily photoperiod 16 h light, 8 h darkness.
b: Complete darkness except for 2 min faint daylight daily.
c: Complete darkness except for 2 min faint green light daily.
d: Complete darkness.
–: Insufficient observations available.

treatment at 5 °C of fresh *P. major* seeds resulted in an increase of germination, whereas only small differences in germinability between prechilled and fresh seeds of *P. lanceolata* and *P. media* were found (Fig. 4.4); this was also observed for *P. coronopus* and *P. maritima.*

The influence of temperature on radicle emergence of plantains was also studied by Sagar and Harper (1960). The percentage germination of *P. media*

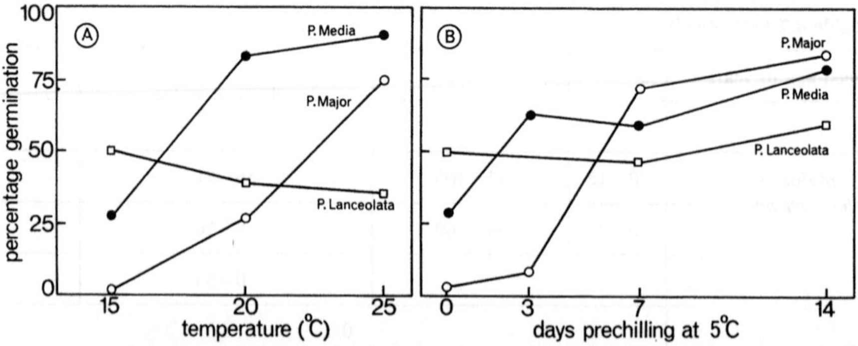

Fig. 4.4. The influence of temperature (**A**) and time of prechilling at 5 °C (**B**) on the germination of seeds of *Plantago media* (●), *P. major* (○) and *P. lanceolata* (□)

and *P. major* increased with a rise in temperature from 15 to 25 °C, especially between 15 and 20 °C (*P. media)* and 20 to 25 °C (*P. major).* Observations on *P. lanceolata* seeds showed no temperature effect; also seeds of *P. coronopus* and *P. maritima* were not affected. These results demonstrate that temperature was important for germination of *P. media* and even more so for *P. major.* The proportion of the seeds of *P. major* capable of germination strongly decreased when the temperature fell below 20 °C; the radicle emergence of fresh to 3-year-old seed did not exceed 3% and for older seeds 7% was obtained. Below 18 °C the radicle emergence of fresh *P. media* seeds was about 5%, and older seeds reached a mean germination of 12% (Sagar and Harper 1960, 1964; Blom 1978). A fluctuating temperature regime (15–20 °C) did not improve germination of seeds of all species, compared with a constant temperature of 25 °C (Blom 1978).

The data of Table 4.7 show that with the exception of *P. lanceolata, Plantago* seed germination was increased by light. Only *P. lanceolata* and, to a lesser extent, *P. media,* were able to germinate in complete darkness. No radicle emergence in *P. coronopus* was observed in the dark or under green light conditions. The results obtained with both subspecies of *P. major* were of particular interest. In complete darkness no radicle emergence took place (25 °C), under so-called safe green light conditions considerable numbers of germinated seeds, especially seeds of *P. major* ssp. *major,* were observed. The influence of light on germination of *P. major* depends on the temperature (Harper and Sagar 1960). At 20 °C, only about 10% of the seeds germinated in light and no germination was found in darkness. At 25 °C, the percentage of germinated seeds in light was raised to about 25% and, contrary to our experiment, in the dark a few seeds germinated (5%). Hawthorn (1974) also found some germination of *P. major* in darkness. Since these results differ from the results found by Blom (1978, Table 1) germination characteristics of *P. major* may differ slightly between different populations.

4.2.2 Germination and the Effects of Moisture

The influence of moisture on germination of *Plantago* species was studied (Blom 1979; Fig. 4.5). At a low moisture content of the soil, dune sand, seeds of *P. lanceolata* and *P. coronopus* germinated. Maximum germination of *P. lanceolata* was found at 21% moisture level (by volume). Maximum germination of *P. coronopus* was observed between 11 and 21%. At a moisture level of 8%, 28% of the *P. maritima* seeds had germinated and a maximum of 91% was found at a moisture content of 21%. *P. major* ssp. *major* and ssp. *pleiosperma* were most sensitive to low moisture levels, followed by *P. media*. Maximum percentages of germinated seeds of these species were obtained at 21% moisture. In water-saturated dune sand all species germinated less and only a few germinated seeds were observed in *P. media* and in both subspecies of *P. major*. It is remarkable that at these circumstances relatively high numbers of seeds from the other species germinated.

4.2.3 Germination and the Effects of Soil and Other Substrates

Experiments on the germination capacity of *Plantago* species strongly suggested that differences in chemical soil conditions had no effects. For example, no important differences in maximum germination were observed when the seeds

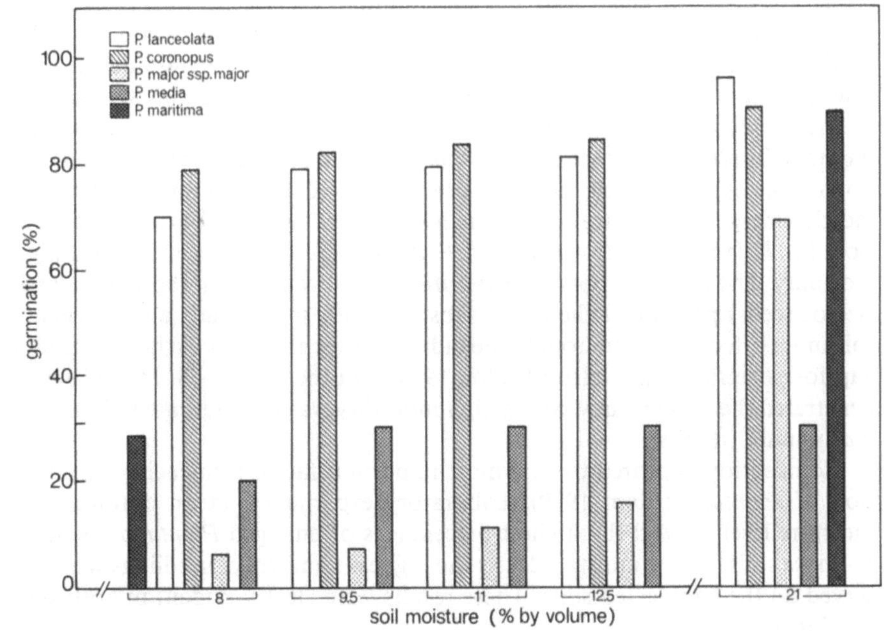

Fig. 4.5. The effects of different soil-moisture conditions on the germination of five *Plantago* species

were sown on pure sand, sand rich in organic matter, sandy clay or clay soils. Besides differences in germination capacity caused by other effects such as seed age, temperature, light conditions and soil-moisture content, physical soil characteristics appeared to be important. Sagar and Harper (1960) demonstrated dramatic differences in germination of *P. lanceolata* between a filter pad, 45% germination, vermiculite, 54% and soil, 72%. Similar results were obtained with seeds of *P. media* and *P. major*. Germination of seeds and seedling establishment in soils were strongly influenced by the size and shape of the seeds and by the microtopography and structure of the upper soil layers (McIntyre 1958; Parker and Taylor 1965; Taylor et al. 1966; Harper et al. 1970; Taylor 1962, 1971; Sheldon 1974; Oomes and Elberse 1976). For this reason a study using glass beads as a more simplified substrate for germination was performed.

Germination and establishment of the five *Plantago* species was studied by varying the size of the glass beads as substrate (Blom 1978). The second important factor in seed/substrate particle interaction was seed size. It decreased in the order *P. maritima, P. lanceolata, P. media, P. major* ssp. *major, P. major* ssp. *pleiosperma* and *P. coronopus*. Experiments were carried out in boxes filled with round glass beads of a diameter varying between 0.01 and 0.05 mm, 0.17 and 0.25 mm, and 0.89 and 1.23 mm. As a reference, boxes were filled with compacted dune sand, the particles varying between 0.05 and 1.0 mm with a median value of 0.19 mm (Table 4.8). Radicle emergence of *P. maritima, P. lanceolata* and *P. media* significantly increased with increasing size of the glass beads. No effect of glass bead size was found in the species with the smaller seeds. Mortality of the young seedlings was highest on substrate with the smallest bead size. On sandy soil and substrates with the low and intermediate bead size, mortality decreased with seed size. Death of seedlings was caused by the inability of the primary root to penetrate the dense substrate caused by the small beads; with exception of *P. maritima* and *P. lanceolata,* root penetration on the substrate with the small beads followed almost immediately upon radicle emergence. Penetration of the lower layers by the roots was relatively slow for *P. lanceolata, P. media* and *P. major* ssp. *major* resulting in a high mortality. The primary roots of *P. maritima* and *P. lanceolata* (both with large size of seeds) penetrated the dense substrate with small beads and the sandy soil more effectively. The roots "meandered" over the bead surface to find a gap for penetration (see also Sheldon 1974). On the sandy soil, many laterals penetrated the surface; however, a high mortality was not prevented (*P. maritima*, unpubl. results).

Compaction appeared to be a most important factor influencing germination of *Plantago* (Blom 1979). Laboratory experiments at an optimal soil-moisture level showed the highest percentages of emerged *Plantago* seedlings on loose soils; at low moisture levels the highest numbers of seedlings were observed on the compacted soils. These results were clearly evident in both subspecies of *P. major*.

In another experiment seeds were sown in turves collected in the field and brought to the laboratory (Blom 1977). Germination of all species increased

Table 4.8. Effect of substrates of different particle size on the seedling radicle emergence, seedling mortality and root length before penetration of four *Plantago* species

Substrate particle sizes (mm)	I 0.01 – 0.05			II 0.17 – 0.25			III 0.89 – 1.23			Sand 0.05 – 1.0; median: 0.19		
	Radicle emergence (%)	Mortality (%)	Length[b] primary root (mm)	Radicle emergence (%)	Mortality (%)	Length primary root (mm)	Radicle emergence (%)	Mortality (%)	Length primary root (mm)	Radicle emergence (%)	Mortality (%)	Length primary root (mm)
P. lanceolata (2.4 – 1.0 – 0.8)[a]	31.9	83.3	x̄: 6.4 5.7 – 7.0	65.6	40.9	x̄: 12.0 8.6 – 15.5	84.4	0	0	93.7	51.4	x̄: 13.8 10.0 – 17.6
P. media (2.0 – 1.0 – 0.3)[a]	58.2	64.9	x̄: 4.0 3.2 – 4.7	81.8	14.8	x̄: 3.3 2.4 – 4.1	75.3	0	0	74.3	19.1	x̄: 3.4 2.2 – 4.5
P. major ssp. *major* (1.2 – 0.9 – 0.3)[a]	88.7	58.8	x̄: 3.0 2.6 – 3.4	94.5	5.0	x̄: 1.8 1.3 – 2.4	97.4	0	0	98.2	1.5	x̄: 2.1 1.2 – 3.0
P. coronopus (0.9 – 0.4 – 0.1)[a]	77.5	7.3	x̄: 1.5 0.9 – 2.1	79.4	1.4	x̄: 0.9 0 – 1.9	70.6	0	0	80.9	0	x̄: 1.1 0.6 – 1.5

[a] Mean length, width and thickness of the seeds (mm).
[b] Mean length (x̄) and 95% confidence interval.
– All values are determined 14 days after sowing.
– Radicle emergence: percentages of germinated seeds with reference to the numbers of sown seeds.
– Percentage mortality with reference to numbers of germinated seeds.

with increasing compaction of the soil, especially in species with large seeds and with *P. major.* This may be probably ascribed to an increased water availability of compacted soils. The increase in water-holding capacity by soil compaction was also observed by Liddle and Greig-Smith (1975) for dune sand an by Krüger (1970) for diluvial loamy sand and alluvial meadow soil. Since seedlings emerged on the turves under competition, a high mortality should be attributed to competition from the neighbouring plants.

In this experiment high percentages of dead *P. major* and *P. media* seedlings were observed and the other *Plantago* species were less affected. Also, in the experiment with the turves, a trampling treatment was applied (Blom 1977). Due to a light trampling treatment an increase in germination occurred in *P. lanceolata* and *P. media.* Apparently, the large seeds were pressed into the soil by the light trampling, which resulted in a better seed-soil contact and more seeds being germinated. Such effects were absent in species with smaller seeds.

4.2.4 Emergence and Establishment Under Field Conditions: P. lanceolata, P. coronopus, P. major, P. media

Laboratory tests are necessary to understand germination, emergence and seedling establishment in the field. To bridge the gap between laboratory conditions and the field, experiments in artificial plots in the field are useful (Blom 1976). Germination experiments were performed in open plots in an experimental garden. The original soil was removed and replaced with dune sand, poor in organic matter, which was compacted to various degrees. Soil moisture depended on rainfall.

Seedlings of *P. lanceolata* and *P. coronopus* emerged in highest numbers on the least compacted soils; seedlings of *P. major* and *P. media* were found in the highest numbers on the more compacted soils and after two years no *P. major* plants survived the loose soil habitat (Fig. 4.6). The opposite reactions between *P. lanceolata* and *P. coronopus* on the one hand and *P. major* and *P. media* on the other were probably caused by the combined action of mechanical resistance and moisture. The soil moisture of these barren sand plots in general was very low. Since *P. lanceolata* and *P. coronopus* germinate better at low soil moisture than *P. major* and *P. media,* soil moisture probably is more limiting for germination for the latter two species. This conclusion also explains the fact that the emergence of *P. major* and *P. media* seedlings was higher on the compacted soils. Mechanical resistance was an important limiting factor for the emergence of *P. lanceolata* and *P. coronopus* seedlings. The low number of seedlings on the intermediate compacted soils was possibly due to relatively large resistance in combination with a relatively low water availability.

Factors interrelated with soil compaction are mechanical resistance, availability of moisture, aeration and temperature of the soil (Eavis 1972; Scott Russell and Goss 1974; Liddle and Moore 1974). Shifts in these factors are due to changes in total pore volume (bulk density) and in pore size distribution.

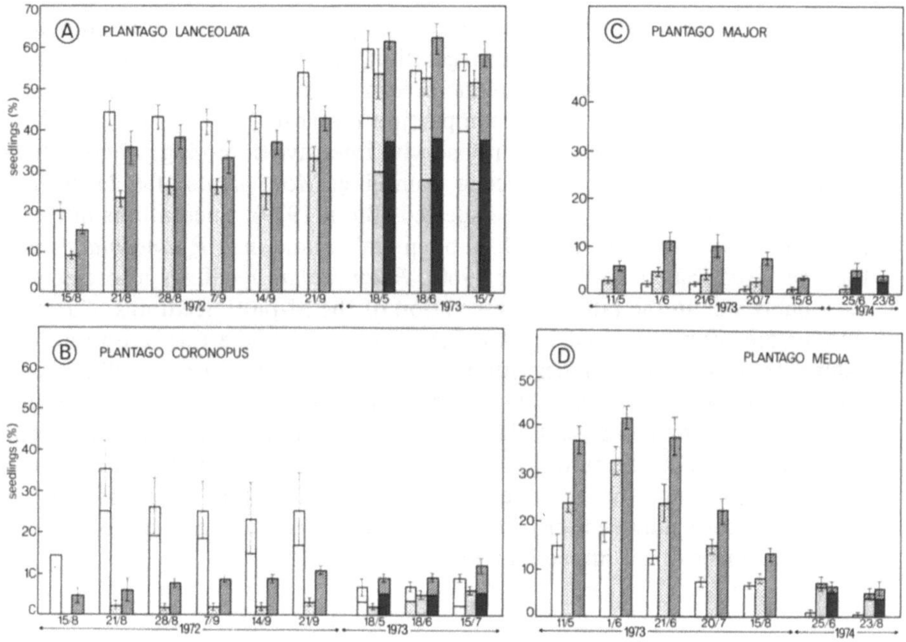

Fig. 4.6. The influence of soil compaction on seedling emergence from buried *Plantago* seeds in experimental plots in the field (mean values of four comparable plots). For the columns on the *right* a distinction was made between newly emerged plants and plants of the preceding year, the *upper part* representing the former and the *lower part* the latter. The *vertical bars* represent twice the standard deviation. ◻ Loose soil; ▨ moderately compacted soil; ▩ compacted soil

Structure of the upper soil layers, mechanical resistance and soil moisture are considered as most important for germination.

To analyze the effect of the environment on establishment of *Plantago*, Sagar (1959) and Sagar and Harper (1960) carried out a series of experiments in which seeds of *P. major*, *P. media* and *P. lanceolata* were sown in a range of habitats which represented plant communities from which plantains were absent and communities in which they normally occurred. In acid heath and valley bogs only few seeds germinated and emerged seedlings succumbed. In beech and oak woodland communities germination was low and seedlings died quickly, probably due to lack of light. In open arable land, without naturally occurring plantains, all species germinated in relatively high numbers, but only *P. lanceolata* seedlings survived. In open grassland on exposed limestone, without naturally occurring plantains, all species germinated and after 1 year all seedlings had succumbed to drought. No germination was found in tall grass habitats, which was ascribed to the absence of light for *P. media* and *P. major*. The fact that *P. lanceolata* was unable to establish itself in its own habitat remains unexplained. Possibly seeds from a meadow ecotype were used. In a similar experiment emergence and establishment were studied by sowing seeds in a dune area (Blom 1977) on paths with compacted soils and partly covered by vegetation (V-plots, Table 4.9), on the edges of paths with moderately compacted soils (M-plots) and a tall, dense vegetation, and on loose soils with an open vegetation (O-plots). In general, an increase in emergence was found at increasing trampling intensity and soil compaction (Table 4.9).

Seedlings of *P. coronopus* did not survive intensive trampling, whereas fullgrown plants had a high resistance to trampling. More plants of *P. lanceolata* survived on paths than would be expected (Blom 1979); fluctuations in trampling probably increased survival. In open arable land only *P. lanceolata* seedlings survived (Sagar and Harper 1960). *P. lanceolata* seedlings also emerged reasonably well in the tall dense vegetation of the M-plot. Seedlings of *P. coronopus* are often observed in gaps between stones where they are protected against trampling. The highest numbers of surviving *P. major* seedlings were found on open, trampled sites. On the plots with a relatively dry soil, *P. major* and *P. media* remained absent. The occurrence of seedlings on wet compacted soils will depend on the germination capacity and the capacity to penetrate the soil. *P. major* clearly met these requirements and the failure in germination by the other species may be ascribed to a low oxygen content, and inhibited root growth by mechanical impedance of the soil.

4.2.5 Emergence and Establishment Under Field Conditions: P. maritima

Survival of seedlings of *P. maritima* was studied by Jerling (1981) on a Baltic seashore meadow; it was related to a topographical gradient from 15 to 54 m from the seashore and to topographical changes in the microhabitat.

Seedling mortality in pre-summer was probably due to desiccation. The seedlings near to the sea suffered more from drought than those found more inland. This may be due to a reduction in vegetation density at the seaside. Later in the year, flooding caused death of the seedlings, especially in the seaside zone. Mortality occurred especially after flooding, but differences in the length of the inundation period due to the differences in microtopography could not explain the observed differences in mortality. Jerling (1981) suggested that light deficiency due to the deposition of iron and salts (FeS, Fe^{3+}) and toxic effects of H_2S caused mortality.

Further studies on *P. maritima* seedlings along the gradient in a seashore meadow by Jerling (1982, 1985a) showed that survival of seedlings was closely related to vegetational characteristics such as total cover, height of vegetation and amount of litter, all factors resulting in reduced light. Superimposed on these characteristics was the duration of flooding. Furthermore, the mortality was positively related to a high density of populations of *P. maritima* seedlings and juvenile plants. It can be concluded that establishment and survival were closely dependent on the availability of "safe" sites as spots lacking high vegetation and spots with a large amount of litter. The number of "safe" sites depended on factors as grazing pattern, deposition of litter by the sea, etc. The number of open and "safe" spots increased approaching the seaside together

Table 4.9. Some environmental factors relating to the emergence and mortality of *Plantago* seedlings from seed sown in plots in the dune area on Voorne (The Netherlands)

Plot	Mean height vegetation (cm)	Cover (%)	Mean penetrometer values ± S.E. at a depth of 5 cm (kg/cm²)	Trampling intensity[a]	Soil moisture 0–5 cm (% dry weight)			Emerged seedlings and dead plants[b]			
					Lowest values	Highest values	Mean values	*Plantago lanceolata*	*Plantago coronopus*	*Plantago major*	*Plantago media*
O_1	0	0	0.9 ± 0.4	–	1.1	6.5	4.4	21.5 (32.5)	16.5 (54.5)	5 (100)	2 (100)
O_2	5	50	5.6 ± 2.1	–	2.0	7.5	4.2	24.5 (24.5)	7 (35.7)	0	0
O_3	0	0	1.7 ± 0.9	–	3.0	16.1	9.9	17.5 (22.9)	16 (18.7)	2 (25)	0.5 (0)
M_1	15	100	10.8 ± 2.3	±	4.4	39.4	13.7	17.5 (51.4)	0	1 (100)	3 (50)
M_2	15	90	14.4 ± 2.4	±	11.6	46.4	27.1	17 (47.1)	6.5 (84.6)	1 (100)	0
M_3	10	100	6.6 ± 1.3	±	42.8	104.2[c]	70.8	19.5 (76.9)	16 (100)	2 (100)	0.5 (100)
M_4	10	100	4.5 ± 1.5	±	50.8	136.8[c]	93.5	42 (72.6)	45 (100)	11.5 (100)	6.5 (100)
M_5	10	100	5.2 ± 1.7	±	34.6	95.5	60.1	12 (70.8)	2.5 (100)	8.5 (94.1)	2.5 (100)
V_1	7	85	17.1 ± 2.0	+	0.1	15.2	7.7	10.5 (42.8)	5.5 (100)	1 (100)	1.5 (66.6)
V_2	5	70	26.8 ± 6.2	+ +	20.5	60.4	43.4	22.5 (64.4)	25.5 (76.4)	11.5 (69.6)	2 (100)
V_3	5	70	31.1 ± 8.8	+ +	11.9	75.0	39.2	10.5 (38.1)	4.5 (100)	21.5 (60.5)	1 (100)
V_4	2	40	21.0 ± 5.8	+ + +	20.5	76.4	38.7	52 (74)	89 (87.6)	99.5 (95.5)	24.5 (79.6)
V_5	3	50	22.5 ± 3.6	+ + +	23.2	41.4	31.4	50 (61)	95 (90.5)	85.5 (86.6)	20 (80)
V_6	3	40	17.8 ± 1.9	+ + +	45.3	104.3[c]	93.3	24 (83.3)	51 (97.1)	71 (80.3)	5.5 (90.9)

[a] Symbols: trampling: – absent; ± sometimes; + moderate; + + rather frequent; + + + very frequent.
[b] Emerged seedlings expressed as percentages of sown seed. In brackets: dead plants (%) calculated with reference to the total numbers of seedlings which had emerged during the test (April – October 1974).
[c] Water table + 1 cm above soil level at the end of October 1974.

Table 4.10. The optimal occurrence of seedlings of five *Plantago* species in relation to some environmental factors

Environmental factor	Low 1	2	3 Ranking	4	High 5
Temperature	*lanc.*	*coron.*	*mar.*	*media*	*major*
Light	*coron.*	*major*	*media*	*lanc.*	*mar.*
Soil moisture	*coron.*	*lanc.*	*mar.*	*media*	*major*
Soil compaction	*lanc.*	*mar.*	*media*	*coron.*	*major*
Trampling	*coron.*	*lanc.*	*mar.*	*media*	*major*
Competition	*major*	*media*	*mar.*	*coron.*	*lanc.*
Flooding	*media*	*lanc.*	*major*	*coron.*	*mar.*

1 = lowest degree of the condition mentioned, 2,3,4,5 = increasing degrees.
coron. = *P. coronopus*; *lanc* = *P. lanceolata*; *major* = *P. major* ssp. *major*; *media* = *P. media*; *mar.* = *P. maritima.*

with the occurrence of flooding (frequency and duration). During years with long-lasting high water, floods will recede the seaside border of plantain establishment from the sea; simultaneously open and "safe" sites with sufficient light will be produced (Ericson 1980).

In a sowing experiment with *P. maritima* in a salt marsh (Blom 1983), it was observed that the seedlings emerged relatively quickly, and that the number of surviving seedlings was low. Heavy trampling and, to a lesser extent, daily flooding were important for seedling mortality. A relatively low mortality was observed in dense vegetation moderately trampled by cattle. The highest survival of seedlings was found on open and sandy soils. Seedling emergence increased markedly after rainfall, which undoubtedly was due to a decrease in soil salinity. In a germination experiment in the greenhouse it was observed that seedling emergence decreased strongly at a salinity higher than 25% of the salt concentration of seawater. Clearly, regularly flooding and heavy trampling caused mortality of *P. maritima* seedlings. The susceptibility of these seedlings to trampling is intermediate between *P. media* and *P. lanceolata.*

The occurrence of the *Plantago* species in relation to the important environmental factors were discussed in this section and summarized in Table 4.10.

4.3 Photosynthesis of *Plantago lanceolata* Populations in the Field

This section compares growth characteristics and photosynthetic response to light intensity of plants of *Plantago lanceolata.* Two contrasting habitats were chosen: the dry dune grassland, Westduinen, and the wet hayfield, Merrevliet. The habitats represent extremes in the distribution of *P. lanceolata* (Sect. 3.1).

The dune grassland microclimate can be classified as dry and warm (Sect. 3.3): during the day it is strongly influenced by convection as a result of heating of the soil by radiation. The hayfield microclimate was humid and

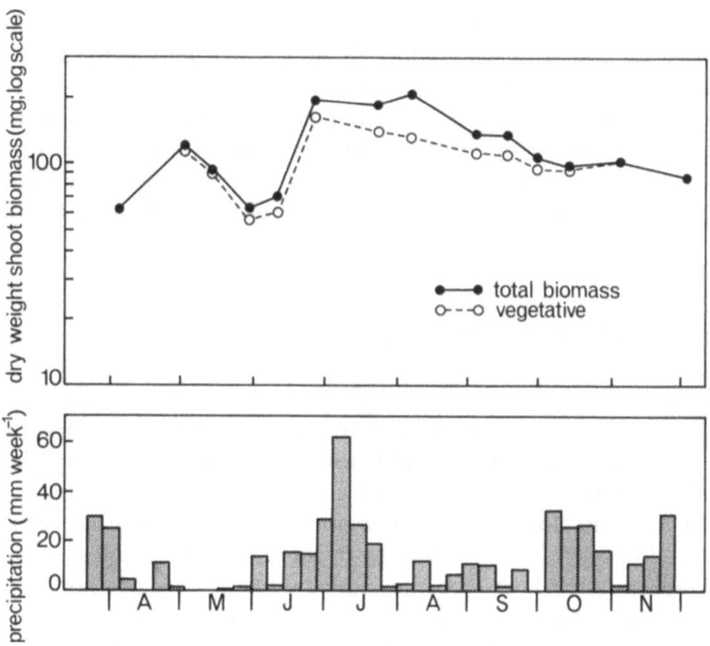

Fig. 4.7. Estimated shoot biomass of *P. lanceolata* in relation to precipitation at the dune grassland. The *columns* represent the summation of the weekly precipitation

temperatures within the canopy were close to the air temperature above the canopy. The moisture-retaining capacity of the dune grassland soil was small (Sect. 3.2). Above-ground biomass production was closely related to the amount of rainfall during the growing season (Fig. 4.7). Above-ground biomass production of *Plantago* at the hayfield was exponential in the first half of the growing period. After the vegetation had reached its final height, shading of *P. lanceolata* by neighbouring plants limited further growth.

4.3.1 Experimental Details

Net photosynthesis (Pn) and (micro-)meteorological parameters were measured using a mobile laboratory. An open system was used to measure CO_2 exchange of the youngest and fully expanded leaf of a *P. lanceolata* plant in its habitat; a differential infrared gas analyzer (URAS 2T, Hartmann & Braun) was used. Air temperature inside and outside the cuvette was measured with small wire thermocouples; outside air humidity was measured with a thermocouple psychrometer (Stoutjesdijk 1980). To maintain air temperature and humidity as *in situ*, the leaf was inside the chamber only for short intervals, 5 to 10 min/h, in order to imitate the present microclimate as close as possible. Stomatal conductance and transpiration were measured with a LiCor LI-1600 steady-state diffusion porometer. Photosynthetic active radiation (PAR) at plant height was recorded with a Si-photocell. Light distribution inside the canopy was measured with small Si light cells at different heights. Leaf water potentials were measured with a pressure chamber on single leaves from rosettes growing in the vicinity of rosettes used for photosynthesis measurements. Shoot biomass was estimated using nondestructive measurements of standard plants with a parallel destructive sampling for calibration (Ondok 1971).

4.3.2 Effects of Drought on Photosynthesis

Temporary effects of water shortage in the plant on Pn were detected in the dune grassland on clear days. *P. lanceolata* plants showed a "midday depression" of photosynthesis (Fig. 4.8), as often found in certain desert species (Lange et al. 1975; Stocker 1976; Schulze and Hall 1982). The depression has been ascribed to stomatal closure, since the evaporative demands of the environment were high (Stalfelt 1955); overheating of the leaf could also play a role (Seibert et al. 1980). Plants which were selected for measurement of photosynthesis along a moisture gradient, viz. the slope of a sand dune, exhibited a different degree in midday depression (Fig. 4.8). This indicates a major effect of the plant's water status on midday depression caused by water availability of the soil. The start of the depression could be explained by hydro-active closure of stomata, which in turn was induced by an internal low water potential (Stalfelt 1955; Hsiao 1973). As a consequence, transpiration was limited and caused an increase in diffusion resistance through the stomatal pores, thereby

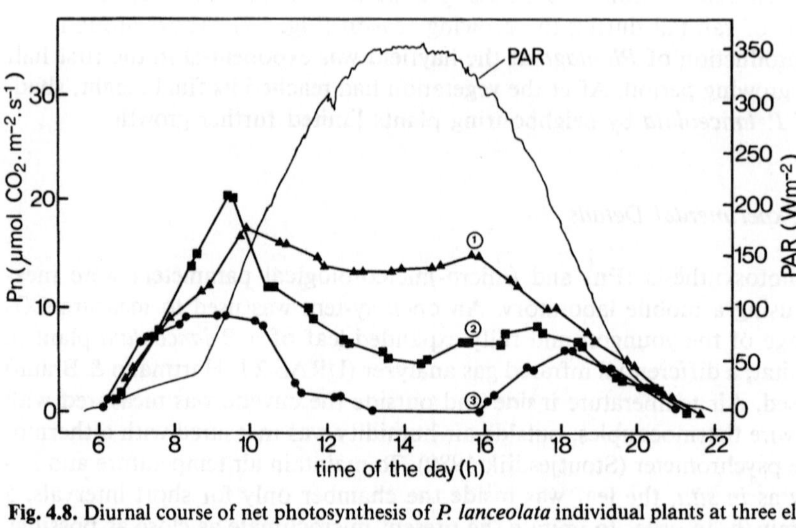

Fig. 4.8. Diurnal course of net photosynthesis of *P. lanceolata* individual plants at three elevations along the slope of a sand dune at the dune grassland site on a clear day. *1* Low elevation; *2* intermediate elevation; *3* high elevation

limiting photosynthesis. Stomata opened again in the afternoon when the leaf water potential still was low. The re-opening of the stomata cannot be solely explained as a reaction to a changed leaf water potential. Schulze and Hall (1982) explained the two peaks in daily photosynthesis by the combined effect of the stomata response to the atmospheric environment close to the leaf surface and an inhibitory effect of high leaf temperature on net photosynthesis. In *Plantago*, the stomata re-opened at 15.00 h and Pn and E were at least partly restored (Fig. 4.9). The evaporative demand of the leaf environment at this mo-

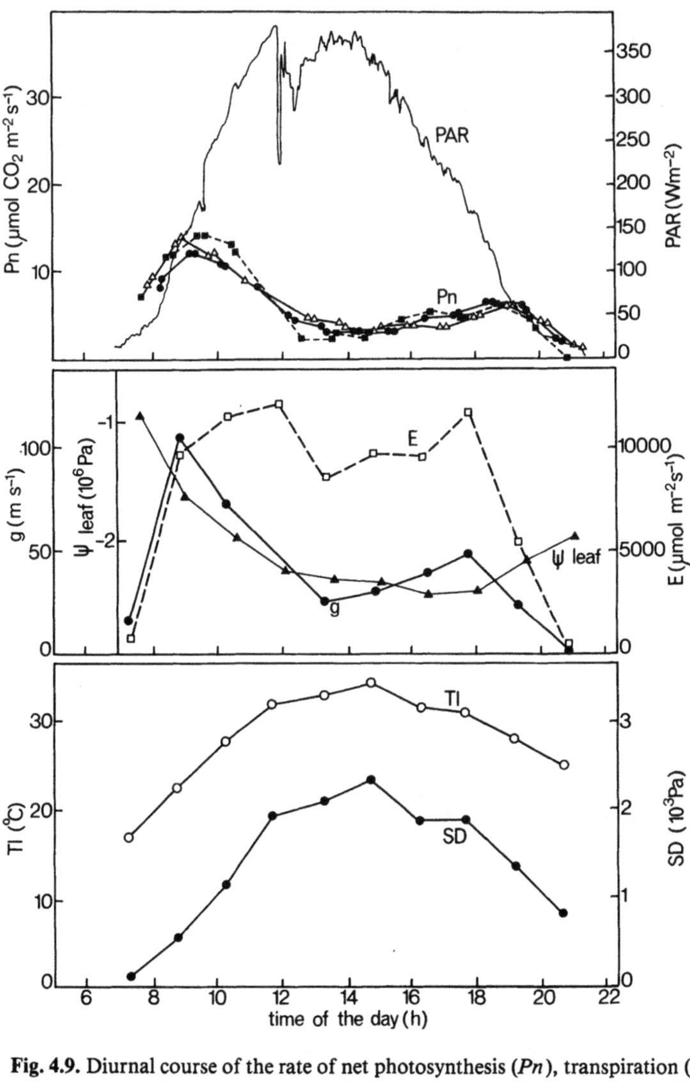

Fig. 4.9. Diurnal course of the rate of net photosynthesis (*Pn*), transpiration (*E*), leaf water potential (ψ), stomatal conductance (*g*) and leaf temperature (*Tl*) of *P. lanceolata* in relation to light intensity (*PAR*) and saturation deficit (*SD*) at the dune grassland site

ment was still high and diminished soon afterwards. This observation is consistent with an optimization theory, proposed by Cowan and Farquhar (Cowan 1978, 1982). The theory suggests that stomatal reactions are controlled by a "feed-forward" system, allowing a maximum photosynthetic gain, with a minimum cost in water loss. Other factors which could be responsible for the daily stomatal action include the hormonal system (Farquhar and Sharkey 1982).

4.3.3 Effects of Shading on Photosynthesis

In the hayfield, *P. lanceolata* survived the winter period as a prostrate rosette; the latter was characterized by numerous relatively short and broad leaves. As weather conditions became favourable for growth, the plant was faced with a diminishing light intensity, due to shading by the surrounding vegetation. *P. lanceolata* plants then formed erect rosettes with fewer and more elongated leaves, up to 35 cm. The specific leaf weight decreased throughout the growing period and the maximum photosynthesis of the newly formed leaves was lower (Fig. 4.10): the latter points towards shade adaptation of the plant (Bjoerkman

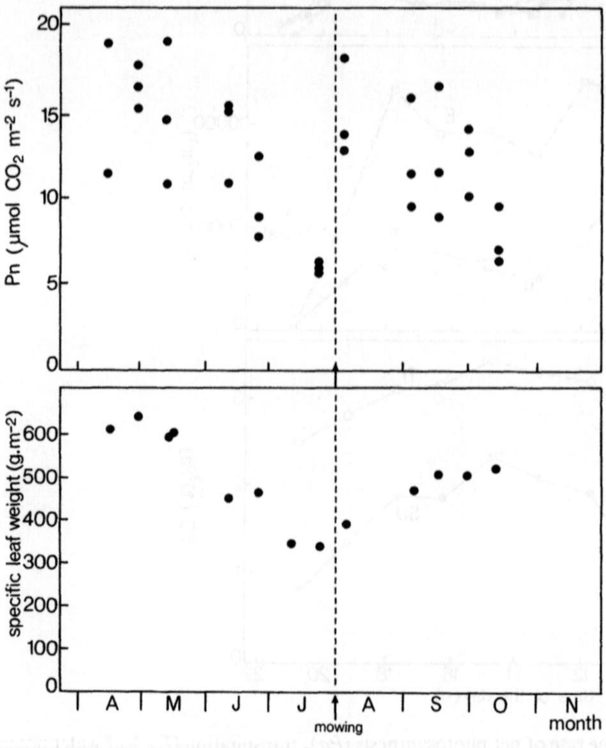

Fig. 4.10. Photosynthetic capacity (Pn-max) and specific leaf weight (SLW) of the youngest fully expanded leaf of *P. lanceolata* rosettes during the growing season at the hayfield site

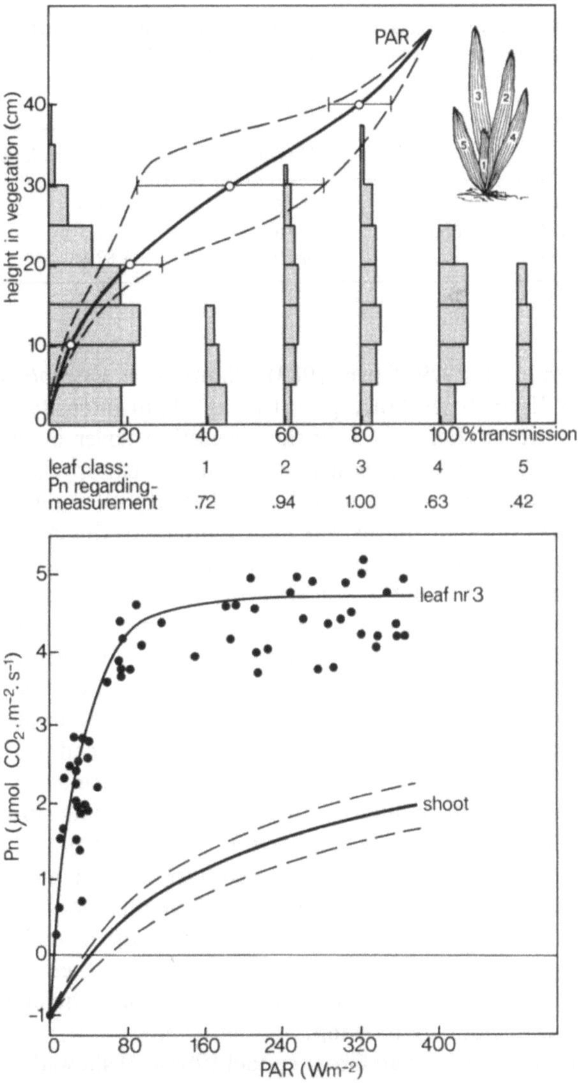

Fig. 4.11. The calculated rate of net photosynthesis (*Pn*) for a standard hayfield rosette in an undisturbed situation in relation to the light intensity above the canopy. The *columns* represent the total leaf area of the rosette per 5-cm canopy layer (*left*) and the area of the subsequent leaves (*right*). The light intensity inside the vegetation is given as percentage transmission of the photosynthetic active radiation. Pn activity of the leaves was related to the most active leaf (nr. *3*). No attention was paid to the leaf temperature, which had only a small influence on Pn in the range from 15 to 32 °C in laboratory experiments

Table 4.11. Net carbon gain of a *P. lanceolata* shoot in an undisturbed situation at the hayfield before mowing and the dune grassland (carbon gain in mg carbohydrate/day)

Weather conditions	Carbon gain	
	Hayfield	Dune grassland
Overcast, less than 20% full sunlight	−6.3	+31.9
Half overcast, between 20 and 70% full sunlight	+6.0	+71.3
Clear, more than 70% full sunlight	+10.4	+82.9

and Holmgren 1963; Clough et al. 1979; Corré 1983). Plants of *P. lanceolata* with its rosette leaves up to 35 cm could not exploit the high light intensity in the upper canopy layer of the hayfield. In July the vegetation cover in the hayfield reached its maximum height; the light intensity in the canopy layer with the leaves of *P. lanceolata* only was 5 to 10% of that of full sunlight (Fig. 4.11). Even on clear days photosynthesis was light limited. In the closed vegetation cover of the hayfield it was not possible to determine the rate of photosynthesis of single leaves, without disturbing the canopy structure. Therefore, the rate of net photosynthesis was calculated from measurements on single leaves, which received full sunlight, in combination with our recordings of light distribution in the undisturbed vegetation (Ondok and Gloser 1978). From the constructed light Pn-curve for the whole shoot in the undisturbed situation, it was calculated that the carbon balance only remained positive on clear days. On days with a heavy overcast weather the balance always was negative (Table 4.11). With the prevailing weather conditions during summer the carbon balance of the shoot may be regarded as close to zero. As it was virtually impossible to establish the rate of root respiration of plants in the hayfield, we conclude that the carbon balance for the whole plant was still less than that represented in Table 4.11. We assume that *P. lanceolata* is unable to persist in such a high vegetation for an extended period; mowing then will be essential for survival of plants in the hayfield. Indeed *P. lanceolata* quickly recovered from removal of the shoot by mowing.

Demographic studies in the hayfield (Van Groenendael 1985a, b) showed an increased death risk for *P. lanceolata* plants after flowering; also flowering in the following year was decreased. Clearly, the precarious C balance of *P. lanceolata* in hayfields contributes to decreased flowering since flowering and seed set will require energy resources. After flowering it will take at least a year to recover enough energy resources to induce flowering again.

The carbon balance for *P. lanceolata* individuals in the dune grassland (Table 4.11) remained positive, even during heavy overcast. The occurrence of the midday depression in net photosynthesis on clear days with a high evaporative demand will cause a reduction in the overall carbon gain in excess of 65%; this reduction never will set the diurnal carbon balance to negative values.

4.4 Nitrogen Nutrition in *Plantago* Species

4.4.1 Introduction

The form of mineral nitrogen delivered by the soil will largely depend on C/N ratio, soil moisture content, soil temperature and soil pH (Focht and Verstraete 1977; Haynes and Goh 1978; Beck 1979). Mineralization will be enhanced by low C/N ratios, e.g., at high nitrogen levels; a high C/N ratio will cause N-immobilization (Paul and Juma 1981). At a low C/N ratio, nitrification will proceed rapidly and its rate will be limited by the rate of ammonification. Nitrification will be affected by oxygen availability and temperature; wet soils are poor in oxygen and have a low temperature, both factors decreasing the rate of nitrification.

The growth sites of *Plantago* species differed markedly in the level of available nitrogen (Sect. 3.1); they ranged from N-poor dune grasslands, N-intermediate hayfields and chalk grasslands to N-rich road sides and field borders (Sect. 3.1). The range of soil pH was also wide, ranging from alkaline soils with *P. media* and *P. coronopus*, neutral soils with *P. coronopus*, *P. lanceolata*, *P. major* and *P. maritima* to acidic soils with *P. lanceolata* (for pH-mean values see Sect. 3.2, Table 3.3).

Plantago populations occur in soils that predominantly deliver ammonium nitrogen as well as in soils delivering mainly nitrate. The populations were present on soils with a various degree of nitrification (Ellenberg 1977; see Table 4.12). Nitrification degrees for *Plantago* population sites were obtained by incubation of soil samples in the laboratory for 4 weeks at 30 °C (Table 4.12). The five *Plantago* species may utilize ammonium- and nitrate-nitrogen. In *P. media* and *P. coronopus* nitrate will be the predominant nitrogen source and in *P. maritima* ammonium; *P. lanceolata* and *P. major* will have an intermediate position.

The nitrogen nutrition of these species was studied in more detail at two sites: a grazed grassland (Westduinen; *P. lanceolata*, *P. major* and *P. coronopus*) and a hayfield (Merrevliet; *P. lanceolata*).

Table 4.12. Nitrification degrees (% NO_3^- in total N mineralization, after Ellenberg 1977; 0<5% NO_3^-, I 5–25%, II 25–50%, III 50–75%, IV 75–90%, V>90%) in plant communities containing *Plantago* species, after Ellenberg 1977 (column 1) and at *Plantago* sites in The Netherlands (column 2). The mean total nitrogen content of the 10-cm top soil of the *Plantago* sites in The Netherlands is listed in column 3. (Troelstra et al. 1981)

Species	1	2	Total N $\mu g\ cm^{-3}$
P. coronopus		V	2720
P. lanceolata	IV	II	3480
P. major	IV	III	2220
P. maritima	II	II	4360
P. media	IV, V	V	2720

4.4.2 The Grassland Habitat

The site was an old dune grassland, used as a pasture. The organic matter content of the 10-cm top soil ranged from 2.4 to 13%, on dune tops about 3 to 4% and in the lower valleys of the field about 6 to 7% of soil dry matter. The pH (water extract) of the 10-cm top soil ranged between 4.5 and 5.5. In the lower parts, particularly around the drinking pools, pH ranged between 6 and 8. Although the organic nitrogen content of the soil ranged from 1000 to 9000 ppm, only a low amount of mineral nitrogen was present – less than 4 ppm (NH_4^+) and less than 2 ppm (NO_3^-). After 5 weeks of incubation in the field a rather low mineralization rate was observed: less than 15 ppm (NH_4^+) and less than 5 ppm (NO_3^-).

The mineralization of nitrogen was followed at different sites during 1978 and 1979 (Fig. 4.12). The highest rates of nitrogen mineralization occurred in early summer and at the end of summer. Nitrate and ammonium mineralization followed the same pattern throughout the year. More ammonium than nitrate was produced throughout the season; the ratio of ammonium and nitrate production differed between sites.

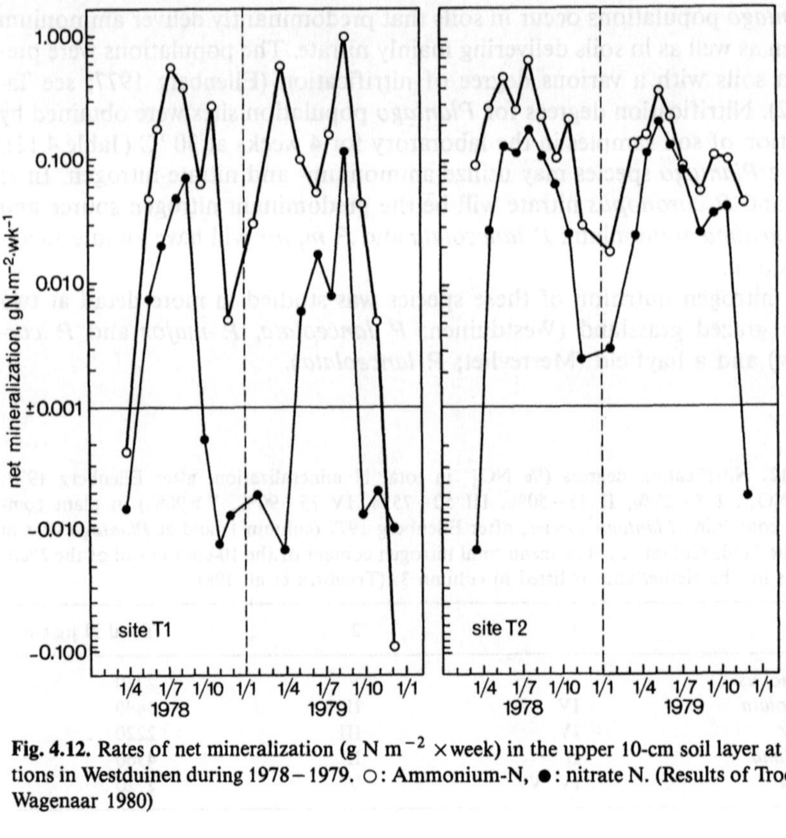

Fig. 4.12. Rates of net mineralization (g N m^{-2} × week) in the upper 10-cm soil layer at two locations in Westduinen during 1978–1979. ○: Ammonium-N, ●: nitrate N. (Results of Troelstra and Wagenaar 1980)

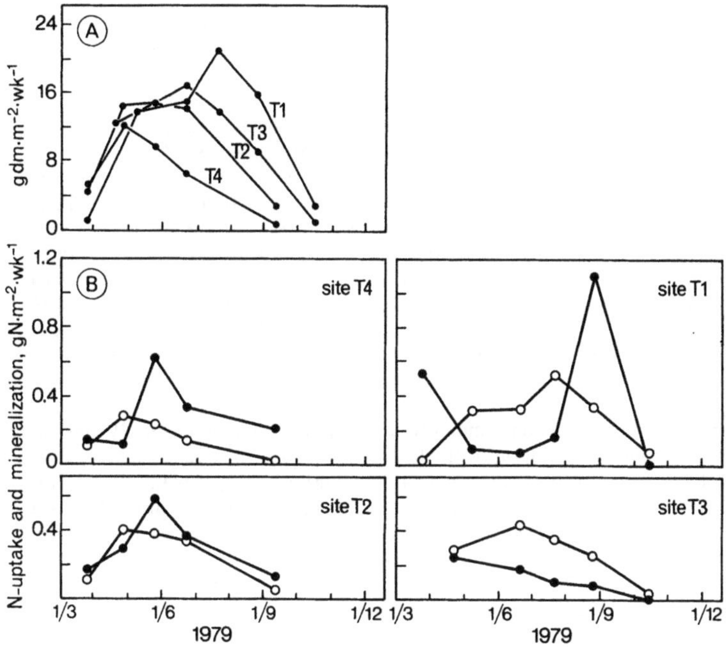

Fig. 4.13. Estimates of above-ground biomass dry matter production of *P. lanceolata* (**A**), and comparison of net production of mineral N with N uptake in the above-ground biomass (**B**) at four sites in Westduinen during 1979. The above-ground N uptake (○) was estimated by periodically harvesting and chemical analyzing. Net mineralization (●) was determined in field incubations. (Results of Troelstra and Wagenaar 1980)

The aboveground biomass of the vegetation was periodically harvested and analyzed (Fig. 4.13). The total nitrogen uptake during March–November 1979 by the above-ground biomass corresponded with the net mineral nitrogen production, observed in field incubation (Table 4.13). The yearly courses of nitrogen uptake and mineral nitrogen production showed considerable agreement (Fig. 4.13). At a particular site, T3, nitrogen uptake was consistently in excess of production, indicating redistribution of nitrogen from roots to shoots or N uptake from deeper soil layers than 10 cm. Plants from that grassland site hardly contained any free nitrate or ammonium.

The organic nitrogen content of the above-ground parts of field grown plants (Table 4.14) was compared with that of plants, grown in solution with an abundant supply of nutrients, and with nutrient-starved plants in a sand culture. The nitrogen concentration was lower in field-grown plants, about half of that of plants grown in nutrient solution. The nitrogen form may influence the concentration of organic nitrogen in the plant: ammonium resulted in a higher concentration than nitrate (Table 4.14). The nitrogen concentration in field-grown plants was considerably higher than that in a nitrogen-starved plant (Table 4.14).

Table 4.13. Comparison of nitrogen uptake by the above-ground biomass and net nitrogen mineralization (0–10-cm layer) in Westduinen at four sites. (Results of Troelstra and Wagenaar 1980)

Site	N taken up from March–Nov. 1979 ($g\,m^{-2}$)	Net mineral N production ($g\,m^{-2}$) in field incubation from March–Nov. 1979		
		NH_4-N	NO_3-N	Total
T_1	8.76	10.08	0.54	10.62
T_2	6.10	5.74	2.90	8.64
T_3	8.44	4.39	0.20	4.59
T_4	3.42	6.85	2.39	9.24

Table 4.14. The reduced nitrogen content (mmol N kg^{-1} dry matter) of shoots of *P. lanceolata* and *P. major* grown on: nutrient solution (with NH_4^+ or NO_3^- (4 mM) as the N source, data of S.R. Troelstra; column 1 and 2), pots with sand (+N: 1.5 mEq NO_3^- per plant at the start of the experiment and 0.188 mEq NO_3^- weekly from week 6–20; –N: only 1.5 mEq NO_3^- at the start). The other nutrients were supplied weekly; data refer to week 18 (data of T. Blacquière), and to plants harvested in Westduinen and in Merrevliet

Species	Nutrient solution		Sand culture		Westduinen	Merrevliet
	NH_4^+	NO_3^-	+N	–N		
P. lanceolata	3933	3005	1404	796	1527	1701
P. major	3464	2514			1557	–

Leaf nitrate reductase activity (NRA) of *Plantago* leaves (Jaworski 1971) throughout the season was considerable (Fig. 4.14). Its activity in roots was about half of that observed in leaves. The NRA of the intact plant was about 5.9 µmol per gram fresh weight per day (30 °C), using a shoot to root ratio of 2, a daylength of 16 h, and assuming that root NRA occurred continuously and shoot NRA only in the light period. The nitrogen concentration of the plants was about 200 µmol N per gram fresh tissue, allowing a relative growth rate (RGR) of 0.029 g g^{-1} day^{-1} in summer (at 30 °C). A RGR of 0.015 was calculated at 20 °C. The RGR of the above-ground part of plants of *P. lanceolata* ranged from zero to 0.02 g g^{-1} day^{-1} in summer (Sect. 4.3) with an extreme value of 0.03. There were also periods in which negative growth rates occurred, since leaves died off in dry periods. The NRA was mostly sufficient for realization of the actual growth; there was enough nitrate supplied by the soil to support NRA activity. At RGR higher than 0.015 g g^{-1} day^{-1}, uptake of ammonium was required or the plant nitrogen content would decline.

Figure 4.15 shows the numbers of nitrifiers, counted in the rhizosphere of the plants used for field NRA measurement. The highest numbers were found in spring and autumn. Nitrate production by 1.5×10^4 *Nitrobacter* cells in a steady-state (Ardakani et al. 1974) was 162 µmol N (g rhizosphere-root)$^{-1}$ wk^{-1}. With a shoot to root ratio of 2, 54 µmol N per g of whole plants was

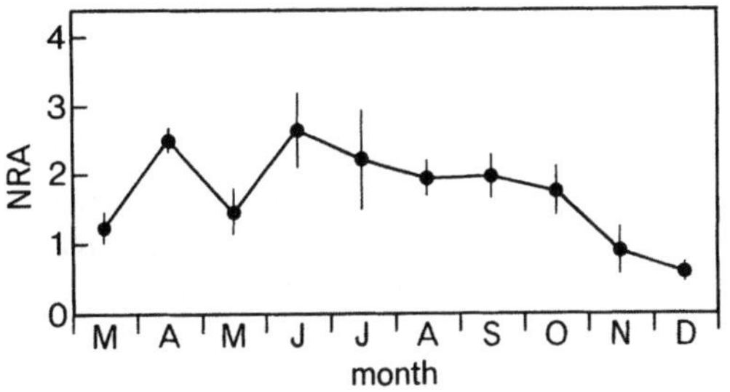

Fig. 4.14. Mean monthly NRA in the leaves of *P. lanceolata* in Westduinen [μmol NO_2^- h^{-1} (g dry weight)$^{-1}$]. Means±SE

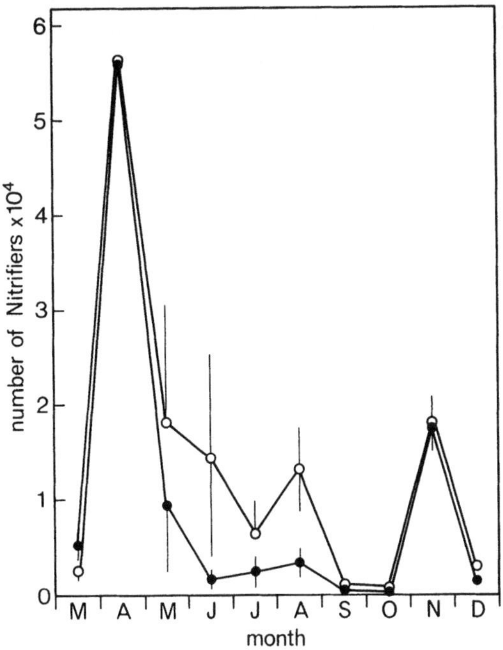

Fig. 4.15. Numbers of nitrifiers in the rhizosphere of *P. lanceolata* in Westduinen. ●: *Nitrosomonas*, ○: *Nitrobacter*. Means±SE

produced per week. Plant total nitrogen concentration was about 200 μmol N (g fresh weight)$^{-1}$. A RGR of 0.27 g g^{-1} wk^{-1} (0.034 g g^{-1} day^{-1}) was realized using nitrate from the nitrifiers in the rhizosphere as an N source. This value was twice as high as RGR permitted by NRA activity. Whether the NRA was underestimated, or the energy requirements of *Nitrobacter*, as estimated by Ardakani et al. (1974) were too high, cannot be decided. Perhaps the NRA of the roots was underestimated due to loss of young roots or parts of roots during isolation of the root system from the swards. Using the data of Ardakani et al. (1974), no correction for temperature could be made. Also a high number of nitrifiers in the soil and rhizosphere may be in a resting state (Beck 1979; Belser 1979), resulting in low energy requirements. Part of the nitrifiers may even grow heterotrophically, and thus not produce any nitrate at all (Smith and Hoare 1968).

4.4.3 Water Limiting Nitrate Utilization in the Grassland Habitat

Plant growth in the studied dune grassland could be limited by summer drought. During such dry periods, growth was strongly hampered. The effect of precipitation after a dry period upon nitrate uptake and reduction was shown in the following experiment. After a long dry period, NRA was detected in the leaves of *Plantago* at a particular site. Two days after the rain had started, NRA was measured again at the same site. After the first NRA measurements were carried out, 22 mm of rain fell (August 2; Fig. 4.16A), and on August 5 another 20 mm followed. The percentage water in the soil samples around the roots of the investigated plants rose steadily from 6% on August 2 to 30% on August 6 (Fig. 4.16B). The NRA of the leaves of *P. major* and *P. lanceolata* rose steadily with increasing soil water content (Fig. 4.16C). The nitrate content of the leaves increased from almost zero to 45 µmol NO_3^- (g dry weight)$^{-1}$ in *P. lanceolata* (Fig. 4.16D).

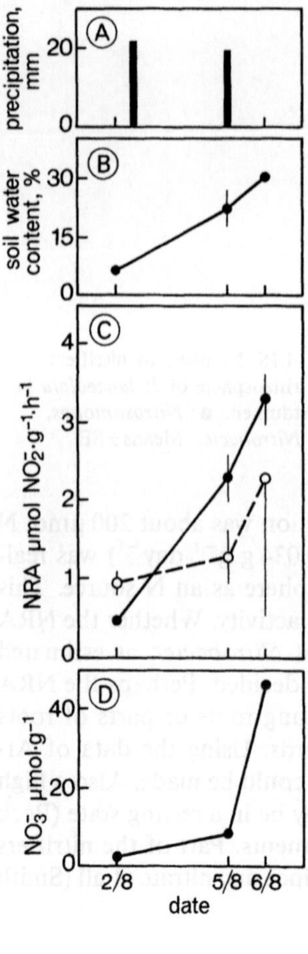

Fig. 4.16. The effect of precipitation after drought on nitrate utilization by *Plantago* in Westduinen. **A** Precipitation during the first week of August 1982. **B** Water content of the upper 10-cm soil layer. Water as a percentage of soil fresh weight (w/w). **C** NRA in the leaves of *P. lanceolata* (●) and *P. major* (○) in µmol NO_2^- (g dry weight)$^{-1}$ h^{-1}. Each *point* represents the mean ± SD of three plants. NRA of each plant was assayed in triplicate. **D** Nitrate concentration in the same leaves of *P. lanceolata* [µmol NO_3^- (g dry weight)$^{-1}$]

The results strongly suggest that the nitrate availability was limited by water stress. Nitrate reserves in the soil were present but could not be utilized by the roots. Immediately after the soil water was raised, nitrate entered the plants again and NRA was induced. The induction of NRA was not high enough to reduce all absorbed nitrate, and nitrate accumulated. This situation differs from that described by Havill et al. (1977), and by Leclerc and Robin (1983, 1985). Their results showed a reduction of the NRA during dry periods but the reduction was accompanied by an increase in nitrate content of the leaves. The water and nitrate flows inside the plant do not seem to restrict NRA, the latter being directly drought-dependent. In *Plantago* under drought the restriction in N utilization seems to be located in the root environment, or at the level of N uptake by the roots in dry soil.

4.4.4 The Hayfield Habitat

At the study site, a wet hayfield on pure peat (Merrevliet), only *P. lanceolata* occurred. The groundwater table was almost at the level of the soil surface, and the soil water content was 85 to 90%. The organic matter content of the soil was 85% of soil dry weight, and total nitrogen content was 23000 ppm. The pH was 5.3 to 6.8.

NRA of *P. lanceolata* was low but significant throughout the entire season. Except for a small period in spring (Fig. 4.17), nitrate reduction was very limited. Nitrifiers in the rhizosphere were only detectable in spring; rather low numbers were observed (Fig. 4.18). Percolation of soil samples with nitrite resulted

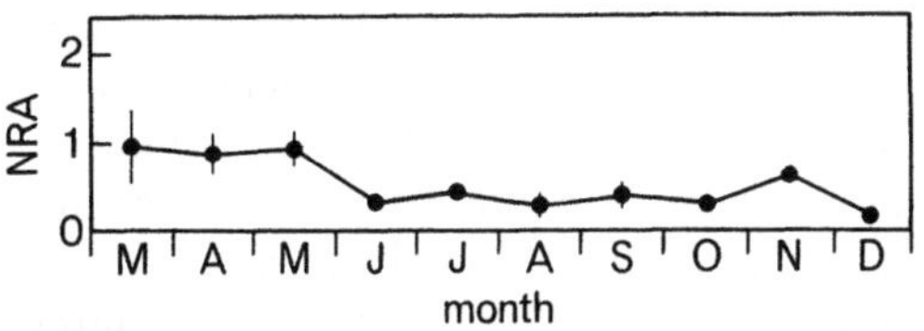

Fig. 4.17. Mean monthly NRA [μmol NO_2^- (g dry weight)$^{-1}$ h^{-1}] in the leaves of *P. lanceolata* in Merrevliet. Means ± SE

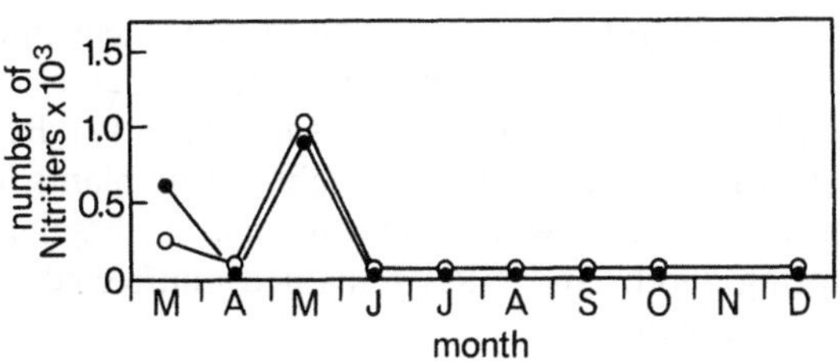

Fig. 4.18. Numbers of nitrifiers in the rhizosphere of *P. lanceolata* in Merrevliet. ●: *Nitrosomonas*, ○: *Nitrobacter*

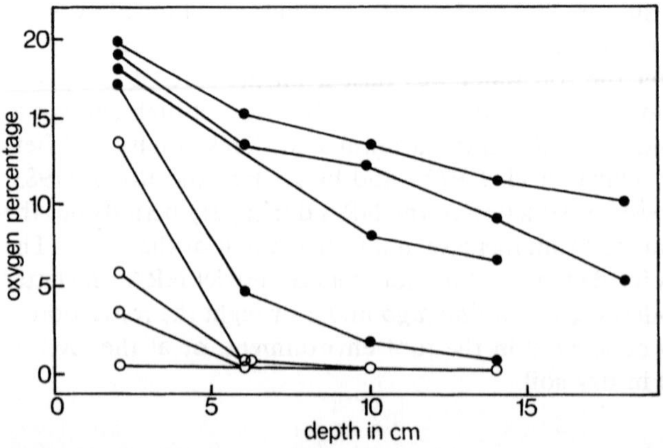

Fig. 4.19. Oxygen profile of the upper soil of the wet hayfield in PVC tubes containing *Plantago lanceolata* alone (○) or with additional vegetation (●). Each *profile* presents the oxygen profile of the soil of a single cylinder

in enrichment of *Nitrobacter*. The numbers were very low and ammonium was the predominant nitrogen source in Merrevliet. No mineralization rates could be measured on samples of Merrevliet soil due to the vulnerability of the hayfield. *P. lanceolata* shoots grew with a RGR of 0.03 g g^{-1} day^{-1} during a great part of summer (Sect. 4.3), rather high in comparison with the plants in the grassland. Growth proceeded without significant fluctuation, in contrast with the grassland habitat. The nitrogen concentration in the plants was rather high, in comparison with grassland plants (Table 4.14). Moreover, the caudices of *P. lanceolata* in the hayfield contained large reserve pools of nitrogen. Together with the observed stable growth rates, this suggests that there was a continuous and sufficient supply of ammonium in the wet hayfield habitat.

To explain the presence of NRA in the leaves the following experiments were conducted. It became very soon clear that nitrate addition by precipitation was present, 4 to 18 μM, but far too small to induce NRA; after a rain shower, NRA of the hayfield plants was not increased. Another alternative for nitrate acquisition by *P. lanceolata* was nitrate production by nitrification. The latter process may occur in wet habitats, provided oxygen is available. Development of aerenchymous roots by *P. lanceolata* itself and by neighbouring plants (other species) is important in this respect. In field tests it soon became clear that roots of *P. lanceolata* and other species in the hayfield contained large air spaces (Blacquière 1988). Such plants added oxygen to the soil (Fig. 4.19). But *Plantago* plants did not benefit from oxygen supply by neighbouring plants in the hayfield. This was demonstrated in plants which were grown in the hayfield in PVC cylinders: survival, rooting depth, and NRA remained unaffected by the enclosure of the roots in a PVC cylinder.

5 Ecophysiology of *Plantago* Species

In this chapter a first analysis of the relation between *Plantago* species and their habitat will be given. In practically all experiments seeds were supplied from a single seed source: the *Plantago* nursery at the Field Station at Oostvoorne. Availability of soil nutrients and functioning of *Plantago* root systems will be evaluated as carboxylate production and proton excretion (Sect. 5.1), mineral nutrient concentration in the soil and in the plant (Sect. 5.2) and nitrogen acquisition and utilization (Sect. 5.3). A more extreme situation, the presence of NaCl in the soil and its effects on functioning of the salt-sensitive and salt-tolerant *Plantago* species will be given in Section 5.4.

Turning to climatic factors and functioning of *Plantago* species, attention is given to frost tolerance of the coastal species and its interaction with NaCl (Sect. 5.5) and light intensity (Sect. 5.6). Finally a *first* synthesis of the phenotypic adaptation of *Plantago* species is given in Section 5.7.

5.1 Carboxylate Production and Proton Excretion

5.1.1 Introduction

Unequal absorption of cations and anions by higher plants, the subsequent transport within the plant and partial utilization of some of these ions invoke the necessity of charge-balance mechanisms to ensure electrical charge neutrality (Raven and Smith 1976; Osmond 1976; Pitman and Cram 1977; Smith and Raven 1979). Nitrogen has the greatest impact upon the ionic balance of plants because (1) it accounts for at least half of the total number of ions absorbed; (2) it is absorbed as an anion (NO_3^-) or a cation (NH_4^+); (3) its ionic form is changed by metabolism (Hiatt and Leggett 1974; Reisenauer 1978). Upon the incorporation of NO_3^- (and SO_4^{2-}) into organic compounds, one OH^- per equivalent reduced is produced; the utilization of NH_4^+ involves the generation of one H^+. The form of N nutrition has a large impact in terms of internal and external pH effects.

The higher plant has several means of preventing internal pH imbalances as active extrusion and biochemical utilizations. The first strategy is applied for NH_4^+ assimilation, which takes place mainly in the root, and for NO_3^- reduction in the roots. For NO_3^- reduction in the shoots, there is a spatial constraint on external disposal of OH^-; OH^- may be removed by carboxylation (the formation of carboxyl groups) via the biochemical pH stat (Davies 1973 a, b, 1986; see also Raven and Smith 1976; Smith and Raven 1979; Raven 1985); the reactions involved in introducing and removing carboxyl groups from metabolites such as, e.g., malate may constitute such a pH stat (Davies 1986). Possible osmotic problems can be overcome by precipitation as with Ca-oxalate or by transport of the organic anions to the roots where OH^- (HCO_3^-) is regenerated by decarboxylation (removal of carboxyl groups) and excreted into the external medium (Raven and Smith 1976).

The overall proton excretion by a plant or vegetation can be expressed in units of mEq $plant^{-1}$ or mEq m^{-2} by the following equation (Troelstra et al. 1985; Troelstra and Blacquière 1986):

$$H^+ \text{ efflux} = (C-A) + 0.946\, N_{org} - 2x - y \, .$$

x and y represent the organic N derived from NO_3^- and dinitrogen fixation, respectively; N_{org} is the total organic N content, and C-A, the difference between accumulated total amounts of cations and inorganic anions, is a measure of the content of carboxylates.

The form of N nutrition causes large differences in plant ion composition. In comparison with NO_3^--fed plants, NH_4^+-fed plants contain relatively low amounts of cations and organic anions, and high amounts of inorganic anions (Mengel 1974). The size of the carboxylate pool will decrease when total anion uptake exceeds cation uptake and when NH_4^+ is utilized. The pool will increase in size when cation uptake exceeds anion uptake and when SO_4^{2-} and NO_3^- are reduced (Houba et al. 1971; Van Egmond 1978).

The concept of an active H^+ extrusion pump can account for all changes in the ionic composition of plants in response to different nitrogen sources; this concept has unified many phenomena in plant ion relations (Raven and Smith 1973). An ATPase-mediated H^+ efflux (proton pump) represents the primary step in the model on ion uptake described by Israel and Jackson (1978); the model adequately accounts for pH regulation and cation-anion balance in plant tissues (Israel and Jackson 1978, 1982; Kirkby 1981; Dejaegere et al. 1981). Organic anions can be generated in response to the operation of the H^+ extrusion pump or processes of nitrate and sulphate reduction.

The concentration of carboxylates, C-A, is in part determined by the form in which N is taken up; it is generally lower for ammonium than for nitrate supply. De Wit et al. (1963) introduced a "normal" carboxylate content, a specific level of organic anions required for optimal growth. However, C-A concentrations may differ considerably for various plant parts and they usually show a decrease during the growing period (Van Egmond 1975). Moreover, increased NO_3^- fertilization will increase organic anion levels in tissues and yield response to N should correlate with organic anion level (Hiatt and Leggett 1974; Kirkby and Knight 1977). The relation between C-A value and yield does not answer the question whether poor growth is due to a low organic anion content per se (Mengel 1974).

It has been suggested that the (C-A)/N_{org} ratio may indicate the extent of ammonium relative to nitrate nutrition. Ammonium assimilation has opposite effects on C-A and organic N accumulation, and the ratio will decrease when more ammonium relative to nitrate is taken up by the plant. However, the picture is made more complicated by the effect of the *level* of nitrate supply: particularly in gramineous plants with their relatively low (C-A)/N_{org} ratios, even under exclusive NO_3^- nutrition (Banwart and Pierre 1975; Argyriadis et al. 1976; Van Egmond 1978; Wollenweber and Kinzel 1988). Therefore, a carboxylate/amide ratio would be more useful to characterize the extent of NH_4^+ nutrition than the (C-A)/N_{org} ratio (Reisenauer 1978; Van Egmond 1978; Wollenweber and Kinzel 1988). On the other hand, relatively high (C-A)/N_{org} ratios (about 1 and higher) will generally be related with a supply of N in the form of NO_3^-.

In this section, results are presented of the composition of *Plantago* species, sampled together with the soil studies of the *Plantago* sites (Sect. 3.2). In addition, comparisons are made with results obtained under more or less controlled conditions in field trials and in greenhouse experiments. Emphasis is on the carboxylate (C-A) metabolism of *Plantago* species; the following section discusses the tissue concentrations of cations and inorganic anions.

5.1.2 Methods

The effect of micronutrients on the ionic balance is often negligible. The quantitatively important inorganic constituents of the balance are NH_4^+, K^+, Na^+, Ca^{2+}, Mg^{2+}, NO_3^-, $H_2PO_4^-$, Cl^-, and SO_4^{2-}. A short description of the procedure for ion determination and total nitrogen in the plant material is given in Section 5.2. The ionic balance of a plant refers to the sum of the accumulated cations (C), the sum of the inorganic anions (accumulated with preservation of their electronegative charge) (A), and the carboxylates (C-A). Organic P is also included in A.

The total of carboxylates, C-A, was calculated from the sums of the inorganic ion concentrations; organic nitrogen was estimated by subtraction of NO_3^- plus NH_4^+ from total N. All values were calculated on the basis of 70 °C-dried material.

5.1.3 Natural Habitats

Plantago species revealed a sequence of C-A levels which decreased in the order *P. media* > *P. major* > *P. lanceolata* > *P. coronopus* > *P. maritima* (Table 5.1, Fig. 5.1). Among the glycophytic species, the sum of cations (C) decreased strongly in the same order, *P. media* > *P. major* > *P. lanceolata*, at about equal levels of inorganic anions (A), resulting in decreasing carboxylate levels in the same order. Both C and A concentrations were high in the halophytic *P. coronopus* and *P. maritima* and carboxylate levels were relatively low, particularly in *P. maritima*.

Table 5.1. Sum of cations (C), sum of inorganic anions (A), carboxylates (C-A), organic nitrogen, and (C-A)/N_{org} ratios in leaves of field-grown *Plantago* species collected at different locations in The Netherlands. Mean values ± S.E.

Species	N[a]	mEq (mmol) (kg dry weight)$^{-1}$				(C-A)/N_{org}
		C	A	C-A	N_{org}	
P. media	16	2650 ± 77	560 ± 61	2090 ± 83	1170 ± 54	1.86 ± 0.13
P. major	25	2280 ± 56	780 ± 50	1500 ± 45	1350 ± 66	1.21 ± 0.09
P. lanceolata	74	1810 ± 30	640 ± 25	1170 ± 25	1240 ± 33	0.98 ± 0.03
P. coronopus	4	2640 ± 267	1610 ± 349	1030 ± 94	1240 ± 130	0.88 ± 0.16
P. maritima	16	3970 ± 134	3190 ± 119	780 ± 65	1360 ± 82	0.62 ± 0.08

[a] N = number of samples.

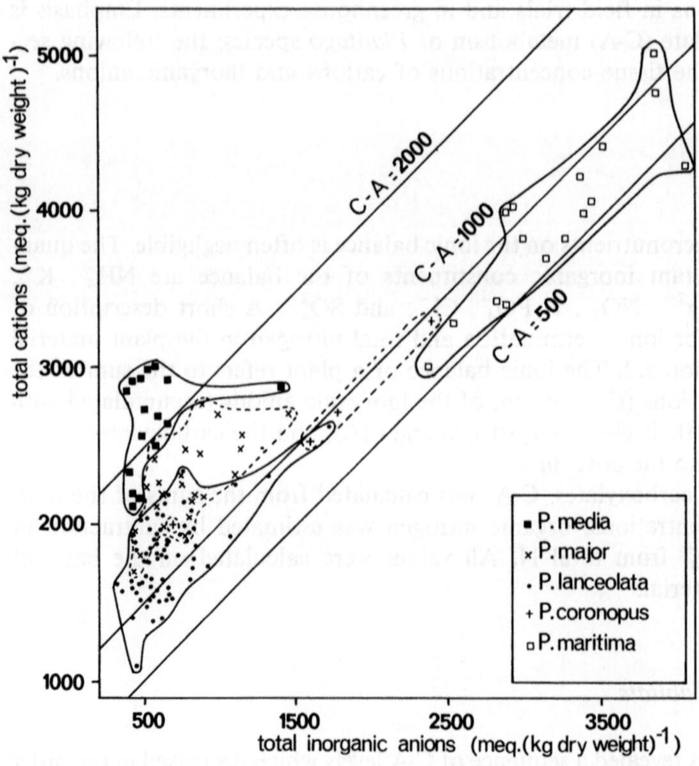

Fig. 5.1. Relation between amounts of accumulated cations (*C*) and inorganic anions (*A*) in leaves of field samples of *Plantago* species collected at different locations in The Netherlands

The high C-A concentration of *P. media* and *P. major* might be related to the generally high pH value of the soil of their habitats: i.e. nitrogen largely available in the form of nitrate (Sect. 3.2). The relatively low C-A level in *P. maritima* might be mainly caused by salinity. Organic-nitrogen concentrations

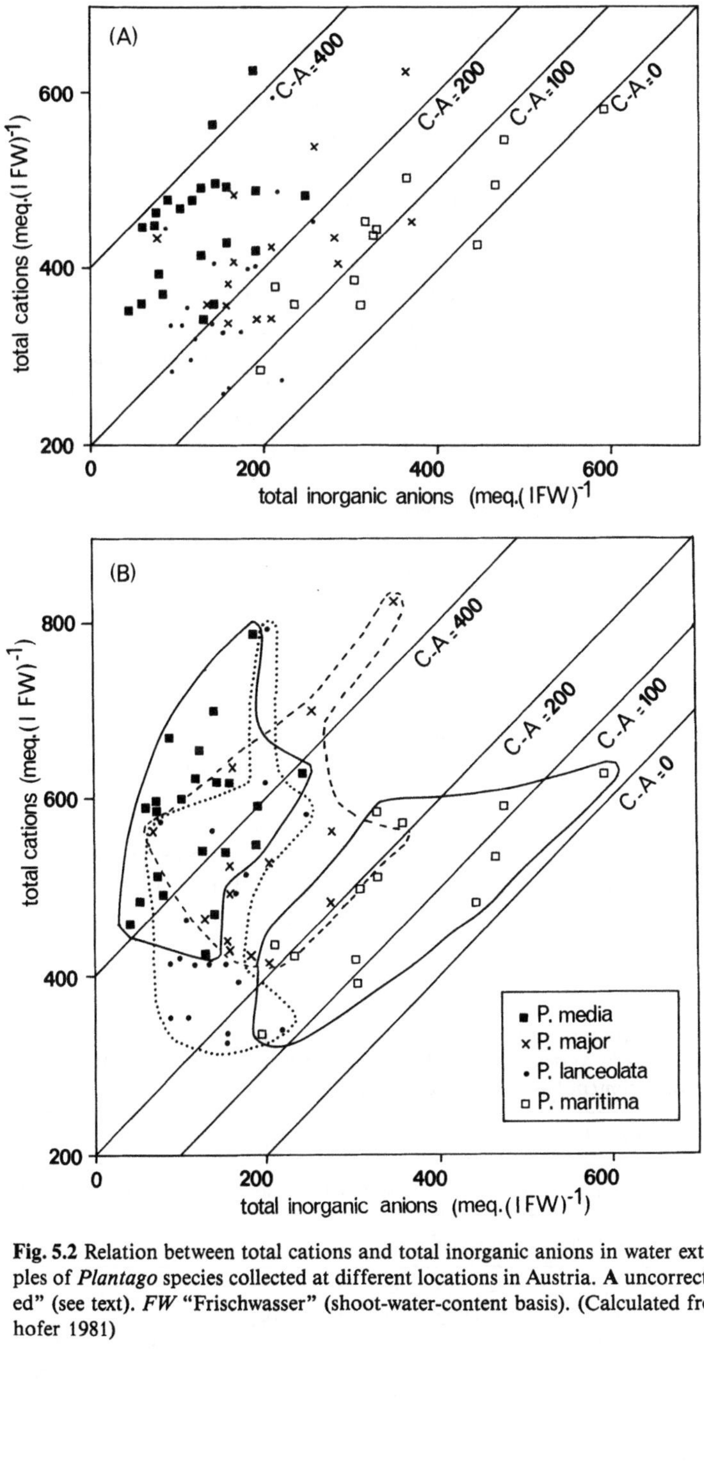

Fig. 5.2 Relation between total cations and total inorganic anions in water extracts of shoot samples of *Plantago* species collected at different locations in Austria. **A** uncorrected, and **B** "corrected" (see text). *FW* "Frischwasser" (shoot-water-content basis). (Calculated from data of Königshofer 1981)

in the field material of the five species were of the same magnitude; the decreasing (C-A)/N_{org} ratio in the order *P. media* to *P. maritima* might indicate the effects of N form and/or salinity (Table 5.1).

The above results were compared with data of Königshofer (1981) for *Plantago* species in Austria: C and A were calculated from data on individual ion concentrations and plotted as in Fig. 5.1, yielding Fig. 5.2A. Assuming a dry matter percentage of 17 in the Austrian field material makes Figs. 5.1 and 5.2A about similar. Further correction is possible since methodical differences were also involved: the single hot-water extraction method (Königshofer 1981), compared with the present digestion/extraction procedures (Troelstra 1983), resulted in similar data for K^+, Na^+, NO_3^-, Cl^-, and SO_4^{2-}, but differences occurred with Ca^{2+}, Mg^{2+} and $H_2PO_4^-$. The hot-water extraction values were approximately 35% (Ca^{2+}), 15% (Mg^{2+}), and 45% ($H_2PO_4^-$) lower. Transformation of the data of Fig. 5.2A into Fig. 5.2B, using corrections for Ca^{2+}, Mg^{2+}, and $H_2PO_4^-$, shows a quantitatively very similar picture for *Plantago* in Austria (Fig. 5.2) or in The Netherlands (Fig. 5.1).

Königshofer (1981; see also Königshofer and Kinzel 1986) analyzed the components of the carboxylate pool, which differed among the species (Table 5.2). Malonate was the main component in *P. media*, whereas the carboxylate pool of *P. major* consisted for almost 80% of iso-citrate. *P. lanceolata* and *P. maritima* accumulated the more commonly occurring malate and citrate. *P. lanceolata* was the only species that contained a considerable percentage of fumarate (average of 20%; not shown in Table 5.2). Malate and citrate were the most important organic anions in the roots of all species. The composition of carboxylates was species-specific and rather constant during the season.

P. major ssp. *major* and *P. lanceolata* were periodically sampled on two sites each (see Sect. 5.2). In general, organic anion levels were highest in *P. major* ssp. *major*, in agreement with the above conclusions. Moreover, carboxylate concentrations were fairly constant for the greater part of the season.

Table 5.2. Carboxylate (C-A) concentrations in shoot material of various *Plantago* species collected at different locations in Austria (calculated from data of Königshofer 1981). Data are expressed on a shoot-water-content basis. Mean values ± S.D.

Species	N[a]	$mEq\,l^{-1}$		% of Σ (C-A)			
		Σ (C-A)		Malonate	Iso-citrate	Malate	Citrate
		Uncorrected	Corrected				
P. media	22	324 ± 69	455 ± 82	80 ± 9	–	12 ± 6	4 ± 1
P. major	15	230 ± 60	329 ± 102	6 ± 4	78 ± 7	8 ± 4	6 ± 2
P. lanceolata	20	195 ± 80	313 ± 110	15 ± 12	–	42 ± 11	18 ± 9
P. maritima	13	79 ± 48	144 ± 72	15 ± 9	–	49 ± 13	27 ± 11

[a] N = number of samples.

5.1.4 Greenhouse and Field Experiments

The analysis of *Plantago* species grown in the greenhouse experiments and on field plots indicated the same sequence of carboxylate concentrations [mEq (kg shoot dry weight)$^{-1}$]: >1000 (*P. media, P. major*); approximately 1000 (*P. lanceolata* and *P. coronopus*); and <1000 (*P. maritima*) (Troelstra and Smant 1979, 1980). These findings together with those of the previous section led to the hypothesis that carboxylate concentrations in *Plantago* could possibly reflect adaptations to habitat factors as pH, N source and salinity. For *P. media* and *P. major* a more or less obligatory preference for nitrate was indicated (Troelstra et al. 1983b).

However, water culture experiments with five NO_3^-/NH_4^+ ratios (at a total N level of 1 to 4 mM and a pH of 6) did not provide evidence for a requirement of different specific organic-anion levels for growth of the five species. Maximum yields for all species were obtained at mixtures of NO_3^- and NH_4^+, which corresponded with intermediate carboxylate concentrations in the shoot (Fig. 5.3). On the other hand, a relatively high C-A concentration in the shoots of *P. media* and *P. major* could still be noticed at all NO_3^-/NH_4^+ ratios, whereas C-A of the roots did not demonstrate any clear difference.

Table 5.3 summarizes results of soil-grown *Plantago* species in pot cultures in the greenhouse and in field plots; nitrate had been the predominant form of N. The field-plot data again suggest the same order of decreasing carboxylate concentration among the species. The picture is somewhat less clear for the pot cultures, but this may in part have been due to differences in N supply, which was higher in the pot cultures. Under the relatively low N conditions of the field plots, *P. media* in particular still accumulated high amounts of organ-

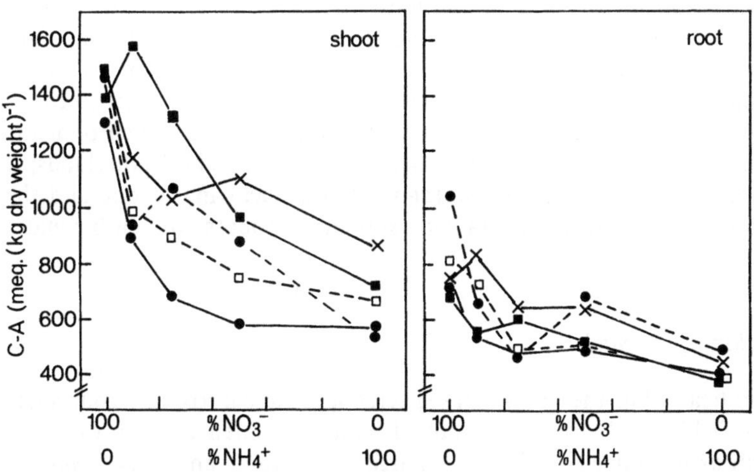

Fig. 5.3. Carboxylate concentrations in *Plantago* species grown on nutrient solutions at different NO_3^-/NH_4^+ ratios (total N, 1–4 m*M*). *P. media* (■——■); *P. major* (×——×); *P. lanceolata* (●——●); *P. coronopus* (● - - - ●); *P. maritima* (□ - - - □)

Table 5.3. Carboxylate (C-A) concentrations and (C-A)/N_{org} ratios in the leaves of *Plantago* species grown in soil under identical experimental conditions

Species	C-A [mEq (kg dry weight)$^{-1}$]			(C-A)/N_{org}		
	Experimental field plot[a]	Pot cultures[b]		Experimental field plot	Pot cultures	
		1983	1984		1983	1984
P. media	1560	2400	2550	2.13	1.60	1.75
P. major	1150	2660	1710	1.89	2.30	2.03
P. lanceolata	960	1730	1480	0.93	1.38	1.58
P. coronopus	790	2000	1740	0.82	1.11	1.82
P. maritima	–	1890	1450	–	1.26	1.29

[a] Study on the influence of trampling and soil compaction (Blom 1979).
[b] Using soil from the locality Westervoort, The Netherlands (% $CaCO_3 > 0.1$, pH-H_2O 6–8; *P. media* and *P. lanceolata* were the naturally occurring species at this site). 1983: mean values of five successive harvests (clippings). 1984: mean values of two to three successive harvests (clippings).

ic anions; in the pot cultures, a similar picture emerged for *P. media*, and to a certain degree also for *P. major*, but distinct differences among the other species were absent.

A summary of pot and water culture experiments in the greenhouse is shown in Fig. 5.4A–D). Differences in organic-anion concentrations among the species in Fig. 5.4A–C were much less pronounced than those found in field material, especially in 100% ammonium nutrition cultures (Fig. 5.4C). The species order of C-A concentrations could be only noticed in Fig. 5.4B, which referred to 100% nitrate nutrition. Figure 5.4D presents an example of the large variation in carboxylate amounts in *P. maritima*. The dry matter yields of plants treated with either demineralized water or sea water were approximately equal, the C-A levels showing an almost fourfold difference. *P. maritima* samples collected at Austrian sites generally had C-A concentrations intermediate between these two extremes, which can be related to environments of minor salt stress compared to the coastal locations in The Netherlands; it is further remarkable that three *Plantago* species, occurring at one and the same location in Austria, contained similar amounts of carboxylates (Fig. 5.4D).

5.1.5 Concluding Remarks

The data indicate that the apparently species-specific carboxylate concentrations in the field mainly were the result of habitat differences with respect to factors such as N source or salinity. A particular organic anion level was not required for normal growth of *Plantago*.

Although the occurrence of *P. media* is restricted to calcareous soils mainly (Sect. 3.2), the species can grow relatively well on 100% ammonium nutrition,

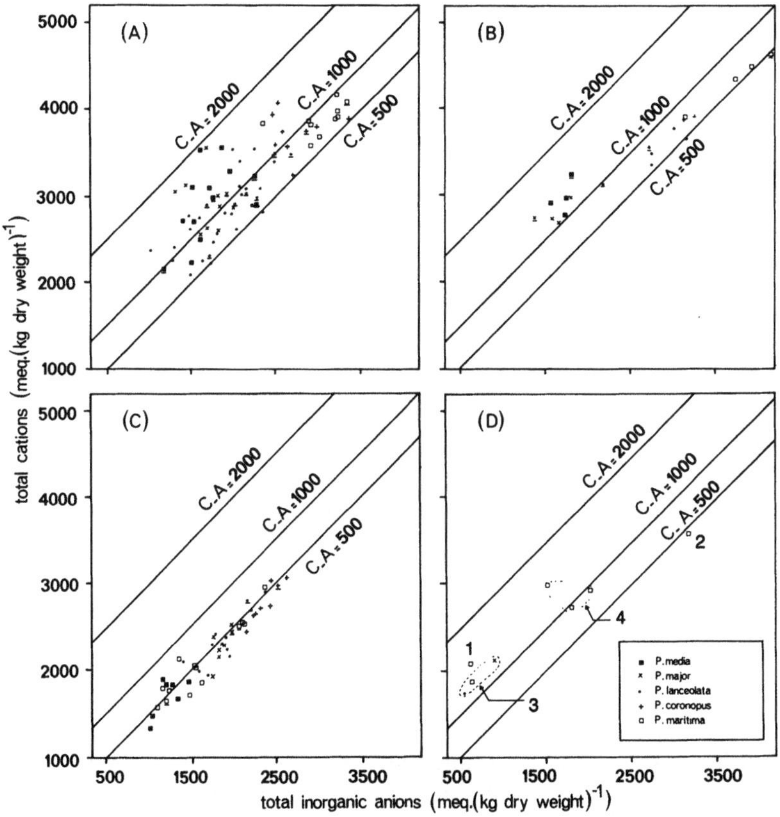

Fig. 5.4. Relation between total cations and total inorganic anions in shoot samples of *Plantago* species grown under identical experimental conditions, unless stated otherwise. **A** Water culture experiments, different NO_3^-/NH_4^+ ratios. **B** Water culture experiments, 100% NO_3^- with different treatments with respect to K-salts (K_2SO_4 or KCl) and NaCl (0 or 10mM). **C** Water culture experiments, 100% NH_4^+ with different ($K^+ + Ca^{2+} + Mg^{2+}$)/Na^+ ratios. **A–C** Maximum yields of species per treatment are underlined. **D** Pot cultures (greenhouse experiment) and field samples (Austria). *1* demineralized-water treatment; *2* sea-water treatment; one (*3*) or different sites (*4*)

provided the acidity and the cation supply, especially Ca^{2+}, were adequately controlled. Perhaps a need for a sufficient cation accumulation, in particular K^+ and Ca^{2+}, in *P. media* and *P. major* may explain the relatively high C-A amounts in these species; they support a direct relationship between carboxylate synthesis and increased cation accumulation (Hiatt and Leggett 1974).

Results obtained by Smouter et al. (1984) suggested a higher activity and capacity of proton pumps in roots of the calcicole *P. media* than in those of the more calcifuge *P. coronopus*. They concluded that this phenomenon could play a role in adaptation of plants to calcareous soils, where the efficiency of the proton pump with regard to ion uptake might be reduced by the presence of a strong buffering system. On the other hand, the low pH limit, tolerated

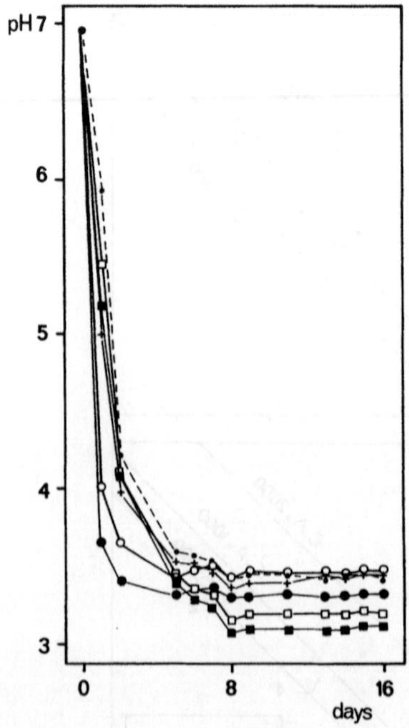

Fig. 5.5. Time courses of nutrient solution pH during growth of *Plantago* species on 100% NH_4^+, without pH adjustment. •, *P. major* ssp. *major*; ○, *P. major* ssp. *pleiosperma*; +, *P. media*; ●, *P. lanceolata*; □, *P. maritima*; ■, *P. coronopus*

by *Plantago*, will also be determined by the ability of the proton pump to cope with increasing external acidity (Smith and Raven 1979). When grown on 100% ammonium and without pH adjustment, a final pH level of approximately 3.3 was reached in all species, with perhaps a slightly lower value for *P. coronopus* (Fig. 5.5).

5.2 Mineral Nutrient Concentrations in the Soil and in the Plant

Apart from recent information by Königshofer (1981), Königshofer and Kinzel (1986) and Wollenweber and Kinzel (1988), there are no quantitative data available on the concentrations of organic nitrogen and individual ions as NH_4^+, K^+, Na^+, Ca^{2+}, Mg^{2+}, $H_2PO_4^-$ NO_3^-, Cl^-, SO_4^{2-} in *Plantago* in their environment. This section summarizes the results of chemical analyses on field-collected plant material, sampled concurrently with soil samples of the same habitat (Sects. 3.2.3 and 3.2.4). In addition, the results of comparative growth studies in pot cultures and in field experiments are discussed.

5.2.1 Methods

Above-ground plant parts were collected in the field and upon arrival at the laboratory, samples were cleaned of impurities and rinsed for a short period with demineralized water, dried at 70 °C for 48 h, and finely ground.

Plant samples were either digested in a concentrated sulphuric acid/salicylic acid mixture with hydrogen peroxide or extracted with demineralized water. In the digests, total N and $H_2PO_4^-$ were determined colorimetrically and K^+, Na^+, Ca^{2+} and Mg^{2+} by atomic absorption. In the water extracts, NO_3^- and NH_4^+ were measured with ion selective electrodes, Cl^- by a potentiometric end point titration, and SO_4^{2-} by turbidimetry. Organic nitrogen was estimated by subtraction of NO_3^- plus NH_4^+ from total N.

Part of the samples was also analyzed for ethanol-soluble carbohydrates and starch (Fales 1951).

5.2.2 The Chemical Composition of Species in the Environment – a Survey

Table 5.4 gives a summary of the ion concentrations of *Plantago* species. Keeping in mind that tissue age differences were involved, the following conclusions may be drawn. Organic nitrogen was similar among the five species. The potassium concentration was relatively high in *P. media* and *P. major* (data refer to ssp. *major* and ssp. *pleiosperma*), and that of magnesium relatively high in *P. coronopus* and *P. maritima*. The average sodium concentration increased in the order *P. media* < *P. major* < *P. lanceolata* < *P. coronopus* < *P. maritima*, whereas calcium showed an increase in the opposite direction.

On an average, nitrate levels were always low, less than 50 mEq kg^{-1} dry matter, in all species; however, some accumulation of nitrate was incidentally observed in the leaves of *P. major* and *P. lanceolata*, 50 to 250 mEq kg^{-1} d.m. The concentration of phosphate was relatively low in *P. media* and *P. lanceolata* and that of chloride was high in *P. coronopus* and even more so in *P. maritima*. Sulphate was high in *P. maritima* and, among the glycophytic species, relatively high in *P. major*. Most of the differences in leaf ion composition were also noticeable, but less pronounced, in the spikes (Table 5.4B).

Königshofer (1981) analyzed four *Plantago* species from several locations in Austria (Table 5.5). There was a good agreement of the results on the Austrian plant material and those mentioned above. Table 5.6, summarizing analyses of different *Plantago* species originating from the same location in Austria, illustrates some of the differences between *Plantago* species listed above.

Plotting the P and K concentrations of the leaves of the field material against the corresponding Olsen-P and exchangeable K values of the 0 to 10 cm soil layer of the natural sites revealed characteristic differences among the glycophytic *Plantago* species (Fig. 5.6). While ion levels generally increased with higher ion availability values in the soil, the P concentration in the dry matter of *P. major* was relatively high at low levels of soil Olsen-P; this species was also more often found at relatively high values of Olsen-P (Fig. 5.6A). On the other hand, P levels in *P. media* and *P. lanceolata* reached very low values, approximately 30 mEq kg^{-1} d.m., which were much lower than the more or

Table 5.4. Ionic contents and organic nitrogen in leaves (A) and spikes (B) of field-grown *Plantago* species collected at different locations in The Netherlands. Means ± S.E.

Species	mEq (mmol) kg^{-1} dry weight								
	K^+	Na^+	Ca^{2+}	Mg^{2+}	$H_2PO_4^-$	NO_3^-	Cl^-	SO_4^{2-}	N_{org}
(A)									
P. media (n = 16)[a]	734 ± 52	36 ± 16	1661 ± 39	221 ± 11	65 ± 6	23 ± 2	310 ± 60	163 ± 10	1170 ± 54
P. major (n = 25)	844 ± 49	39 ± 14	1148 ± 57	246 ± 10	111 ± 6	19 ± 6	274 ± 48	378 ± 18	1347 ± 66
P. lanceolata (n = 74)	397 ± 23	347 ± 31	883 ± 27	184 ± 6	79 ± 4	13 ± 4	298 ± 24	251 ± 7	1236 ± 33
P. coronopus (n = 4)	234 ± 33	963 ± 257	916 ± 153	525 ± 109	126 ± 29	16 ± 9	1145 ± 271	320 ± 68	1239 ± 130
P. maritima (n = 16)	387 ± 28	2280 ± 112	715 ± 43	580 ± 17	102 ± 5	10 ± 3	2488 ± 134	587 ± 49	1358 ± 82
(B)									
P. media (n = 12)	605 ± 32	18 ± 8	521 ± 36	122 ± 8	81 ± 10	8 ± 1	126 ± 30	84 ± 5	1024 ± 95
P. major (n = 4)	736 ± 86	24 ± 19	451 ± 66	177 ± 14	157 ± 17	33 ± 18	126 ± 76	129 ± 25	1455 ± 74
P. lanceolata (n = 63)	375 ± 15	152 ± 14	315 ± 10	115 ± 3	82 ± 4	6 ± 2	90 ± 10	89 ± 2	972 ± 31

[a] n = number of samples analyzed.

Table 5.5. Ionic contents in leaves of field-grown *Plantago* species collected at different locations in Austria. (Calculated from data of Königshofer 1981). Mean ± S.E.

Species	No. of locations	No. of samples analyzed	mEq l^{-1} FW[a]							
			K^+	Na^+	Ca^{2+}	Mg^{2+}	PO_4^{3-}	NO_3^-	Cl^-	SO_4^{2-}
P. media	12	22	166 ± 7	2 ± 0.8	222 ± 9	59 ± 12	8 ± 0.8	1 ± 0.6	73 ± 11	45 ± 4
P. major	11	15	149 ± 10	2 ± 0.5	184 ± 15	87 ± 15	18 ± 2	3 ± 0.8	89 ± 13	105 ± 43
P. lanceolata	14	20	141 ± 7	3 ± 5	163 ± 14	56 ± 8	18 ± 2	3 ± 0.9	76 ± 9	59 ± 6
P. maritima	8	13	83 ± 7	188 ± 29	77 ± 14	90 ± 12	10 ± 2	1 ± 0.3	182 ± 14	163 ± 29

[a] "Frischwasser" (shoot-water-content basis).

Table 5.6. Ionic contents in leaves of *Plantago* species occurring at the same location in Austria (Königshofer 1981; S.R. Troelstra unpubl. results)

Site	Species	K^+	Na^+	Ca^{2+}	Mg^{2+}	PO_4^{3-}	NO_3^-	Cl^-	SO_4^{2-}	Σ OS organic anions)
		mEq l^{-1} FW (hot-water extraction)								
Neusiedlersee, saline	*P. maritima*	59	102	74	50	5	1	139	54	60
	P. lanceolata	159	7	122	52	9	1	84	44	164
Pulkautal, saline	*P. maritima*	40	224	49	47	18	1	225	71	76
	P. lanceolata	145	24	95	66	14	1	98	62	173
	P. media	182	15	182	108	12	1	88	48	407
Pulkautal, saline	*P. maritima*	94	153	67	114	9	1	252	186	28
	P. major	184	1	105	115	33	1	143	112	151
Maiersdorf, non-saline	*P. maritima*	126	1	246	68	–	1	218	111	70
	P. lanceolata	148	2	145	35	2	1	91	61	169
	P. media	179	1	257	40	1	2	71	47	308
		mEq kg^{-1} d.m. (digestion + extraction)								
						($H_2PO_4^-$)				(C-A)
Neusiedlersee[a]	*P. maritima*	247	257	888	458	60	0	391	166	1247
	P. lanceolata	341	32	925	420	83	0	294	148	1205
	P. major	339	12	1330	430	92	0	347	435	1247

[a] Plant material collected at one site, analyzed in The Netherlands.

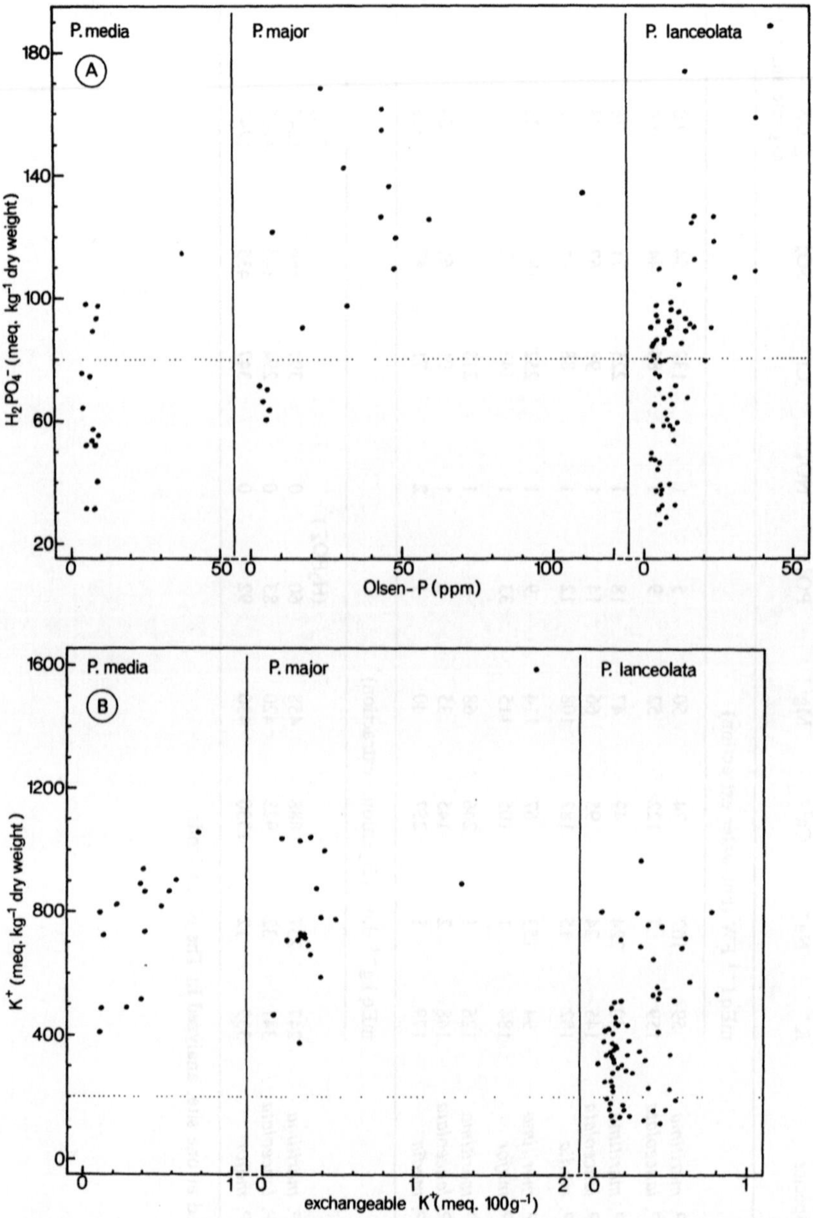

Fig. 5.6. The relationship between the P concentration in the leaves and the Olsen-P value of the 0–10-cm soil layer **(A)** and the relationsphip between the K content in the leaves and the exchangeable K value of the 0–10-cm soil layer **(B)** for *Plantago media*, *P. major* and *P. lanceolata* at various sites in The Netherlands. The ***dotted lines*** are indications of an adequate P supply and a supposedly required minimum K level, respectively, as suggested by De Wit et al. (1963) for grassland plants

less optimum value of 70 to 80 mEq kg^{-1} d.m., mentioned by De Wit et al. (1963) for grassland plants.

In water culture experiments with varying levels of P supply, P concentrations in both *P. major* and *P. lanceolata* reached minimum values of about 30 mEq kg^{-1} d.m. at the lowest level of P supply, where dry matter yield was clearly limited by P supply; visible symptoms of P deficiency hardly occurred (Troelstra et al. 1983a). Even values <30 mEq kg^{-1} d.m. were commonly found in field material of *P. media* and *P. lanceolata*, especially in material from chalk grasslands (Fig. 5.6A). De Jager and Posno (1979) observed a minimum value of about 20 mEq kg^{-1} d.m. for the glycophytic *Plantago* species in water culture experiments.

As far as potassium was concerned, it was noteworthy that the field material of *P. media* and *P. major* almost showed no concentrations less than 400 mEq kg^{-1} d.m., a level which may be considered to be adequate (De Wit et al. 1963; Fig. 5.6B). However, more than half of the samples of *P. lanceolata* were below this level and even K levels below 200 mEq kg^{-1} d.m. occurred, which may create K shortage (De Wit et al. 1963; Van Tuil 1965). Plotting the K and Na levels against each other, the glycophytic species showed a certain replacement of K by Na in the leaves of *P. lanceolata*; no such relationship and very low Na levels were found in the other two species (Fig. 5.7). The ability of *P. lanceolata* to substitute Na for K in growth may play a role in its tolerance

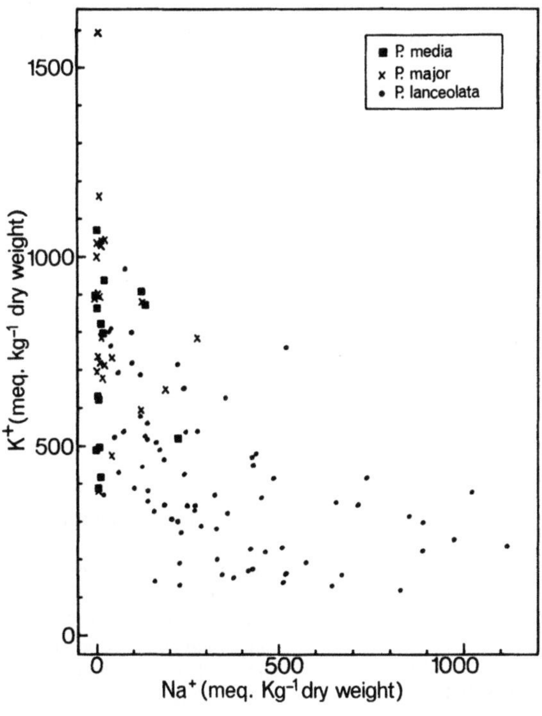

Fig. 5.7. The relationship between potassium and sodium contents in field samples (leaves) of *Plantago media*, *P. major* and *P. lanceolata* collected at different locations in The Netherlands

of low K supply, as indicated by the relatively high yields of *P. lanceolata* at low K supply (Troelstra et al. 1983a); in these experiments, K contents generally decreased to 100 to 200 and 70 to 150 mEq kg^{-1} d.m. in shoots and roots at the lowest K supply, with distinct symptoms of K deficiency (Troelstra et al. 1983a). According to De Wit et al. (1963) plantain does not show a major preference for any of the four cations. Their results, however, referred to *P. lanceolata* and it seems likely that this conclusion cannot be regarded as valid for all *Plantago* species; it probably does not hold for *P. major* ssp. *major* (Troelstra et al. 1983a).

The relatively high P and K levels of *P. major* in the field material could indicate a possible higher ion requirement for *P. major* or they could be the result of the solitary occurrence of *P. major* on compacted sites with less competition and better soil transport conditions for P and K. The present data and those of Troelstra et al. (1983a) with varying P and K supply support the data of Kruijne et al. (1967): *P. major* apparently shows a preference for a relatively high P and K status of the soil.

In the water culture experiments with varying K supply, the shoot: root ratios were relatively constant, those of *P. major* ssp. *major* generally being higher than those of *P. lanceolata* (Troelstra et al. 1983a); at increasing P supply the shoot: root ratio of *P. major* ssp. *major*, starting at the low level of 2 to 3, increased to a much higher value that that of *P. lanceolata*, approximately 7 to 8 versus 4 to 5 (Troelstra et al. 1983a; see also De Jager and Posno 1979).

Carbohydrate levels of *P. media*, *P. major*, and *P. lanceolata* are shown in Fig. 5.8: ethanol-soluble carbohydrates were relatively high in *P. lanceolata* and ethanol-insoluble carbohydrates, starch, were relatively high in *P. major*. These

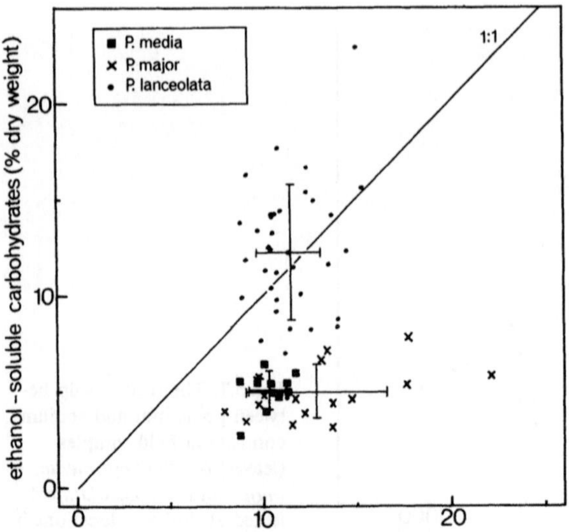

Fig. 5.8. The relationship between ethanol-soluble and ethanol-insoluble carbohydrates (starch) in leaves of *Plantago media*, *P. major* and *P. lanceolata* collected at various sites in The Netherlands. +, standard error of mean

Table 5.7. The chemical composition of leaves and roots of *Plantago lanceolata*, sampled along a 50-m transect in the Westduinen (relatively old stabilized undulating dune grassland complex with maximal differences in height of about 2–4 m). At each sampling point bulk samples of several individual plants were collected. Means ± S.E.

	mEq (mmol) kg^{-1} dry matter	
	Leaves (n = 15)	Roots (n = 12)
K^+	399 ± 20	204 ± 8
Na^+	282 ± 12	144 ± 9
Ca^{2+}	632 ± 19	207 ± 8
Mg^{2+}	224 ± 6	175 ± 6
$H_2PO_4^-$	72 ± 4	49 ± 3
NO_3^-	9 ± 0.3	3 ± 0.9
Cl^-	402 ± 22	87 ± 5
SO_4^{2-}	235 ± 10	41 ± 3
C-A	817 ± 13	550 ± 18
N_{org}	1557 ± 34	810 ± 30

results are in agreement with those of Königshofer (1981), who found the highest accumulation of water-soluble carbohydrates in *P. lanceolata.*

It was concluded from a separate study on the spatial variability in chemical composition of *P. lanceolata* that the chemical composition along a 50-m transect in a dune grassland revealed a relative constancy (Table 5.7), in spite of considerable variation in the 0 to 10 cm soil layer of various soil chemical properties (Troelstra 1978; transect II).

5.2.3 Seasonal Analysis of the Chemical Composition of Plantago Species

As part of a detailed soil chemical study at selected *Plantago* sites (Sect. 3.2.4), above-ground plant parts were collected and analyzed frequently (generally at weekly intervals) during the growing season. The data on the ion balance and organic nitrogen are summarized in Tables 5.8 and 5.9.

The levels of NH_4^+ and NO_3^- in the plant generally were extremely low throughout the season. Possibly the mineralization rates, ammonification and nitrification, and the rate of transport of mineral N to the roots never exceeded the uptake rate. On an average, the highest NH_4^+ levels occurred in the plant material of *P. major* ssp. *major* (De Bruuk), in agreement with NH_4^+ levels in the soil; a notable accumulation of nitrate was only observed at the Angeren site, where organic N levels also were relatively high (Table 5.9).

K levels were low at the acidic locations (Achterberg and De Bruuk); an extremely low K level throughout the season in *P. lanceolata* (Achterberg) was remarkable (Fig. 5.9 A) and was also in agreement with the soil analysis. Relatively low K levels at the end of July in *P. major* ssp. *major* (De Bruuk; Fig. 5.9 A)

Table 5.8. Ionic contents, sum of cations (C), sum of inorganic anions (A), organic anions (C-A), and organic nitrogen in leaves and spikes of *Plantago lanceolata* and *Plantago major* ssp. *major* collected during the period April 1983 – May 1984 at four locations in The Netherlands. [Ranges of values; n = 23 – 28 (leaves) or 5 – 16 (spikes)]

	mEq (mmol) kg^{-1} dry matter							
	Leaves				Spikes			
	P. lanceolata		*P. major* ssp. *major*		*P. lanceolata*		*P. major* ssp. *major*	
	Heteren hayfield	Achterberg hayfield	Heteren path	De Bruuk path	Heteren hayfield	Achterberg hayfield	Heteren path	De Bruuk path
NH_4^+	1 – 10	1 – 5	3 – 29	3 – 30	2 – 8	2 – 7	1 – 4	3 – 20
K^+	448 – 970	83 – 267	438 – 1113	213 – 820	373 – 653	102 – 326	175 – 560	463 – 616
Na^+	39 – 427	175 – 546	2 – 36	1 – 57	26 – 187	95 – 294	3 – 15	10 – 62
Ca^{++}	448 – 1256	577 – 1078	269 – 1943	263 – 1710	193 – 506	222 – 428	315 – 452	386 – 545
Mg^{2+}	134 – 333	96 – 218	135 – 312	139 – 582	110 – 171	68 – 128	131 – 167	148 – 224
$H_2PO_4^-$	77 – 202	45 – 124	67 – 192	37 – 144	83 – 180	60 – 152	67 – 154	67 – 117
NO_3^-	1 – 11	1 – 5	0 – 5	0 – 32	0 – 3	0 – 2	0 – 3	1 – 9
Cl^-	215 – 1193	183 – 467	87 – 566	144 – 919	108 – 412	38 – 211	13 – 215	167 – 415
SO_4^{2-}	92 – 244	146 – 411	46 – 590	66 – 621	43 – 87	33 – 136	29 – 126	100 – 196
C	1131 – 2431	1096 – 1814	1166 – 2886	1096 – 2732	882 – 1233	654 – 958	706 – 1185	1078 – 1467
A	410 – 1600	408 – 951	325 – 1197	302 – 1500	269 – 626	161 – 408	185 – 389	391 – 717
C-A	561 – 1164	588 – 1021	841 – 1912	730 – 1450	519 – 806	447 – 581	521 – 819	600 – 750
N_{org}	769 – 2507	880 – 2295	771 – 2398	996 – 2552	628 – 2007	650 – 1845	976 – 1445	987 – 1575

Table 5.9. Ionic contents, sum of cations (C), sum of inorganic anions (A), organic anions (C-A), and organic nitrogen in leaves of *Plantago major* ssp. *pleiosperma* collected during the June – September period in 1983 at three locations in The Netherlands. [Ranges of values; n = 3 (Angeren) or 6]

	mEq (mmol) kg^{-1} dry matter		
	Oostvoornse Meer, embanked beach plain	Kwade Hoek, salt marsh	Angeren, riverbank
NH_4^+	4 – 12	5 – 9	12 – 24
K^+	359 – 583	393 – 647	339 – 644
Na^+	43 – 99	593 – 1227	211 – 430
Ca^{2+}	1764 – 2597	838 – 1174	1398 – 2188
Mg^{2+}	272 – 482	459 – 650	233 – 323
$H_2PO_4^-$	41 – 55	59 – 108	111 – 168
NO_3^-	1 – 4	1 – 3	4 – 114
Cl^-	564 – 1450	1041 – 1777	1071 – 1387
SO_4^{2+}	408 – 971	205 – 462	363 – 396
C	2444 – 3748	2568 – 3424	2498 – 3304
A	1025 – 2049	1436 – 2298	1606 – 2007
C-A	1382 – 1699	1116 – 1275	892 – 1297
N_{org}	671 – 950	740 – 1303	1987 – 2066

were the lowest reported for field material of this species. At the Achterberg site, K levels decreased below the critical value of 200 mEq kg^{-1} d.m. The ability of *P. lanceolata* for replacement of K by Na is clearly illustrated in Fig. 5.10. Such a relationship was not found in *P. major* ssp. *major* (see also Sect. 5.2.2). Calcium levels of the leaves of *P. major* increased steadily to values higher than in *P. lanceolata,* and the highest levels were reached at the end of the growing season (Fig. 5.9B); only at Kwade Hoek, where Na concentration in *P. major* ssp. *pleiosperma* was relatively high, accumulation of Ca was slightly less (Table 5.9; see also Sect. 5.3). As contrasted with *P. major* ssp. *major* the ability to accumulate Na in the shoot in *P. major* ssp. *pleiosperma* might contribute to its salt tolerance (Tables 5.8 and 5.9). Magnesium levels were high in *P. major* at De Bruuk and at both coastal locations, and low in *P. lanceolata* at Achterberg.

The relatively low P concentrations in the leaves of *P. major* ssp. *major* (De Bruuk) and *P. major* ssp. *pleiosperma* (Oostvoornse Meer) were the lowest determined for this species. Chloride levels showed considerable fluctuations and highest values were recorded in *P. major* ssp. *pleiosperma* at the saline site (Table 5.9). Strong increases in $H_2PO_4^-$ and Cl^- concentrations were observed in *P. lanceolata* following mowing of the vegetation at the end of June (Fig. 5.9C, D), particularly at the Heteren site. Increase in these ion levels might be related to the removal of the sink for nutrients by mowing or by increased N mineralization (nitrification) due to a high soil temperature after mowing. Nitrate reduction in the root will generate OH^-(HCO_3^-) in the root

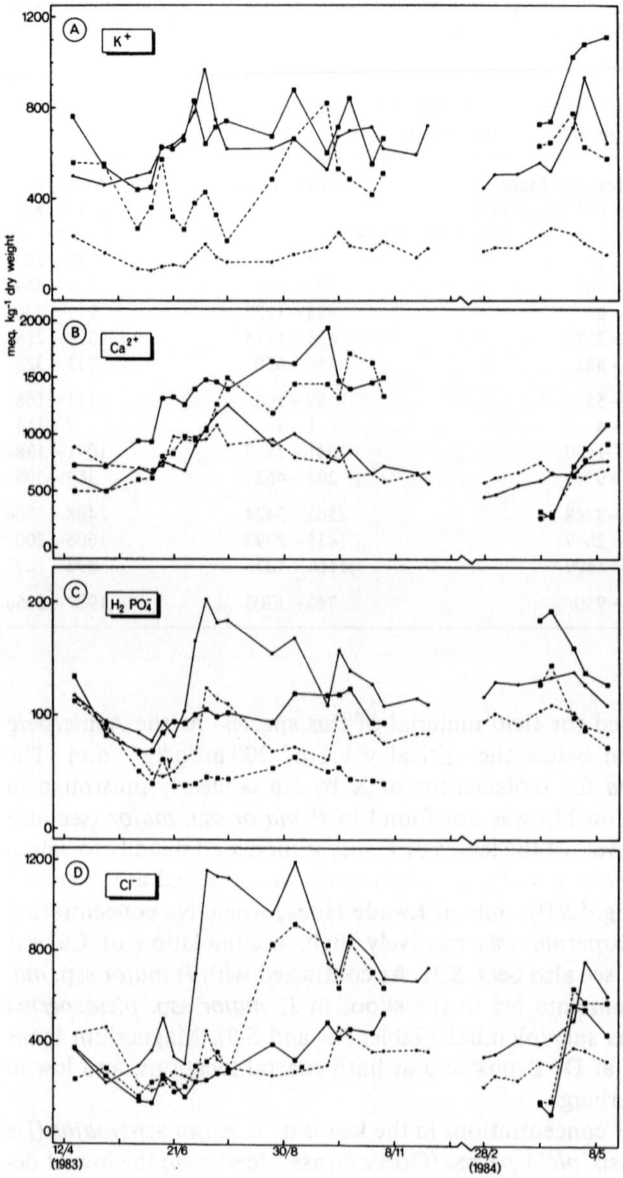

Fig. 5.9. Time course of potassium (**A**), calcium (**B**), phosphate (**C**), and chloride contents (**D**) in field samples (leaves) of *Plantago lanceolata* at Heteren (●———●) and Achterberg (● – – – ●) and of *P. major* ssp. *major* at Heteren (■———■) and De Bruuk (■ – – – ■)

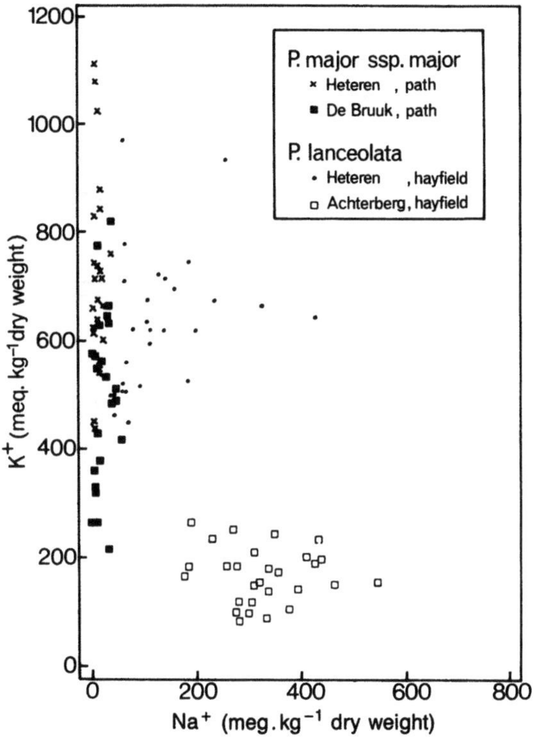

Fig. 5.10. The relationship between potassium and sodium contents in field samples (leaves) of *Plantago lanceolata* and *P. major* ssp. *major* collected during the period April 1983–May 1984 at four locations in The Netherlands

cytosol and will result in OH^-(HCO_3^-) extrusion; the latter may be responsible for increased uptake of inorganic anions other than NO_3^-. Sulphate accumulated to relatively high levels in the leaves of *P. major* (particularly at Oostvoornse Meer; Table 5.9), whereas the levels in *P. lanceolata* were highest immediately after mowing.

Generally, the total cation level (C) reached its highest value in *P. major* at the end of the season. In *P. lanceolata,* C was relatively constant at one site (meadow: Achterberg), whereas at Heteren (hayfield) there was a strong increase in C immediately after mowing, which was then followed by a steady decrease. Total inorganic anions (A) followed more or less the same pattern as C, but at a lower concentration. Thus, C-A levels were in most cases fairly constant for the greater part of the season (Fig. 5.11 A); they were higher in *P. major* than in *P. lanceolata* (Sect. 5.1). The relatively low C-A level in *P. major* ssp. *major* at the acidic site, De Bruuk, may have been caused by availability and uptake of NH_4 at this site (see Sect. 3.2 and 5.1).

A remarkable finding was the rather similar time course of organic N in *Plantago* at the four inland locations (Fig. 5.11 B). Low levels were observed at the end of June, which coincided with the time of mowing at the *P. lanceolata* sites. Following mowing, N_{org} strongly increased in *P. lanceolata,* whereas a peak value was also found around the end of August in the leaves

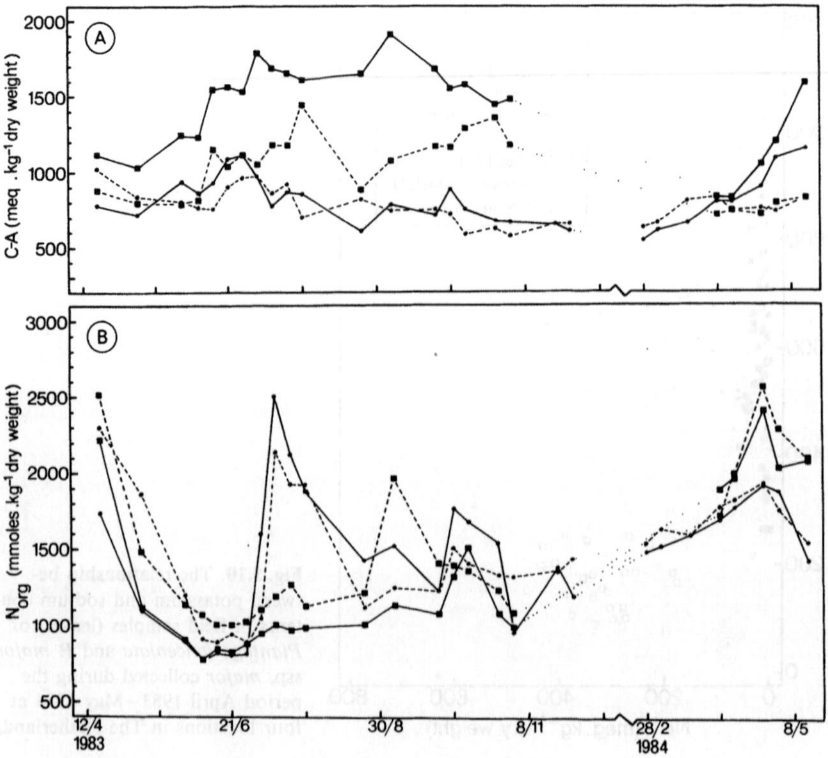

Fig. 5.11. Time course of organic anions (**A**) and organic nitrogen (**B**) in field samples (leaves) of *Plantago lanceolata* at Heteren (●——●) and Achterberg (● - - - ●) and of *P. major* ssp. *major* at Heteren (■——■) and De Bruuk (■ - - - ■)

of *P. major* ssp. *major* at the acidic site (De Bruuk); the latter increase might be due to mowing (Troelstra et al. 1984a). In general, N_{org} was somewhat higher at the acidic sites, probably due to partial uptake of N in the NH_4^+ form (Sect. 3.2.). The time course of N_{org} in *P. lanceolata* was comparable to other results for a hayfield site on a marine clay soil (Berendse 1981). The N_{org} data of *P. major* ssp. *pleiosperma* clearly indicated the relatively low N supply of the coastal site (Oostvoornse Meer) and the relatively high N supply of the inland site (Angeren; Table 5.9).

On two occasions, *P. major* ssp. *major* and *P. lanceolata* samples were taken from the same site (De Bruuk), where plants occurred at a few cm distance from each other. Several species effects, observed at different sites, were also revealed at this single site (results not shown): in *P. lanceolata* a high Na content; in *P. major* ssp. *major* high levels of Ca, Mg, SO_4, C, A (to a lesser extent), and C-A; levels of N_{org} were higher in *P. major* ssp. *major* at the start of the growing season, which is due to differences in age of the tissue.

Ethanol-soluble carbohydrates showed a distinct seasonal fluctuation, which was most pronounced in *P. lanceolata*: relatively low levels in early spring and mid-summer, and peak values in between and during late summer–early autumn (Troelstra et al 1984a). In general, ethanol-soluble carbohydrate levels were higher than those of starch in *P. lanceolata* leaves, whereas the reverse held for the leaves of *P. major* ssp. *major.* Starch levels were always of the same order of magnitude, but ethanol-soluble carbohydrates were much higher in *P. lanceolata* (Troelstra et al. 1984a), which agrees with data of the initial survey (Sect. 5.2.2) and with data of Königshofer (1981).

5.2.4 Comparative Studies on Growth and Mineral Nutrition of Plantago

The overall impression from a large number of greenhouse experiments with water and with pot cultures was that *Plantago* species reacted strikingly similarly for ion levels in relation to a variation in experimental conditions (Troelstra, Brouwer et al. unpubl. results), except for a few notable differences; the latter were already recognized in the field material. Among the glycophytic species, cation levels were relatively high in *P. media* and *P. major*; an exception was the very low Na levels in the shoots of *P. major* ssp. *major,* which was due to limitation of Na transport to the shoot: the root Na level was quite high (Troelstra et al 1983a). *P. coronopus* and *P. maritima* had relatively high ion concentrations, especially Na, Cl, and the total cations (C) and inorganic anions (A). These differences may originate in part from differences in growth rate among the species.

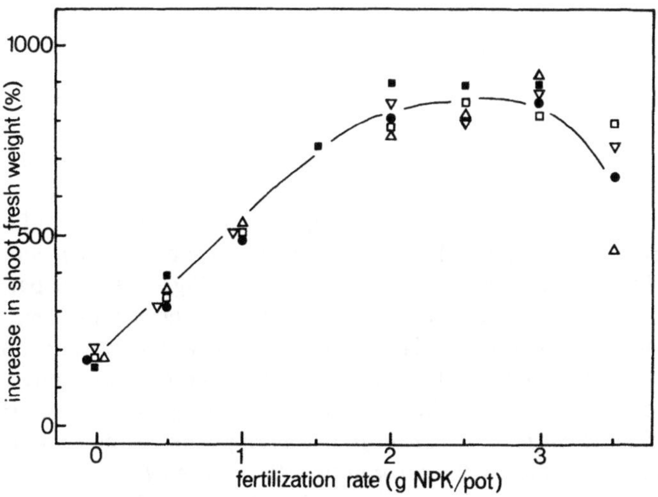

Fig. 5.12. The effect of nutrient level on the relative increase in shoot fresh weight of *Plantage media* (●), *P. major* ssp. *major* (▽), *P. major* ssp. *pleiosperma* (■), *P. lanceolata* (△), and *P. coronopus* (□) during a 30-day growth period on the respective treatments, after a precultivation period of 30 days without any fertilizer addition. Plants remained in the vegetative stage for the duration of the experiment . NPK: 12% N–10% P_2O_5–18% K_2O

Table 5.10. Survival percentage of *Plantago* seedlings following transplantation into pots filled with soil and fertilized at different nitrogen levels

Species	N-supply (g NH_4NO_3 pot^{-1})				
	0	0.4	0.8	1.6	3.2
P. media	93	75	25	0	0
P. major	100	100	83	43	13
P. lanceolata	100	100	65	30	3
P. coronopus	100	100	100	88	68

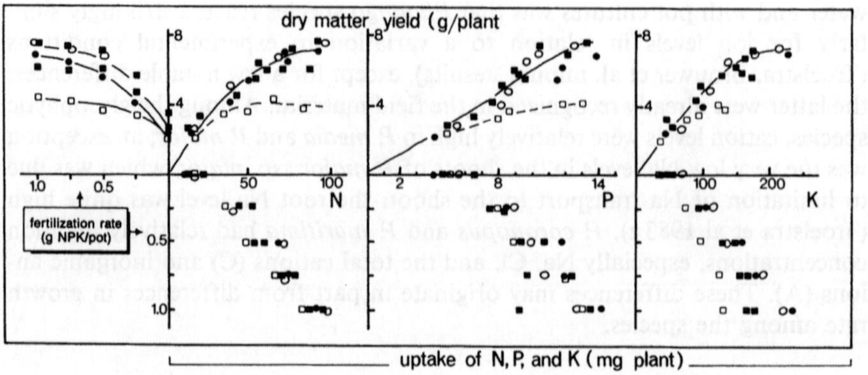

Fig. 5.13. Influence of the fertilization rate on dry matter production and nutrient uptake of four *Plantago* species; NPK: 12% N – 10% P_2O_5 – 18% K_2O. ●, *P. media*; ○, *P. major*; ●, *P. lanceolata*; □, *P. coronopus*

Pot experiments with varying fertilizer rates of either N + P + K, N alone or P alone, showed positive effects on vegetative and generative growth at increasing nutrient supply in all species (see Fig. 5.12). Seedling survival appeared to be highly negatively correlated with fertility level, especially in *P. media* (Table 5.10). Growth of *P. coronopus* in the reproductive phase slowed down considerably, resulting in high levels of N and P in this species (Fig. 5.13). The pattern of dry matter distribution was strongly influenced by the fertility level, with high shoot: root ratio at high levels of nutrient supply. The proportion of the leaves surprisingly remained constant in the reproductive phase (Fig. 5.14). The order of increasing shoot: root ratio among the species, *P. lanceolata* < *P. media* < *P. major* ≪ *P. coronopus*, was opposite to the ranking order of increased leaf area, *P. coronopus* ≪ *P. major* < *P. media* < *P. lanceolata*. Because of these opposite rankings, differences in relative growth rates often were comparatively small: *P. lanceolata* ~ *P. media* ~ *P. major* ≥ *P. coronopus*.

When the species were grown together in mixtures, the similarity in response to increasing fertilization, as found in monoculture, completely disappeared

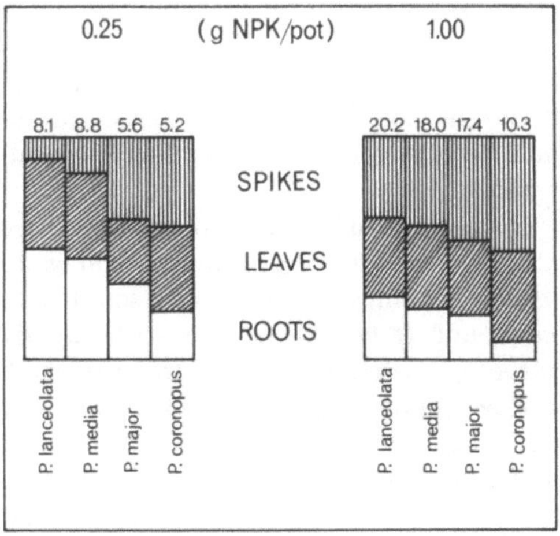

Fig. 5.14. Distribution of the dry matter of four *Plantago* species at two levels of nutrient supply; ***figures on top of the columns*** indicate the total dry matter yields (g $plant^{-1}$). NPK: 12% N – 10% P_2O_5 – 18% K_2O

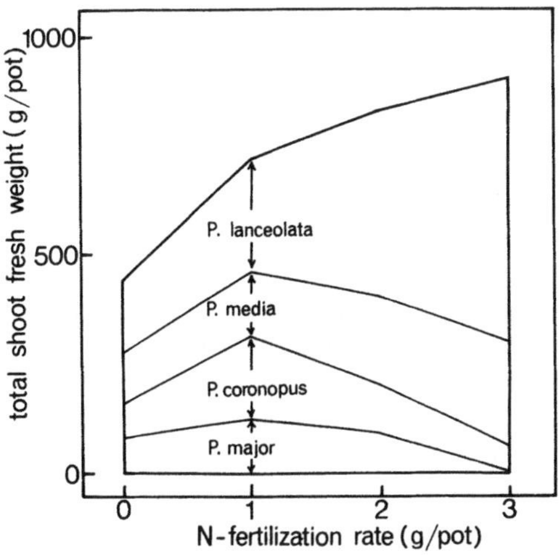

Fig. 5.15. Shoot fresh weights of four *Plantago* species when grown together in a mixture for a total period of 5 months at four levels of nutrient supply; total of four plants per pot, i.e. one individual per species

due to morphological differentiation. At high levels of nutrient supply, the more erect growing *P. lanceolata* and *P. media* demonstrated a better competitive ability than *P. major* and *P. coronopus* (Fig. 5.15). Such effects refer merely to above-ground competition and do not yield information on below-ground competition for nutrients.

5.2.5 Concluding Remarks

This section provides information on the ion composition of the different *Plantago* species in their environment, and on the response to varying levels of nutrient supply. It is unlikely, however, that the distribution of the species can be simply explained in terms of habitat differences for availability of nutrients. Differences, e.g. in depth and pattern of root systems, in competitive ability above and below ground, and with respect to non-*Plantago* species, and in salt tolerance, may be needed to explain the distribution of the species. Moreover, the complete growth cycle should be considered for evaluation of the adaptation of species to a particular habitat.

Apart from salt tolerance, some conclusions on nutritional aspects can be drawn. In contrast to the apparent indifference of *P. coronopus* seedlings to a wide range of nutrient levels, *P. media* seedlings were extremely sensitive to high nutrient concentrations. *P. media* and *P. lanceolata* seem to be clearly adapted to low-nutrient environments, as indicated by the low P status of chalk grasslands, where *P. lanceolata* and *P. media* occurred together with very low P concentrations in the dry matter, and the low K and Mg status of *P. lanceolata* habitats. The low availability of K at the Achterberg site, for example, was possibly only tolerated by *P. lanceolata* by its ability for replacement of K^+ by Na^+; *P. major* ssp. *major* was unable to exchange Na^+ for K^+. On the other hand, K^+ levels in *P. lanceolata* at this site, which had not been fertilized since 1970, were clearly marginal and the dry matter production of *P. lanceolata* might still be limited by the K supply, and not by N or P. Analysis of large *P. lanceolata* individuals just outside the hayfield plot and adjacent to fertilized farmlands revealed a large response in K concentration: an approximately fivefold higher level when compared with the hayfield plants. P and N concentrations were comparatively little affected, an 8% decrease and a 23% increase, respectively.

Although less clear, there seems to be some evidence that *P. major* ssp. *major* has a preference for a relatively high level of nutrient supply, particularly of K and P. Little information is available about the actual shoot:root ratios in the field situation. However, *P. major* ssp. *major* had a tendency toward relatively high shoot:root ratios in the field situation, when compared to *P. lanceolata*. The former species will require a comparatively high P status in order to maintain an adequate supply of P (Wild 1981) towards a limited root system. Both *P. major* and *P. media* showed a relatively high cation requirement, K and Ca, which explains their relatively high organic anion levels (C-A) and their occurrence at sites with a high pH.

5.3 Acquisition and Utilization of Nitrogen

Nitrogen is available to the roots of higher plants as nitrate or ammonium. The availability of the ions in the environment of the *Plantago* species has been discussed in Section 4.4. This section deals with aspects of nitrogen metabolism, which might be seen as adaptions to the nitrogen availability of the environment, in particular in *Plantageo lanceolata* and *Plantago major* ssp. *major* (for convenience we use the name *P. major* below.).

5.3.1 Nitrate Uptake

5.3.1.1 Kinetic Parameters

Nitrate uptake of several species shows the characteristics of Michaelis-Menten kinetics, and in some species at least two systems for nitrate uptake are present (Doddema 1978). Nitrate uptake can be characterized by the parameters V_{max} and K_m, the maximum uptake capacity and the affinity for nitrate, respectively (Epstein 1972). On account of his experiments with *Arabidopsis thalinana* mutants, Doddema (1978) assumed that system I transported nitrate ions over the plamalemma into the cytoplasm and that system II was responsible for transport of nitrate directly into the vacuole.

Van de Dijk et al. (1982) studied the kinetic parameters of *P. lanceolata* and *P. major*, and also of *Urtica dioica, Hypochaeris radiacata*, ssp. *radicata* and ssp. *ericetorum* Van Soest. *Urtica dioica* is a eutrophic species and *H. radicata* ssp. *ericetorum* is a species from nutrient-poor heath soil. The kinetic parameters of uptake system I were measured during a growth period of several weeks. The objectives were to determine the effect of nutrient availability on V_{max} and K_m, and to investigate a possible correlation between the values found for the kinetic parameters of these species and the nutrient availability in the environment. Such a correlation can be expected according to Lancaster (cited by Huffaker and Rains 1978). Van de Dijk et al. (1982) grew their plants in tanks with nutrient solution at optimum availability of nutrients (1/4 strength Hoagland, containing 3750 μM nitrate) or at a limiting supply (2% of a 1/4 strength Hoagland, containing 75 μM nitrate), with periodical depletion and replacement of the latter solution. The V_{max} of *Urtica dioica* was significantly higher at optimum nutrient availability than when depletion occurred; for the other species no significant difference was found. Table 5.11 shows that there was no clear correlation between the values of the apparent V_{max} and the nutrient availability of the environment of the species investigated. A high shoot to root ratio and a high RGR imply a high demand for nitrate so that a high uptake capacity can be expected in eutrophic species (Chapin 1980). Although *P. major* is found in habitats with the highest nutrient supply (after *U. dioica*), it has the lowest V_{max} in combination with the low shoot to root ratio and a low RGR when grown under the conditions of Van de Dijk et al. (1982). Table 5.11 shows in addition that there is no correlation between the K_m values and the distribution of the five species in the field as regards the nutrient availability.

In a second investigation into the kinetic characteristics of the *Plantago* species, the nutrient variable was the nitrate concentration in the strict sense of the word since the plants were grown in a flowing solution system, containing half-strength Hoagland with 100 or 5 μM nitrate. Under these conditions the apparent V_{max} of *P. major* was significantly higher than that of *P. lanceolata* and there were no differences in K_m value (Table 5.11). Apparently the uptake capacity for nitrate of *P. major* was higher than that of *P. lanceolata* when grown at 100 and at 5 μM nitrate. Because *P. major* forms less root

Table 5.11. Values of V_{max} [$\mu M\ NO_3^-$ (g dry root)$^{-1}$ h^{-1}] and K_m (μM) for NO_3^- absorption, RGR (g g^{-1} day^{-1}) and shoot to root ratio (dry weight basis) of several plant species grown either in a nutrient solution at 3750 $\mu M\ NO_3^-$ or 75 $\mu M\ NO_3^-$ (data from van de Dijk et al. 1982) or in a flowing nutrient solution at 100 $\mu M\ NO_3^-$ or 5 $\mu M\ NO_3^-$

Plant species	Nutrient regime	V_{max}	K_m	RGR shoot	Shoot/root ratio
Urtica dioica	3750	150	11	0.160	4.0
	75[e]	94	9		1.6
Hypochaeris rad. ssp. *rad.*	3750	70	13	0.120[b]	6.3[c]
	75[e]	82	13		3.8–3.0[c,d]
Hypochaeris rad. ssp. *ericetorum*	3750	75	17	0.110[b]	4.6[c]
	75	90	15		2.8–2.0[c,d]
Plantago lanceolata	3750	150–20[a]	20	0.134	2.5
	75[e]	60	23		1.9–1.1[d]
	100	84	3	0.230	3.7
	5	74	3	0.220	1.9
Plantago major ssp. *major*	3750	45	15	0.120	2.6
	75[e]	39	15		2.7–1.1[d]
	100	143	7	0.280	3.4
	5	167	5	0.240	2.3

[a] V_{max} was gradually declining during the 3-week experimental period.
[b] RGR was gradually declining during the 3-week experimental period. An average value has been presented.
[c] Shoot to lateral root ratio.
[d] Shoot to root ratio was gradually decreasing during the 3-week experimental period.
[e] Periodical depletion occurred.

length from the same amount of dry matter, the difference in uptake capacity between the two *Plantago* species was greater when expressed per unit root length.

The aim of the above-mentioned experiments was to investigate the significance of the uptake parameters in explaining the distribution of *Plantago* species in the field. The plants used, however, were grown in nutrient solution of various composition prior to the determination of the uptake parameters of system I, which were measured under the same experimental conditions. Comparison of the results shows that the K_m values found in the second series of culture experiments for both *Plantago* species were lower than those found by Van de Dijk et al. (1982). Experiments with *Urtica dioica* have shown that the K_m value for nitrate was affected by the composition of the nutrient solution and that the K_m was markedly reduced when the plants were grown in the presence of Cl^-. The difference between the results of Van de Dijk et al. (1982) and the second set of experiments, therefore, probably can be ascribed to differences in Cl^- content. Under the conditions applied in the latter experiments, viz. a flowing solution system, more Cl^- was accumulated in the plants than under those applied by Van de Dijk et al. (1982), where hardly any Cl^- accumulation was found. The second set of experiments probably yields the most appropriate values for the K_m of system I, because system II (which is supposed to be responsible for the filling of the vacuole; Doddema 1978), is not operative when vacuoles were filled with Cl^-.

Also the V_{max} values found for *P. major* were markedly different. Those of the second series of experiments were markedly higher than those of Van de Dijk et al. (1982). Furthermore, a higher RGR was found by the first author. The two approaches differed with respect to concentration as well as to supply of nitrate. It is unlikely that the difference was caused by the fact that the nitrate concentration in the nutrient solution of Van de Dijk et al. (1982) per se was too high. *P. major* grown in a flowing nutrient solution at 7500 μM (Freijsen and Otten 1984) exhibited the same RGR as when grown at 9.5 μM. However, a nitrate concentration of 3750 μM was probably too high for *P. major* but not for *P. lanceolata* under light-limiting conditions. Unpublished experiments with *P. major* have shown that in some cases a switch from 3750 μM to 75 μM resulted in a higher RGR. In the experiments of Van de Dijk et al. (1982) self-shading of *P. major*, resulting in a lower light intensity at the plant level, was probably important since the plants were grown more densely than in the second series of experiments.

Summarizing, we conclude that the experimental conditions during growth can strongly influence the values of the uptake parameters. Conclusions on the adaptive value of K_m of V_{max} values have to be drawn carefully, taking into account differences in experimental procedures.

5.3.1.2 Specific Uptake Rate

The relationship between the nitrate uptake rate of *P. lanceolata* and *P. major* and the external nitrate concentration was also investigated by calculation of the specific uptake rate (SUR), i.e. the mean N absorption, as measured by the N increment in the plant, over several days, with the equation of Williams (1948). The SUR values were determined for plants either grown in 1/4 Hoagland solution containing 3750 μM nitrate (Van de Dijk et al. 1982) or in a flowing solution system containing 1/2 strength Hoagland with 7500 or 9.5 μM nitrate (Freijsen and Otten 1984) or 100 or 5 μM nitrate (Table 5.12).

The SUR of *P. lanceolata* was lower at 9.5 μM than at 100 and 7500 μM nitrate. The effect of the ambient nitrate concentration on the uptake rate of *P. major* in the flowing solution of 9.5, 100 and 7500 μM nitrate was small. Under these conditions *P. major* attained a high SUR at a low concentration. For both species SUR at 100 μM nitrate was higher than at 5 μM and no difference between the species was found. Probably many plant species can attain maximum uptake rates at low or even very low external nitrate concentration and they also grow optimally under these conditions (Clement et al. 1978; Bhat et al. 1979). The experiments of Van de Dijk et al. (1982) yielded contrasting results for the SUR value at an unlimited nutrient supply (Table 5.12): the SUR of *P. major* was lower that that of *P. lanceolata*. For both species, V_{max} of system I apparently was a good measure for actual nitrate uptake since V_{max} and

Table 5.12. Values of V_{max} [$\mu M\ NO_3^-$ (g dry root)$^{-1}$ h^{-1}] and specific uptake rate for NO_3^-, SUR, [$\mu M\ NO_3^-$ (g dry root)$^{-1}$ h^{-1}] of *Plantago lanceolata* and *P. major* ssp. *major* grown either in a nutrient solution at 3750 μM NO_3^- (data from van de Dijk et al. 1982) or in a flowing nutrient solution at various NO_3^- concentrations. (Freijsen and Otten 1984)

Nutrient regime ($\mu M\ NO_3^-$)	*Plantago lanceolata* V_{max} (1/2 V_{max})		*Plantago lanceolata* SUR	*Plantago major* V_{max} (1/2 V_{max})	*Plantago major* SUR
7500	–		154	–	175
3750	104		100	45	63
100	84		165	143	140
9.5	–		92	–	163
5	74	(37)	69	167 (84)	81

SUR values were approximately similar in experiments with unlimited nitrate supply. For plants grown at the lowest concentration, 5 μM, SUR was compared with 1/2 V_{max}, since the concentration of 5 μM was close to the K_m value. SUR values of *P. lanceolata* exceeded half V_{max} and full V_{max} at 5 and 100 μM, respectively. Apparently both *Plantago* species use the total absorbing capacity of system I during growth on nutrient solution; the measured value V_{max} of *P. lanceolata* apparently was lower than the actual nitrate absorption during reduced growth.

5.3.1.3 Ecological Significance

The aim of the uptake experiments was to investigate the relevance of the uptake parameters K_m and V_{max} of system I in explaining distribution of plant species in the field. We concluded from the present experiments that the K_m value was influenced by the experimental conditions prior to its determination and has to be interpreted carefully. The data of Van de Dijk et al. (1982) were obtained for five plant species, grown under the same experimental conditions; they showed that the K_m was not important for the distribution of these species in the field, although a change of the K_m at extremely low nitrate concentrations could still not be excluded.

As regards the V_{max} values of system I we conclude that *P. major* could better absorb nitrate from a low (9.5 μM) nutrient concentration than *P. lanceolata* and this was also the case for a high concentration (Freijsen and Otten 1984). However, the experimental conditions influenced the V_{max} value and at the same time SUR and RGR of *P. major*: all three parameters were lower under the conditions applied by Van de Dijk et al. (1982). In all experiments the V_{max} value and SUR were in general agreement with each other.

According to Chapin (1980), natural selection in nutrient-rich environments promoted species with high rates of uptake and of growth to arise. In his opinion, under conditions of suboptimal nutrient availability, this type of plant was more productive than the oligotrophic one due to a higher V_{max}.

Table 5.13. Fresh weight (g) of shoots of four *Plantago* species grown together for 5 months in pots with soil. At the start of the experiment various amounts (in g) NPK fertilizer (12% N) were added to the pots

Species	Fresh weight shoot NPK (g)			
	0	1	2	3
P. lanceolata	1.6	2.5	4.2	5.9
P. media	1.1	1.4	2.1	2.4
P. coronopus	0.8	1.9	1.0	0.6
P. major	0.8	1.2	0.9	–

From the present V_{max} data we cannot conclude whether *P. major* is better adapted to conditions of rich nutrient supply than *P. lanceolata*. Van de Dijk et al. (1982) concluded that for the interpretation of V_{max} data, other factors such as RGR and shoot/root ratio have to be taken into account. Both RGR and shoot/root ratio of *P. major* were lower under the conditions of Van de Dijk et al. (1982) than when grown in a flowing solution system at 100 μM nitrate (Table 5.11).

Brouwer determined the fresh weight of the shoots of four *Plantago* species grown in containers with various amounts of NPK fertilizer. Differences in shoot weight between the species were found when the four species were grown together in a competition experiment. *P. lanceolata* and *P. media* were predominant at harvest time, esspecially at the highest NPK addition (Table 5.13). This result can be explained by the fact that the leaves of *P. lanceolata* and *P. media* were less affected by poor light conditions from shading. We therefore conclude that the shoot morphology of a species is probably a very important factor in explaining the distribution of *Plantago* species in the field and could well be more important than the nutrient uptake parameters.

5.3.2 Nitrate Reduction

The reduction of nitrate was measured in plants of *P. lanceolata* and *P. major* grown in tanks with either 1/4 strength Hoagland solution containing 3750 μM NO_3^- or 2% of that solution, containing 75 μM NO_3^- (Stulen et al. 1981 a,b). The in vivo nitrate reductase activities of both shoots and roots of *P. lanceolata* depended on the nutrient supply, but the activities of both shoots and roots of *P. major* were rather independent (Table 5.14). A similar pattern was obtained for the activity of the enzymes, involved in the assimilation of reduced NO_3^-, glutamine synthetase (GS) and glutamate dehydrogenase (GDH), and for the total amount of reduced nitrogen (Stulen et al. 1981 a,b). These results lead to the hypothesis that the maintenance of high enzyme activity levels as found for *P. major* might be advantageous under circumstances in which the nutrient supply is limited during a relatively short period, so that

Table 5.14. Comparison of in vivo nitrate reductase activity [μM NO_3^- h^{-1} (g starch free dry weight)$^{-1}$] and NO_3^- concentration [μM NO_3^- (g starch free dry weight)$^{-1}$] of leaves and roots of *P. lanceolata* and *P. major* grown at either 3750 μM or 75 μM NO_3^-. (Data from Stulen et al. 1981 a, b). Mean ± SD; number of independent determinations, 4–5

Nitrate conc. μM	Plant part	NRA		Nitrate content	
		P. lanceolata	*P. major*	*P. lanceolata*	*P. major*
75	Leaves	10 ± 1	18 ± 2	0.46 ± 0.25	0.23 ± 0.17
3750	Leaves	17 ± 2	17 ± 4	1.44 ± 0.25	0.72 ± 0.23
75	Root	28	47	0.53 ± 0.29	0.66 ± 0.10
3750	Root	53	52	1.36 ± 0.29	1.11 ± 0.10

the plant would not have to readjust its enzyme activity levels to an altered supply of nitrate. Whether this hypothesis holds has to be established in experiments in which the nitrate supply is interrupted, with emphasis on periods shorter than 1 day. For the conditions applied by Freijsen and Otten (1980), the growth of *P. lanceolata* was less reduced by an intermittent supply of NO_3^- (1 day with 25 μM NO_3^- alternating with 1 or 2 days without) than that of *P. major.*

5.3.3 Nitrate Accumulation

If nitrate uptake exceeds nitrate reduction, nitrate will accumulate in the vacuole. Nitrate accumulation of *P. lanceolata* and *P. major*, grown on nutrient solution, was compared (Table 5.15). *P. lanceolata* accumulated more nitrate than *P. major* under all conditions. The greatest species difference occurred at optimal supply. *P. lanceolata* maintained a high accumulation when its specific uptake rate was reduced by low external concentrations. Finally, *P. lanceolata* accumulated more nitrate than *P. major* when the nutrient medium was subjected to periodical depletion (2 days out of 3 at the end of the experimental period).

P. lanceolata clearly tended to accumulate nitrate. Three physiological aspects should be considered. Firstly, *P. lanceolata* had a higher requirement for internal nitrate for nitrate reductase activity than *P. major* (Table 5.15). Apparently, *P. lanceolata* had a high demand for minerals as an osmoticum. At ample supply, nitrate was used to build up a high vacuolar concentration; at restricted supply it was replaced in the vacuole by chloride (Clement et al. 1978), or by glucose in case chloride was absent (Stulen et al. 1981 a). Finally, increased accumulation implied increased storage. *P. lanceolata* and *P. major* were grown in nutrient solution with intermittent nitrate supply, days with 225 μM NO_3^- alternated with days without NO_3 (Freijsen and Otten 1980): growth of *P. lanceolata* was less reduced than that of *P. major.* Probably, *P. lanceolata* maintained a higher growth rate in the intervals without nitrate, due to more nitrate stored on days with nitrate. The experiments of Stulen et al. (1981 a, b) showed that when the plants were switched from high to low nutrient conditions, the NO_3^- concentra-

Table 5.15. Comparison of shoot NO_3^- concentration [μM NO_3^- (g starch free dry weight)$^{-1}$] of *P. lanceolata* and *P. major* ssp. *major* grown in nutrient solutions at various regimes of NO_3^- availability. (Stulen et al. 1981 a, b); Freijsen and Otten 1984)

Nutrient regime μM NO_3^-	NO_3^- content	
	P. lanceolata	*P. major*
7500	1530	890
3750	1440	730
100	1454	753
75[a]	460	230
9.5	710	660
5	710	520

[a] Periodical depletion occurred.

tion of the roots of *P. lanceolata* decreased more rapidly than in *P. major*. In a period of 4 days, the NO_3^- concentration of the roots of *P. lanceolata* decreased by 60%, and root growth was not reduced. In the same period the NO_3^- concentration of the root of *P. major* was reduced by only 11% and root growth was decreased. The storage function of accumulated nitrate, found in the present culture experiments, could also be significant in the field. The ability to store minerals was regarded as an adaptation of oligotrophic species to soils with interrupted supply (Chapin 1980). A capacity to store nitrate could indeed be more significant for *P. lanceolata* than for *P. major* in view of the fact that the former is more widely found on temporarily desiccated soils: in dry soil transport of nitrate to the roots will be hampered. In addition, stored nitrate could be used in hibernating organs (Pate 1983).

Speculation on the relevance of nitrate accumulation by *Plantago* in the field seems idle as long as this accumulation has not been established. We know that there is a large discrepancy between the high nitrate concentrations in shoots of plants in culture experiments and the low concentrations in field samples: mature plants of *P. lanceolata* and *P. major* contained negligible amounts of nitrate (Sect. 4.4). Shoots of juvenile plants of both species grown in the experimental garden, possessed NO_3 concentrations lower than 1 mmol kg^{-1} dry matter.

5.3.4 Nitrate Starvation

The influence of a long period of nitrate starvation on growth and nitrate metabolism of *P. lanceolata* and *P. major* was examined by Blacquière and Koetsier (1988). Plants of both species were grown separately in pots on sand, watered with nutrient solution containing either nitrate (+N treatment) or no nitrate (−N treatment) (Table 5.16): root growth of *P. lanceolata* was stopped whereas root growth of *P. major* continued.

Table 5.16. Comparison of root weight (fresh weight in g) of *P. lanceolata* and *P. major* ssp. *major* grown in pots with silversand with an initial supply of nitrate (1.50 mmol per plant per pot). The plants were watered with either 1/4 strength Hoagland solution containing nitrate (50 ml per plant per week) (+N) or with 1/4 strength Hoagland solution without nitrate (50 ml per plant per week, NO_3^- was replaced by Cl^-)

Weeks after sowing	Root fresh weight			
	P. lanceolata		*P. major* ssp. *major*	
	+N	−N	+N	−N
7	0.56	0.53	0.43	0.64
9	1.4	2.1	1.2	1.6
16		5.8 ± 1.2		6.1 ± 1.2
18	8.2 ± 0.3	5.8 ± 1.1	10.3 ± 1.8	8.3 ± 0.9

At the moment it is highly speculative to assume that this difference was caused by the fact that *P. major* could absorb nitrate better at a low nitrate concentration (see Sect. 5.3.1.1). *P. major* forms a compact and shallow root system, which may be regarded as efficient in a moderately rich and wet soil habitat, whereas *P. lanceolata* has the tendency to form a long main root axis (Freijsen and Otten 1984). The capacity to continue root growth during a prolonged period without nitrate could be of competitive value under field conditions, especially in the upper layer of the soil, where N content tends to be high (Sect. 3.2).

On the whole the results of Blacquière and Koetsier (1988) showed no major differences in the response towards nitrate depletion between the two species.

5.3.5 Ammonium Assimilation

Most of the experiments on nitrogen metabolism of *Plantago* species were carried out with nitrate as the nitrogen source. Although ammonium nitrogen could be important in the field (Sect. 4.4), comparatively few experiments as regards its influence on nitrogen metabolism were carried out with ammonium. In those experiments the influence of the nitrogen source on the level of two ammonium assimilating enzymes, glutamate dehydrogenase (GDH) and glutamine synthetase (GS), was investigated.

Firstly the effect of ammonium on the level of GDH in the root was investigated. *P. lanceolata* and *P. major* were grown in nutrient solution containing 3.75 mM NO_3^-. After 48 h on a nitrogen-free solution the plants were transferred for 24 h into a solution containing varying concentrations of ammonium. Roots of *P. lanceolata* and *P. major* reacted with an increased level of GDH (Table 5.17) and the response of the two species clearly depended on the age of the plants. Enhancement of GDH levels in roots by ammonium was also found by Kanamori et al. (1972), Taylor and Havill (1981) and Gasparikovà

Table 5.17. Influence of a 24-h treatment on nutrient solution containing various concentrations of NH_4^+ (in mM) after a 48-h period in nutrient solution −N on the level of GDH (glutamate dehydrogenase) [in μmol $NADH^-$ (g dry weight)$^{-1}$ h^{-1}] of roots of *P. lanceolata* and *P. major* ssp. *major* of various age (1 and 3 weeks on nutrient solution)

Species weeks of nutrient treatment	*P. lanceolata*					*P. major*				
	Concentration of NH_4^+ (mM) in solution									
	0	0.1	1	10	100	0	0.1	1	10	100
1	13	30	40	55	40	12	12	21	39	16
3	13	10	9	12	16	30	30	30	60	17

et al. (1976). No stimulation by ammonium was found after 33 weeks (*P. lanceolata*); stimulation at a high ammonium concentration was still found in *P. major*.

In experiments of Blacquière et al. (1988), the levels of both GDH and GS in shoots and roots were determined in plants grown on nitrate, ammonium or ammonium nitrate (0.5 mM N; Table 5.18). Ammonium enhanced the level of GDH and GS in *P. lanceolata* and *P. major*. An increase in GDH and GS levels by ammonium was also found for *Lemna* by Rhodes et al. (1976). The

Table 5.18. Influence of the nutrient solution on the levels of glutamate dehydrogenase (GDH) and glutamine synthetase (GS) [in μmol product (g dry weight)$^{-1}$ h^{-1}] of shoots and roots of *P. lanceolata* and *P. major* ssp. *major*. The nutrient solution contained 0.5 mM as NO_3^-, as NH_4^+ or as NH_4NO_3

	Day	*P. lanceolata*				*P. major*			
		Nutrient regime							
		NO_3^-	NH_4NO_3	NH_4^+	NH_4^+/NO_3^- ratio	NO_3^-	NH_4NO_3	NH_4^+	NH_4^+/NO_3^- ratio
GDH									
Shoot	32	14.8	22.5	46.6	3.1	6.3	6.8	8.1	1.3
	35	26.2	31.3	50.2	1.9				
	36					9.6	11.9	18.4	1.9
Root	32	24.5	25.3	64.9	2.6	9.5	6.7	18.8	3.4
	35	18.2	37.3	69.6	3.8				
	36					7.3	11.4	34.8	4.8
GS									
Shoot	32	67.0	57.3	123.6	1.8	139.4	158.6	141.2	1.0
	35	116.0	130.6	123.5	1.1				
	36					112.3	84.3	114.7	1.0
Root	32	158.2	474.7	968.5	6.1	21.3	120.8	226.5	10.6
	35	289.4	533.3	637.6	2.2				
	36					37.0	68.8	299.2	8.0

level of both enzymes was highest in *P. lanceolata*; the relative increase by ammonium was highest in the roots of *P. major*.

Summarizing, we conclude that the levels of GDH and GS were clearly enhanced by ammonium in *Plantago*, especially in the roots. But differences in ammonium/nitrate nutrition, as found in the field, and nitrate nutrition were less clear, especially for GDH (see also Blacquière 1988). Roots cannot be easily removed from the soil without severe damage. Therefore, the activity of the two ammonium assimilating enzymes cannot be used as an indicator for the nitrogen source in the field. Whether GDH is really involved in the assimilation of NH_4 is still doubtful in view of the high K_m for NH_4 (Wallsgrove et al. 1983). However, an increase in GDH activity could well be of physiological significance when for both *Plantago* species ammonium is available in high concentrations.

5.3.6 Conclusions

The aim of the present experiments was to investigate in how far differences in nitrogen metabolism play a role in the distribution of *Plantago* species.

The major part of the experiments was carried out under various conditions in the laboratory and they made clear that experimental conditions can strongly influence the results, in particular K_m and V_{max} values for uptake of nitrate. The uptake experiments showed that it is unlikely that either K_m or V_{max} of system I are significant in adaptation of *P. lanceolata* and *P. major* to its habitat. These *Plantago* species differed with respect to accumulation and use of accumulated nitrate but it remains doubtful whether *P. lanceolata* shows an advantage in this respect in the field regarding the nitrogen concentration of material collected in the field (Sect. 3.2). Nitrogen starvation experiments showed that *P. major* endures nitrogen limitation during a prolonged period as well as *P. lanceolata*.

Summarizing, we conclude that differences in shoot morphology and possibly also root morphology were more important than the differences in parameters of nitrogen metabolism found in explaining the distribution of *P. lanceolata* and *P. major* in the field.

5.4 Salinity and *Plantago*

The genus *Plantago* contains species that differ widely in salt tolerance. *P. media* seems to avoid saline soils, *P. maritima* is a halophyte and *P. coronopus* occurs in habitats which may vary in salinity (Sect. 3.2). Most of our studies on salt tolerance concern the three species and less attention has been paid to *P. major* and *P. lanceolata*.

5.4.1 Growth and Transpiration

In several experiments growth of *P. media* and *P. major* ssp. *major* already was reduced at as little as 25 mM NaCl; leaf expansion was reduced and necrosis occurred at the leaf edges. At this salt level the leaves of *P. major* spp. *major* were succulent in appearance. *P. coronopus* could easily grow at 150 mM NaCl and *P. maritima* maintained growth at 300 mM. Both species were succulent and the dry weight percentage of the leaves generally decreased with increasing salt concentration (Erdei and Kuiper 1979).

In *P. maritima* shoot growth was stimulated when the plants were exposed to 75 or 100 mM NaCl, a common observation for halophytes (Poljakoff-Mayber 1975). It has been suggested that such a growth stimulation is due to a water deficit present in no-salt plants (Munns et al. 1982). Possibly halophytes have a need for salt accumulation to enable the plants to acquire water, despite a low hydraulic conductivity of the roots for water. Preliminary experiments with a pressure bomb showed that the hydraulic conductivity of *P. maritima* roots was 50% of that of *P. media* roots, viz. 45 and 90 µl h^{-1} $(10^5\ Pa)^{-1}$ (g fresh weight of $roots)^{-1}$ (De Boer 1985). Growth of roots seemed to be reduced at lower concentrations of NaCl than that of shoots, but no simple relation between shoot/root ratio and salinity was evident. The rate of transpiration will influence the flux of ions into the shoot (Greenway 1965). At 95 to 100% relative humidity of the air, the glycophyte *P. media* and the halophyte *P. maritima* accumulated equal amounts of Na^+ in the roots, but no translocation to the shoot was evident in either species. At 60% relative humidity Na^+ was transported to the shoot in *P. maritima* but in *P. media* only after a certain lag period (De Boer 1985). This lag period in Na^+ translocation seems to be due to a greater reduction in transpiration of *P. media* by salt stress than in *P. maritima*.

5.4.2 Ionic Balance of Plantago Under Salt Stress

Sodium. *Plantago maritima* is a typical Na^+-includer (Fig. 5.16). Plants of *P. maritima* accumulated about 120 mEq (kg dry $weight)^{-1}$ of roots at 1 mM NaCl applied. Also, transport of Na^+ to the shoot was high at this NaCl concentration, viz. about 400 mEg NaCl (kg dry $weight)^{-1}$ (Tánczos et al. 1981; De Boer 1985). At 10 mM NaCl the level of Na^+ in the shoot already exceeded 100 mEq (kg dry $weight)^{-1}$ and it further rose to 6000 mEq (kg dry $weight)^{-1}$ at 150 mM NaCl (Erdei and Kuiper 1979). Addition of Ca^{2+} (10 mM) to the 10 mM NaCl solution did not affect Na^+ uptake by the roots, but translocation to the shoot was significantly reduced.

Plantago media is an effective Na^+-excluder. At 1 mM NaCl virtually no Na^+ was observed in roots and shoots (Tánczos et al. 1981; De Boer 1985). At 10 mM NaCl, Na^+ uptake and translocation are evident; but for short periods *P. media* still was able to exclude Na^+ at this concentration.

Potassium. Also for K^+ uptake and translocation the halophytic *Plantago* species (*P. maritima* and *P. coronopus*) and the salt-sensitive *P. media* showed different reactions to salinity (Fig. 5.17). In *P. maritima*, K^+ uptake and translocation to the shoot were already reduced at 10 mM NaCl and a further

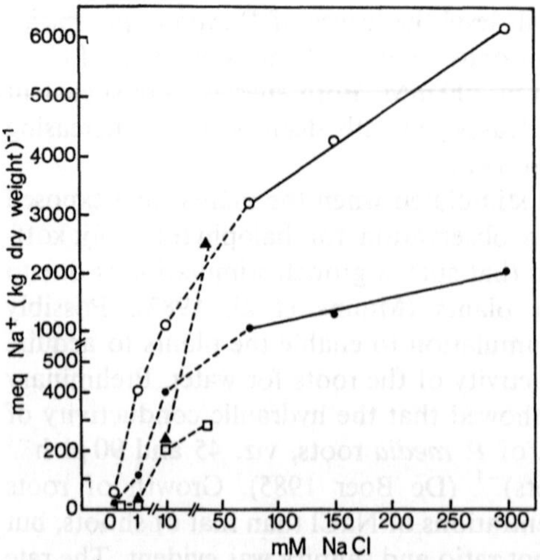

Fig. 5.16. Uptake and translocation of Na^+ by the halophyte *P. maritima* and the glycophyte *P. media* as affected by salinity. Experimental data on water cultures by Erdei and Kuiper (1979), Tánczos et al. (1981), De Boer (1985). ●, *P. maritima*, roots; ○, *P. maritima*, shoot; □, *P. media*, roots; ▲, *P. media*, shoot

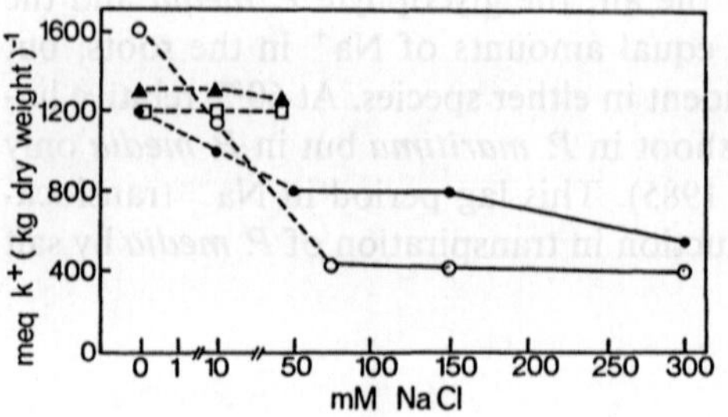

Fig. 5.17. Uptake and translocation of K^+ by *P. maritima* (halophyte) and *P. media* (glycophyte) as affected by salinity (NaCl). Experimental data by De Boer (1985) and Erdei and Kuiper (1979). Symbols as in Fig. 5.16

reduction in uptake and translocation of this ion was noted up to 300 mM NaCl. The data suggest competition between Na^+ and K^+ for uptake and translocation in halophytic *Plantago* species. This suggestion is confirmed in a salinity experiment under nutrient-poor conditions; at 75 mM NaCl the Na^+ content of the roots had already reached the same level as that of plants grown at 150 mM NaCl under nutrient-rich conditions (Erdei and Kuiper 1979).

In halophytes a decrease in K^+ at increasing Na^+ is often found: Na^+ ions may substitute for K^+ in some of its metabolic functions (Rush and Epstein 1976) and replace K^+ in the vacuole as an osmoticum (Stelzer and Läuchli 1978). Therefore the K^+ content of halophytes may be reduced to a large extent before growth reduction becomes manifest (Neales and Sharkey 1982). In *P. maritima* addition of extra Ca^{2+} to the root environment restored K^+ uptake and translocation to the level observed in control plants (no salinity). In *P. media* uptake and translocation of K^+ to the shoot were practically unaffected up to 50 mM NaCl. Only at 150 mM NaCl was K^+-content of the shoot slightly reduced, but at this concentration severe symptoms of salinity damage were visible. Clearly, *P. media* is a species with a strong preference for

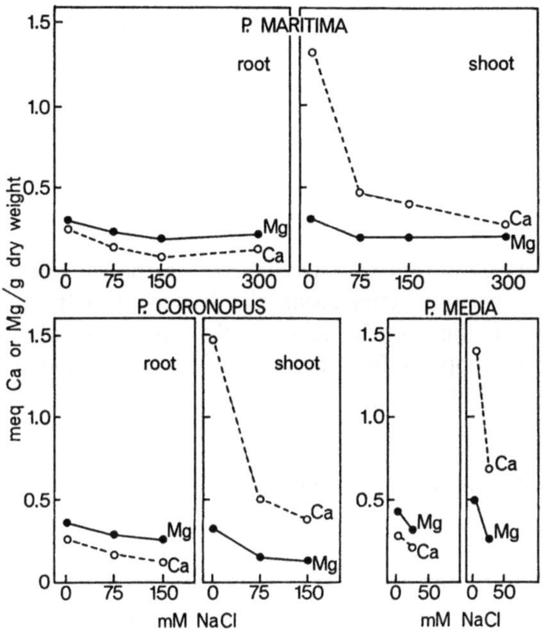

Fig. 5.18. Ca and Mg levels of *P. maritima* and *P. coronopus* (both halophytes) and *P. media* (glycophyte) as affected by salinity. (Experiment by Erdei and Kuiper 1979)

K^+ above Na^+. Interestingly, the reaction of *P. media* to excess Ca^{2+} also differed from that of *P. maritima:* stimulation of K^+ uptake (10 mM NaCl) above the control level (no salinity), at the expense of K^+ translocation to the shoot.

Calcium. In *Plantago,* Ca^{2+} content of the roots was much lower than that of the shoot. Salinity did not affect Ca^{2+} content of the roots up to 10 mM NaCl concentration (De Boer 1985). Above this concentration a gradual decrease in Ca^{2+} level of the roots with salinity was observed. Salinity drastically decreased Ca^{2+} translocation to the shoot (Fig. 5.18). Addition of extra Ca^{2+} to the root medium restored the level of Ca^{2+} in the shoot (P. maritima) or even exceeded the level observed in control plants (*P. media*) (De Boer 1985).

Magnesium. The levels of Mg^{2+} in roots and shoots of *Plantago* were about the same. In all studied species Mg^{2+} uptake and translocation were gradually reduced with salinity, this in contrast to the response of Ca^{2+} in the shoot to salinity.

5.4.3 Effect of Diethylstilbestrol (DES) on Uptake and Translocation of Na^+

DES, an inhibitor of plasmamembrane ATPase of plant root cells (Cheeseman et al. 1980) reduced Na^+ uptake of *P. maritima* roots at low NaCl concentration (De Boer 1985). At the same time it stimulated Na^+ uptake and translocation in *P. media* (Tánczos et al. 1981). Obviously, DES affected energy dependent Na^+ intake in *P. maritima* roots, while it blocked the Na^+ extrusion pump in the roots of *P. media.*

5.4.4 Conclusions

P. media seems to lack the ability to store toxic ions like Na^+ in the vacuoles. It attempts to exclude Na^+ at various steps along the transport pathway through the roots. *P. maritima* has the capacity to store Na^+ in its vacuoles. It seems to take advantage of this property for regulation of its water balance and turgor under conditions of salinity.

5.4.5 Characterization of the H^+ Translocating ATPase in Plantago Roots

It is now widely assumed that the energy for ion transport in plant cells is obtained by establishment of a proton gradient across the plasma membrane by a H^+ translocating ATPase. This ATPase has been characterized in a number of halophytic plant species: *Beta vulgaris* (Hansson and Kylin 1969), *Avicennia* (Kylin and Gee 1970), *Halocnemum* (Vakhmistrov et al. 1982), and *Plantago maritima* and *P. coronopus* (Erdei and Kuiper 1980; Brüggemann and Janiesch 1987).

Purified plasma membrane fractions of roots of *P. maritima, P. coronopus, P. lanceolata* and *P. major* ssp. *major* were obtained by separation with the aqueous two-phase system (Brüggeman and Janiesch 1987). Purity of the preparation was checked by marker enzymes for mitochondria, tonoplast and endoplasmic reticulum and by the reaction of the ATPase towards various inhibitors. Tests proved to be satisfactory. The pH-dependency of the H^+ ATPase from the roots showed narrow pH optima for three *Plantago* species, pH 6.0–6.5. For *P. coronopus* a wider pH activity profile was noted. All preparations showed a high affinity for ATP and contamination by non-specific phosphatases was negligible. Mg^{2+} was needed for activation, Ca^{2+} acted in inhibitory fashion. Monovalent cations further activated ATPase activity, especially at the optimal pH, 6.0. No synergistic effects between Na^+ and K^+ were observed. The high similarity of the preparation of the roots of the four species do not indicate a relation with salt tolerance. In *P. maritima* and another *Plantago* species, *P. crassifolia*, the plasma membrane ATPase remained unchanged when the plants had adapted to saline conditions (Brüggemann and Janiesch 1988). The function of the H^+-ATPase seems to be limited to energize the membrane for ion transport and it obviously does not play a direct role in ion flux control.

5.4.6 Membrane Lipids

Membrane lipid composition is an important factor in salt resistance of higher plants, since structure and permeability characteristics of the lipid matrix of plant cell membranes strongly depend on it. Structure and permeability characteristics of the lipid matrix will determine ionic composition and organic osmoregulator concentration of the cell cytoplasm under saline conditions. In addition, lipids are necessary for the functioning of the H^+-translocating membrane ATPase (Hansson et al. 1973; Erdei et al. 1977; see Sect. 5.4.5). A short review of the effect of salinity on various lipid components, with emphasis on *Plantago*, will be presented.

Free sterols, which are found in plasmalemma and tonoplast, are very effective in regulation of ionic permeability. Sterols differ in degree of control of permeability of plant cell membranes, cholesterol being most effective and sitosterol being least effective. In roots of *Plantago maritima*, the level of sitosterol was reduced with increasing salinity, indicating that the increased level of cholesterol became more effective in regulation of permeability (Fig. 5.19; Erdei et al. 1980). A similar observation was made in *Citrus*, where sitosterol was replaced by stigmasterol when the roots were exposed to NaCl (Douglas and Walker 1983). Regulation of sterol metabolism may be a factor in the regulation of ion fluxes in plants under saline conditons.

It was shown in *Plantago* that the phospholipid content of shoots and roots apparently decreased with salinity, but, expressed on a total lipid basis, no changes were observed (Stuiver et al. 1982; Fig. 5.20). Evidently, the decreased phospholipid content (per g dry matter) was not due to decreased synthesis or breakdown of phospholipids, but may be caused by changes in the plant cell upon exposure to salt stress, such as, e. g., production of more cell wall material and accumulation of NaCl in the cell. Also, the phospholipid composition of roots and shoots of salt-resistant as well as salt-sensitive *Plantago* species was unchanged by long-term exposure to NaCL stress, indicating that NaCl did not lead to "adaptive" changes in phospholipid composition in *Plantago*. These data are in contrast with data on barley and sugarbeet. In the latter species phosphatides possible play a role in glycine-betaine synthesis (Hitz et al. 1981). *Plantago*, however, produces sorbitol as an osmoregulator under salinity stress,

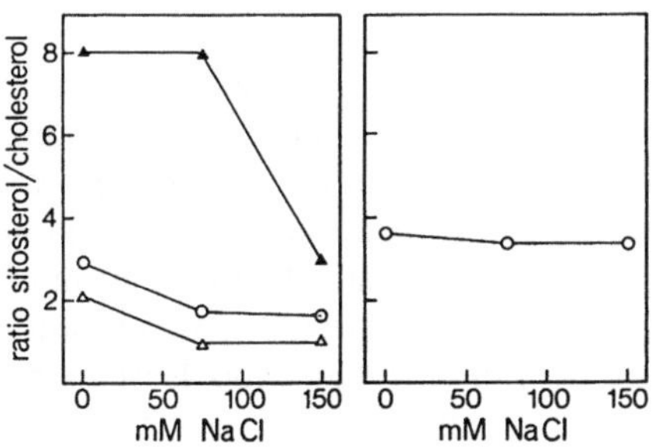

Fig. 5.19. Sitosterol/cholesterol ratio of lipids of the roots of *P. maritima* (*left*) and *P. coronopus* (*right*) as affected by salinity. Roots, nutrient-poor conditions (▲); shoot, nutrient-rich conditions (○); shoot, nutrient-poor conditions (△). (After Erdei et al. 1980)

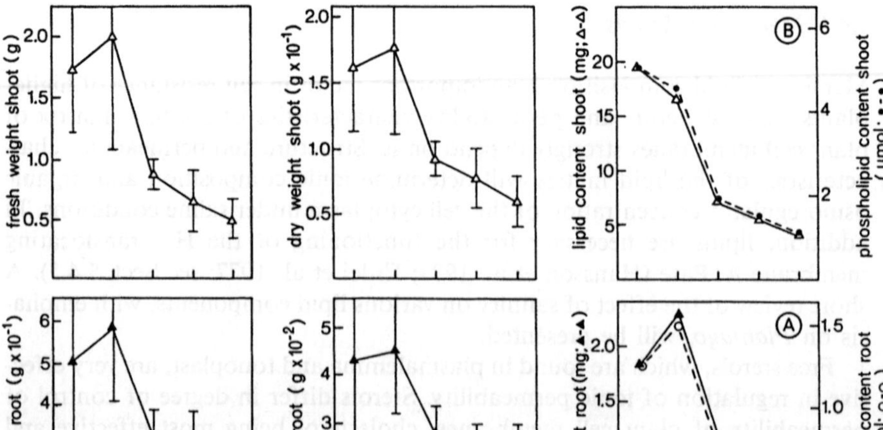

Fig. 5.20. Effect of salinity (NaCl) on growth, total lipid and phospholipid content of *P. media* (glycophyte, **B**). A similar picture was obtained for *P. coronopus* (halophyte, **A**; after Stuiver et al. 1982)

and only traces of betaine were observed (Lambers et al. 1981 a). In *Plantago* phospholipid biosynthesis and degradation evidently are regulated to a high degree.

In contrast to phospholipids, the level of glycolipids was often affected by salinity. In the salt-sensitive *Plantago media*, and in the halophytic *P. coronopus* and *P. maritima* under more extreme saline conditons, the glycolipid level of shoot and roots was decreased by salt stress. Synthesis of chloroplast galactolipids may be reduced by inhibition of the chloroplast envelope enzymes galactosyl transferase and -acylase (Mueller and Santarius 1978).

5.4.7 Electrical Potentials in Roots of Plantago

Differences in Na^+ translocation from root to shoot of the studied *Plantago* species indicate the existence of an ion-selective barrier at the symplast/xylem interface of the roots. It is generally agreed upon that an electrogenic proton pump exists at the plasmalemma of epidermal and cortical cells, regulating ion transport between the root environment and the symplast. The mechanism of ion transport from the root symplast into the xylem has been extensively studied in *Plantago maritima* and *P. media* (De Boer et al. 1983). Models were presented that in roots two active transport mechanisms in fact exist in the form

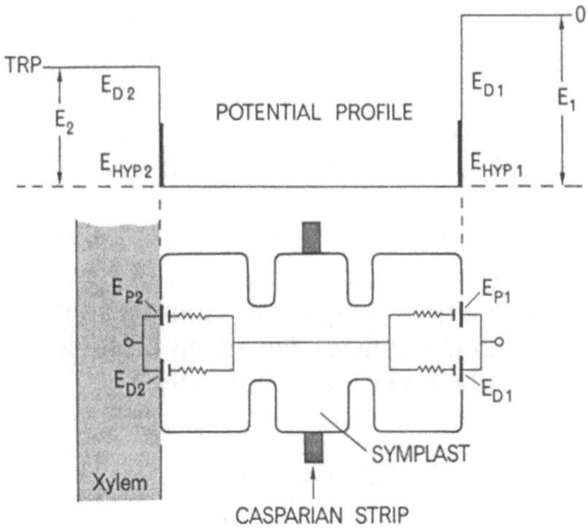

Fig. 5.21. Schematic presentation of the electrophysiology of *Plantago* roots (after De Boer et al. 1983). Trans root potential, *TRP*; potential of epidermal/cortical cells, *E1*; potential of xylem parenchym cells, *E2*; electromotive force of the electrogenic pumps, *Ep1* and *Ep2*, respectively; passive channels, *ED1* and *ED2*, respectively

of two spatially separated electrogenic pumps, one at the environment/symplast interface and the other at the symplast/xylem interface. The two proton pumps work in opposite directions as was first suggested by Okamoto et al. (1978; Fig. 5.21). Measurements on the trans root potential (TRP) under different conditions of O_2 supply of the roots yielded evidence for the above model. Upon gradual exposure of the plant roots to anoxia, a biphasic reaction of TRP was observed (Fig. 5.22). An initial rapid hyper-

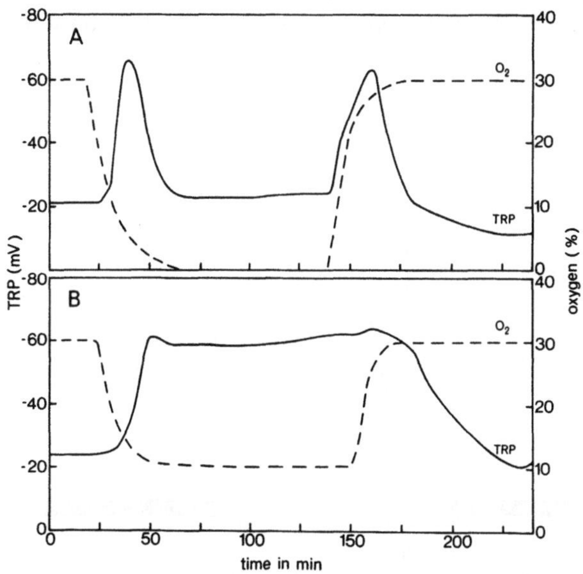

Fig. 5.22. Reaction of the trans root potential of *Plantago* roots under anoxia (**A**) and when pO_2 was reduced to 10% (**B**). (After De Boer et al. 1983)

polarization followed by depolarization was observed, in agreement with the hypothesis that the inner pump was more quickly inactivated by anoxia than the outer pump and the fact that both pumps operated in opposite directions. Restoration of O_2 availability first restored the outer pump, resulting in hyperpolarization, followed by restoration of the inner pump and resulting in depolarization.

Further studies are needed to evaluate the role of both pumps in salt resistance of *Plantago*. Possibly differential hormonal interaction with the electrogenic pumps may explain important differences in Na^+ translocation to the shoot, as observed in halophytes and glycophytes. For recent progress see Maathuis and Puis (1990), Maathuis (1991) and Staal et al. (1991).

5.4.8 Carbon Metabolism of Plantago Under Saline Conditions

It is sometimes suggested that growth reduction of plants by saline conditions is partly due to shortage of energy, because of the energy needed for ion transport and repair of salt-induced damage (Shone and Gale 1983). An energy balance of *Plantago coronopus* grown under non-saline and saline conditions (50 mM) yielded no differences between both groups of plants as far as growth, photosynthesis, dark respiration of shoots, respiration of the roots, including the cytochrome pathway and the alternative route, were concerned (Blacquière and Lambers 1981). Evidently, *P. coronopus* did not require extra energy for growth at 50 mM NaCl.

In short-term experiments interesting interactions between energy metabolism and osmoregulation were observed in the same *Plantago* species. *P. maritima* (Ahmad et al. 1979) and other *Plantago* species produce sorbitol for

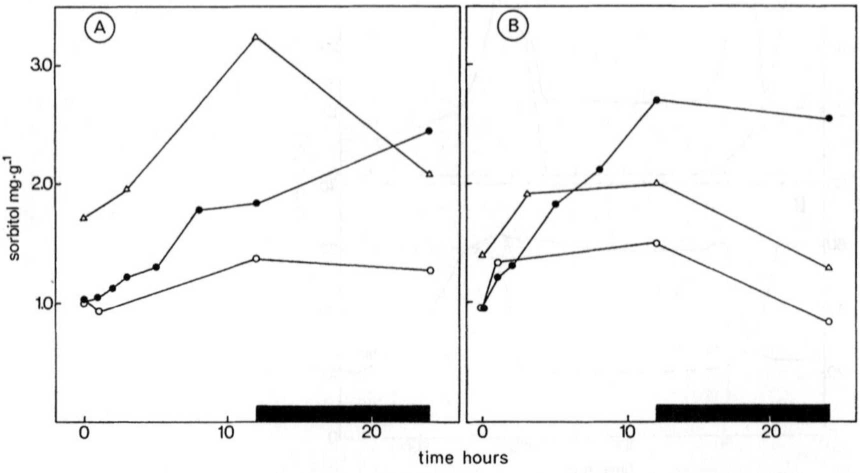

Fig. 5.23. Time course of sorbitol level in shoots (**A**) and roots (*B*) of *P. coronopus* (halophyte). The *black bar* indicates the night period. Plants grown continuously under saline conditions (△); plants exposed to NaCl at time zero (●); control plants (○). (After Lambers et al. 1981a)

osmotic adjustment in a saline environment. As expected, the level of sorbitol in the roots rose quickly after exposure of the roots to 50 mM NaCl. The increase in sorbitol in the shoot started later. As in the long-term experiment, growth, dark respiration of the shoots and photosynthesis were unaffected by transfer of the plants to 50 mM NaCl. Respiration of the roots decreased immediately upon addition of NaCl, which could be attributed to a decreased activity of the cyanide-resistant alternative pathway (Fig. 5.23).

It was calculated that the amount of carbohydrate saved in respiration matched the amount of sorbitol needed for osmoregulation, suggesting a role of respiration via the alternative path in regulation of energy metabolism under changing conditions of salinity (Lambers et al. 1981 a).

Older plants of *P. coronopus* (>55 days) no longer adapted to an increased level of NaCl. Their growth and reproduction was severely reduced upon exposure to 50 mM NaCl. These plants did not increase the level of sorbitol in roots or shoots; the activity of the cytochrome path declined and, instead of a decrease, an increase in the activity of the alternative path was found (Blacquière and Lambers 1981).

5.5 Low Temperature Tolerance of *Plantago coronopus* and *Plantago maritima* as Affected by Salt (NaCl)

5.5.1 Introduction

Low temperature presents an important limiting factor for plant distribution in temperate regions (Larcher 1981). Little information is available about the low temperature tolerance of *Plantago* species. Restriction of the distribution of *P. lanceolata* and *P. media* in Britain by frost was considered unlikely because both species retained green leaves, even after severe frost (Sagar and Harper 1964). The same holds true for *P. major* in Britain. This species overwinters at or below ground level and in herbage an above-ground rosette is maintained in all but severe winters (Bates 1935). In Denmark *P. major* ssp. *major* has been described as more winter hardy than *P. major* ssp. *pleiosperma*, which was killed completely during winter (Mølgaard 1976). Very little is known about the frost tolerance of the salt-tolerant species *P. coronopus* and *P. maritima* and about the role of low temperature as a limiting factor in their distribution.

Towards the northern distribution of its range, as in South Sweden, *P. coronopus* is bound to coastal areas. In more southern regions such as Central and southern France, inland occurrence is frequently observed (Sect. 3.3). *P. maritima* also is mainly restricted to coastal areas. This species occasionally occurs in the inland on gypsum and marl soils in pioneer vegetations in Germany (Oberdorfer 1979). NaCl is not a prerequisite for growth of *P. coronopus* and *P. maritima*; growth in nutrient solution was not enhanced by salt (Erdei and Kuiper 1979).

The question arises why these species in the northern region are mainly restricted to coastal areas with saline soils. Plants generally respond to a high salt concentration with the accumulation of organic molecules in the cytoplasm in order to maintain the osmotic balance between the cytoplasm and the environment (P. J. C. Kuiper 1984). A similar increase of the concentration of organic molecules in the cytoplasm occurs during frost hardening. The resulting higher osmotic value of the cells prevents damaging intracellular ice formation (Levitt 1980).

A possible explanation for the preferential occurrence of *P. coronopus* and *P. maritima* on saline soils could therefore be that these species acquire a higher frost resistance by increasing the osmotic value of their cell sap as a consequence of the presence of NaCl in the soil. The objective of this study was to investigate whether the presence of NaCl in the root medium affected the frost sensitivity of *P. coronopus* and *P. maritima*.

5.5.2 Material and Methods

Seedlings of *Plantago coronopus* L. (Westduinen) and *Plantago maritima* L. (Noordpolderzijl) were placed in a 25% Hoagland nutrient solution (Hoagland and Snijder 1933) and grown in a greenhouse at a minimum day temperature of 22 °C and a minimum night temperature of 16 °C. After 4 weeks, the plants were placed in 30-l tanks with 0, 10, 25, 50 or 75 mM NaCl in 25% Hoagland solution. Youngest fully grown leaves of 10- to 15-week-old plants were used for experiments.

Two pieces of 1 cm length were cut at one third from the leaf base. Roots were blotted with filter paper and cut in 1-cm pieces. Samples of two leaf pieces or 0.5 g root material were put into test tubes. The tubes were closed with rubber stoppers and placed in ethanol in a Kryostat (Lauda UK 30 D) at +4 °C for 1 h. The tubes were than cooled at a rate of 2 °C h^{-1} to −10 °C. To prevent supercooling the samples were seeded with ice cristals at −2 °C (Levitt 1980). By freezing tubes with plant material in liquid nitrogen, 100% damaged controls were obtained (Horvàth et al. 1981); other controls were directly placed at 4 °C (0% damage).

Electrolyte leakage after freezing and thawing was measured with a conductivity meter. After thawing at 4 °C overnight, 10 ml of demineralized water was added to each tube. The conductivity was measured after shaking the tubes during 30 min on a shaking machine at room temperature. Damage was expressed as the percentage conductivity of the 100% (treatment in liquid N_2) minus the 0% control. The temperature at which 50% damage occurred, LT 50, was determined from the percentage of plant damage against temperature.

After low temperature treatment and thawing overnight, leaf pieces were placed on an aluminium plate covered with wet filter paper at 20 °C. After a dark period of 1 h, the rate of chlorophyll fluorescence induction of the leaf pieces was measured with a fluorimeter. The relative capacity of chlorophyll fluorescence induction was determined by measuring the maximal rate of the fluorescence rise (Smillie and Hetherington 1983).

5.5.3 Results

Table 5.19 shows the effect of salt concentration in the nutrient solution during growth on the LT 50 of *P. coronopus* and *P. maritima* shoots and roots. Roots, LT 50 about −3 °C, were more frost-sensitive than the shoots, LT 50 about −5 °C. There was no difference in frost sensitivity between the two species and between the various salt concentrations. The presence of NaCl in the nutrient solution, however, affected the frost sensitivity of leaves of both species during prolonged freezing at −4 °C, above LT 50 (Table 5.20). Damage to *P. coronopus* leaves decreased from 21.6% (without salt) to only 0.6%, when grown at 50 mM NaCl. Damage of *P. maritima* leaves decreased from 7.7% (without salt) to 0, when grown at 75 mM NaCl. Without any NaCl during growth, *P. maritima* leaves were more tolerant to a prolonged period at −4 °C than leaves of *P. coronopus*.

In addition to frost-induced electrolyte leakage, the chlorophyll fluorescence induction capacity after freezing and thawing was determined. Figure

Table 5.19. The effect of NaCl concentration of the nutrient solution during growth on freezing damage, measured as LT 50 (°C) and determined by electrolyte leakage of leaves and roots of *P. coronopus* and *P. maritima*. Mean values of three separate measurements of LT 50 (°C) with their standard deviation (in brackets) are given

NaCl (mM) during growth	LT 50 (°C)			
	P. coronopus		*P. maritima*	
	Leaf	Root	Leaf	Root
0	−5.2 (0.2)	−3.1 (0.3)	−5.6 (0.4)	−3.8 (0.3)
10	−5.2 (0.1)	−3.2 (0.1)	−5.3 (0.3)	−3.1 (0.1)
25	−5.3 (0.3)	−2.8 (0.3)	−6.2 (0.7)	−3.6 (0.1)
50	−5.5 (0.2)	−3.0 (0.3)	−6.2 (1.8)	−3.8 (1.1)
75	–	–	−6.4 (1.9)	−4.4 (1.3)

Table 5.20. The effect of NaCl concentration of the nutrient solution during growth on low temperature damage of leaves of *P. coronopus* and *P. maritima* after cooling to −4 °C for 2.5 h. The results represent means of three separate series of ten samples with their standard deviation (in brackets)

NaCl (mM) during growth	Damage (% electrolyte leakage of the control)	
	P. coronopus	*P. maritima*
0	21.6 (10.1)	7.7 (6.2)
10	5.0 (5.2)	7.7 (5.9)
25	4.6 (3.4)	3.3 (3.1)
50	0.6 (1.0)	1.0 (1.2)
75	–	0.0 (0.6)

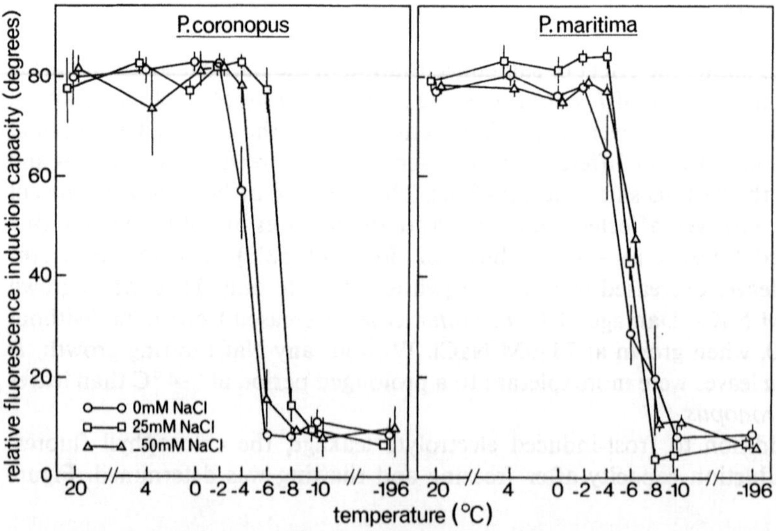

Fig. 5.24. The effect of cooling to the indicated temperatures on the chlorophyll fluorescence induction capacity of leaves of *P. coronopus* and *P. maritima* plants grown at different NaCl concentration. Every point represents the mean of eight separate measurements with their standard deviation

5.24 shows the changes in relative chlorophyll fluorescence induction capacity of leaves of *P. coronopus* and *P. maritima*, grown at different salt conditions, after cooling to the indicated temperatures. An abrupt decrease in fluorescence was observed between −4 and −8 °C leaves of plants grown with NaCl. In the absence of NaCl during growth the decrease in fluorescence started at a higher temperature, resulting in a significantly lower fluorescence capacity at −4 °C (both species). The effect of the salt concentration of the nutrient solution during growth on the chlorophyll fluorescence induction capacity after freezing to −4, −6 and −8 °C is shown in Table 5.21. Optimal protection against frost was observed in leaves of *P. coronopus*, grown in the presence of 25 mM NaCl. In these leaves the induction capacity after cooling to −6 °C was only slightly decreased (undamaged control leaves) while the capacities of leaves from plants grown with 0 or 50 mM NaCl were very small. The frost tolerance of *P. maritima* was less affected by the salt concentration of the growth medium. Leaves of plants grown with 50 or 75 mM NaCl showed a small but significant increase in frost tolerance at −4 °C and a more pronounced increase at −6 °C; at −8 °C salt-treated plants were more damaged. There was hardly any difference between the protective effect of 50 and 75 mM NaCl.

5.5.4 Discussion

Field observations indicate a relatively high frost sensitivity of *P. coronopus* in The Netherlands. A high winter mortality was observed in dune slacks,

Table 5.21. The effect of freezing to the indicated temperatures of *P. coronopus* and *P. maritima* leaves on the capacity for induction of chlorophyll fluorescence. Capacity is expressed in degrees. Plants were grown at various NaCl concentrations. Data represent the mean of four determinations with their standard deviation in brackets

NaCl (mM)	Freezing temperature (°C)					
	P. coronopus			*P. maritima*		
	−4	−6	−8	−4	−6	−8
0	56.9 (9.1)	8.1 (1.3)	7.7 (1.6)	63.7 (8.2)	28.7 (6.1)	19.2 (11.5)
25	83.4 (1.0	77.2 (4.8)	14.0 (3.8)	–	–	–
50	78.4 (6.2)	15.8 (3.5)	8.8 (0.6)	77.8 (1.7)	47.7 (3.2)	10.1 (1.6)
75	–	–	–	84.2 (0.7)	42.6 (3.8)	3.3 (3.3)

which was mainly ascribed to flooding damage (Schat 1982). However, also on non-flooded soils a high, >95%, mortality was observed after severe winters (Haeck). The results of this study do not support the hypothesis that the preferential occurrence of *P. coronopus* and *P. maritima* on saline soils can be primarily ascribed to the acquisition of a higher frost resistance, expressed as LT 50; the LT 50, a common measure to express frost sensitivity, of both species was hardly influenced by the presence of NaCl in the nutrient solution during growth. Nevertheless, a small inhibition of frost damage at freezing temperatures above LT 50 was observed in plants grown in the presence of salt. The inhibition was most evident in *P. coronopus* grown with 25 mM NaCl. Such an increase of frost tolerance could be due to the presence of sorbitol, which is synthesized in *Plantago* in order to counteract the osmotic effects of NaCl taken up by the plants (Ahmad et al. 1979; Jefferies et al. 1979; Lambers et al. 1981 a). Sorbitol could inhibit frost damage in a way similar to sucrose, viz. by its colligative effect (Lineberger and Steponkus 1980). Salt-induced inhibition of damage at mild freezing temperatures could play a role in preventing injury by night frosts during springtime. This factor, combined with the moderating effect of the presence of the sea on the prevailing temperature, may be advantageous for survival of *P. coronopus* and *P. maritima* in coastal areas of the Northern region and may explain the preference for coasts of these *Plantago* species.

5.6 Seed Germination of *Plantago major* ssp. *major* and *Plantago lanceolata*

5.6.1 Introduction

Many seeds, particularly small ones, require light for germination. Perception of light by the phytochrome system is considered to be involved in the light response. A certain threshold level of the far-red absorbing form of phytochrome (Pfr) is required for germination in seeds that are under phytochrome control (Toole 1973). The level of Pfr is determined by the light conditions in the seed, either during ripening, the photostationary state of phytochrome being arrested when the seed dries, or after imbition of the seed. Various light-independent physiological processes complicate this picture (Smith 1982).

Plantago major has rather small seeds; the germination is strongly promoted by light, but not under far-red. The seeds of *Plantago lanceolata* are larger and to a lesser extent influenced by light, if at all (Steinbauer and Grigsby 1957; Sagar and Harper 1964; Van der Veen 1970; Gorski et al. 1977; Blom 1978; Frankland and Poo 1980; Grime et al. 1981; Soekarjo 1982; Pons 1986; Pons and Van der Toorn 1988).

So far the effects of light on seeds of the two species have been studied only under a limited set of conditions, optimal for germination. In order to understand more precisely the effect of light on germination in the field, the following study was carried out. The effect of various pretreatments and germination conditions were studied separately and in combination with light of various spectral energy distribution. The results are discussed from a physiological and an ecological point of view.

5.6.2 Plantago major ssp. major

5.6.2.1 Physiological Aspects

In various seed collections freshly harvested and dry stored seeds of *P. major* only germinated maximally at temperatures of 25 °C and higher. Cold stratifi-

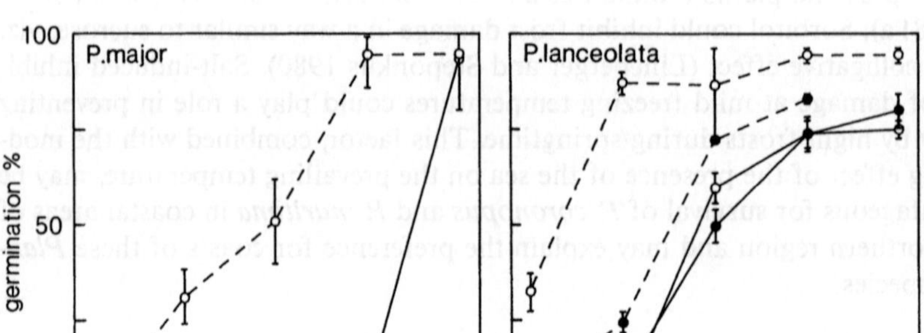

Fig. 5.25. Influence of temperature, stratification and light on seed germination (mean % ± SE) of *Plantago major* (collection Utrecht, 1980) and *P. lanceolata* (collection Maartensdijk, 1980). The stratification treatment included burial of the seeds at 3 cm depth in soil during one winter. (Pons and Van der Toorn 1988)

cation caused a lowering of this temperature threshold, to about 10 °C (Fig. 5.25); often the rate of germination was also increased (Soekarjo 1982). Normally, the seeds do not germinate in darkness. Of the various seed collections tested, a single one gave 5% germination at 30 °C in darkness, after stratification. Temperature fluctuations had no clear effects. Differences from the mean constant temperature could for the larger part be ascribed to specific effects of the maximum or the minimum temperature (Pons 1986).

The germination % of *P. major* seeds decreased with a decrease in red/far-red light ratio (R/FR) to which the seeds were exposed (Fig. 5.26; Frankland and Poo 1980). The relationship dependend on temperature during the test and whether the seeds were stratified. A low R/FR was less inhibitory

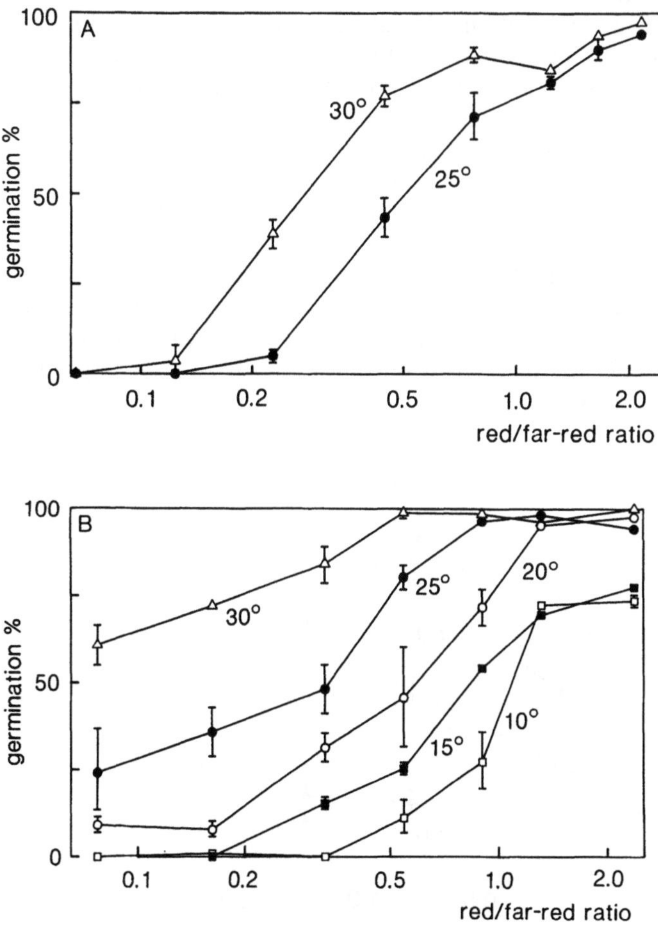

Fig. 5.26. Influence of temperature and stratification on the response of *P. major* seeds (collection Heteren, 1981) to the red/far-red ratio of the light (R/FR). **A** Seeds stored dry at 4 °C. **B** Seeds stratified for about 8 weeks at 4 °C. (Pons 1986)

Table 5.22. Influence of the red/far-red ratio of the light (R/FR) on germination % of *Plantago major* with and without nitrate (10^{-2} M) added

	R/FR	
	0.3	1.2
$-KNO_3$	66	97
$+KNO_3$	87[a]	99

[a] Difference significant at a R/FR of 0.3 for $P<0.05$).

at high temperature and when the seeds were stratified. Apparently there was a decrease in the level of Pfr required for germination, with increasing temperature and when stratification had occurred (Van der Woude and Toole 1980; Pons 1983, 1984). The combination of stratification + high germination temperature made germination even possible at the low level of Pfr, established by far-red (Taylorson and Hendricks 1971). The actual level of Pfr at the site of action in the seeds is difficult to establish, due to pigmentation of the seed coat (Rezk 1974). Germination under green safelight (Blom 1978), especially after a period of burial (Froud-Willams et al. 1984a), was most likely due to the extreme sensitivity of the seeds to the low level of Pfr, established under these ligth conditions.

Steinbauer and Grigsby (1957) reported a stimulating effect of nitrate on germination. We found that nitrate stimulated germination only under suboptimal conditions of temperature and light (Table 5.22). The influence of nitrate on the response of germination to light indicates that nitrate somehow interferes with the action of Pfr.

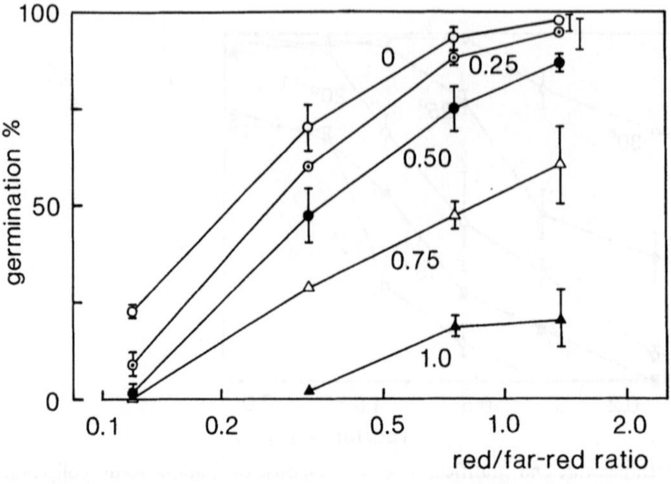

Fig. 5.27. Influence of osmotic strength (MPa) on the germination response of *P. major* seeds (collection Heteren, 1981) to the R/FR ratio at 28 °C. (Pons 1986)

A decrease in osmotic potential decreased germination, particularly at low R/FR ratio (Fig. 5.27), pointing to an increase in Pfr requirement for the generation of a sufficiently large growth potential of the embryo. The latter is needed to counteract the low osmotic potential and break the testa (Scheibe and Lang 1965; Karssen 1970b).

5.6.2.2 Ecological Aspects

P. major is a rather short-lived perennial, which propagates only occasionally by vegetative means; its habitat is typically exposed to disturbance (Sagar and Harper 1964). Survival of populations depends very much on continued establishment of new populations from seeds. Seeds of this species are an important constituent of seedbanks of Dutch grasslands (van Altena and Minderhoud 1972), due to the potentially long lifespan of the seeds in the soil (Kivilaan and Bandurski 1973).

The almost absolute light requirement of the seeds is another important requisite for survival in the soil; fatal underground germination of the small seeds, about 0.2 mg, is avoided. Disturbance of the soil will expose buried seeds to light which can induce germination. Changes of survial of the resulting seedlings are enhanced, due to reduction or absence of competition by established plants (Angevine and Chabot 1979). Mortality of the small seedlings in dense vegetation would otherwise be very high (Sagar and Harper 1961; Van der Toorn and Pons 1988).

Emergence of *P. major* seedlings generally takes place in spring; seeds shedding is in late summer and autumn. Germination of these newly shed seeds is, to some extent, prevented by a high temperature requirement, although low night temperatures may alleviate dormancy (Pons 1986). When the seed has dropped under a leaf canopy, germination will be inhibited by the low R/FR ratio; Blom (1978) reported hardly any emergence from non-stratified seeds, sown in a dense vegetation in spring. In temperate regions dormancy should be completely alleviated in spring. Leaf canopies, if present, are rather open in spring, and the R/FR ratio is only slightly depressed. At low temperatures, seed germination is inhibited when the R/FR ratio decreases only slightly below normal daylight values, about 1.2 (Fig. 5.26). Hence, it may be predicted that the rather open leaf canopies, present in spring, may eventually prevent germination of seeds at the surface due to the prevailing low temperatures. But many of the *P. major* seeds, shed in the previous season, will have entered the soil through cracks and animal activity, and consequently are excluded from light. Later in the season germination of seeds at the surface may be inhibited by the decreasing R/FR ratio in the vegetation due to leaf development, by low water potentials and induction of secondary dormancy (Froud-Williams et al. 1984b; Pons and Van der Toorn 1988).

5.6.3 *Plantago lanceolata*

5.6.3.1 Physiological Aspects

Most freshly harvested seeds of *P. lanceolata* are dormant. The degree of dormancy of the seed population depends on ecotype (Van Groenendael 1985 a, b) and probably also on conditions during ripening of the seeds. This dormancy can be alleviated to some extent during after-ripening (Blom 1978). Further release of dormancy proceeds during cold stratification which causes an increase in germination, especially at lower temperature and in darkness (Fig. 5.25); the rate of germination is also increased. The influence of the various factors on dormancy as described above makes the response of germination rather complicated.

It has been shown that nitrate can stimulate germination of *P. lanceolata* in light and in darkness (Steinbauer and Grigsby 1957; Williams 1983; Pons 1989). The critical concentration being 0.1 mM and the optimum 10 mM (Fig. 5.28).

Various authors report little effect of light on germination of *P. lanceolata* (Sagar and Harper 1964; Gorski et al. 1977; Blom 1978; Grime et al. 1981). However, the light effect varies between seed collections (Van Groenendael 1985 a) and due to pretreatment of the seeds. Seeds of *P. lanceolata* that had after-ripened for 4 months showed a rather pronounced effect of light of different R/FR ratio at a range of temperatures (Fig. 5.29). The effect of R/FR ratio was much reduced after a long period of after-ripening (Fig. 5.30). There was also little response of the seeds to the R/FR ratio at 20 and 15 °C after stratification but at 10 °C there was still an effect (Fig. 5.29 b). The latter indicates that a large proportion of the seeds was still under phytochrome control. *P. lanceolata* seeds showed in many respects a similar response to temperature, stratification and light as the seed of *P. major*, although the germination response was much less strict. Variability between seed collections with regard to the seed responses was larger than in *P. major*.

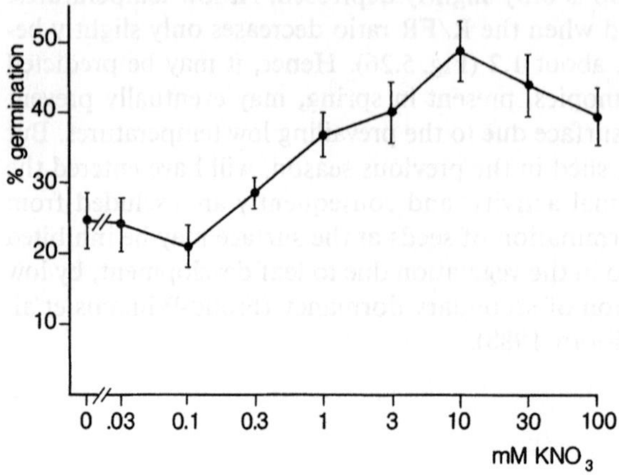

Fig. 5.28. Germination of *P. lanceolata* seeds (collection Vrakelberg 1985) in darkness at various nitrate concentrations. (Pons 1989)

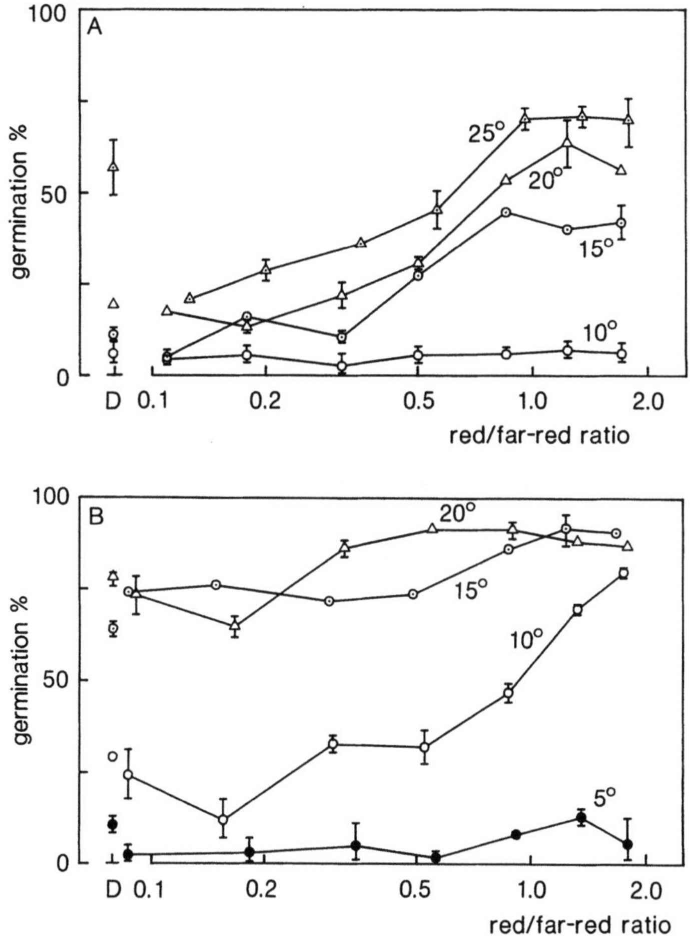

Fig. 5.29. Influence of temperature and stratification on the response of *P. lanceolata* seeds (collection Maartensdijk, 1981) to the R/FR ratio. **A** Seeds stored dry at 20 °C for about 4 months. **B** Seeds stratified at 4 °C for about 2 months after **A**; (4–10% of the seeds germinated during the stratification treatment were excluded from the experiment). (Pons and Van der Toorn 1988)

How the phytochrome system funtions in *P. lanceolata* seeds is not clear. The seeds that were kept in soil of a low water potential for 3 weeks germinated thereafter to nearly the same extent in darkness as the control seeds; there was no difference at all after stratification (Fig. 5.31). Disappearance of Pfr should have proceeded during the incubation period (Toole 1973). Possibly the seeds that germinated in darkness were no longer under phytochrome control anymore.

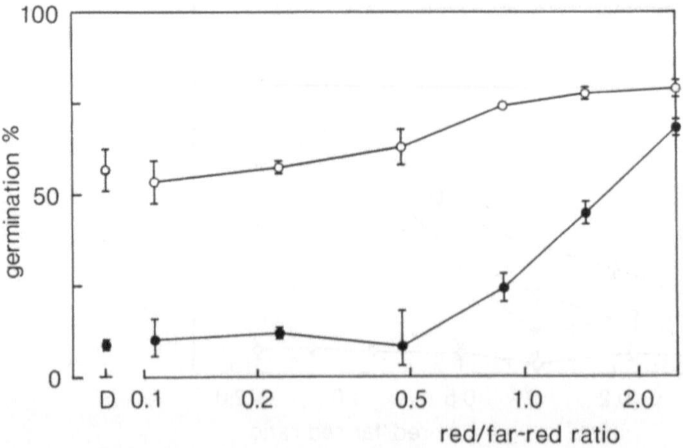

Fig. 5.30 Germination of *P. lanceolata* (collection Maartensdijk, 1980) at various R/FR ratios. ○, seeds stored dry for 9 months at 20 °C and 6 months at 4 °C; ●, seeds stored dry for 6 months at 20 °C and buried in soil at 20 °C for 4 weeks; 52% of the seeds germinated during burial and the remaining seeds were used for the experiment. (Pons and Van der Toorn 1988)

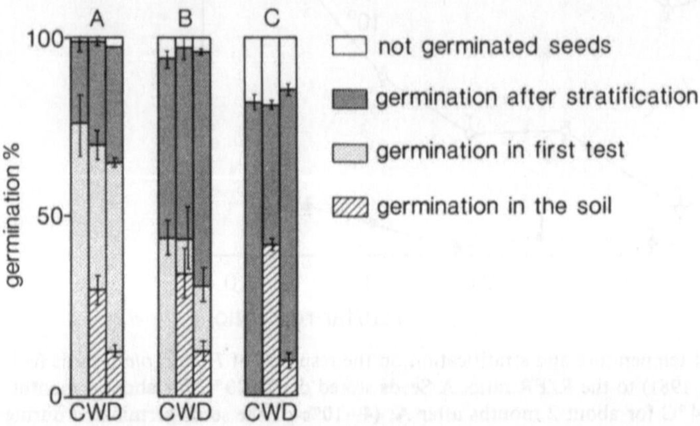

Fig. 5.31. Influence of soil moisture, light and stratification on dormancy and germination of *P. lanceolata* seeds (collection Maartensdijk). Seeds were buried in soil for 3 weeks at 22 °C. Germination was tested thereafter at an alternating temperature regime of 12/22 °C in light (**A**), darkness (**B**) or not tested at all (**C**). The seeds of treatment **B** were exposed to light during recording of the germination %. Subsequently the seeds were stratified during 6 weeks at 4 °C and germination was tested (again) at 12/22 °C in light (**A** and **B**) or darkness (**B**and **C**). Soil moisture treatments (löss from South-Limburg): *W* wet soil (36% on dry weight basis, approx. $-2\cdot10^{-3}$ MPa); *D* dry soil (21% approx. −1.0 MPa); *C* control, not incubated in soil

5.6.3.2 Ecological Aspects

P. lanceolata in general is not a very long-lived perennial under competitive conditions; it shows little vegetative reproduction (Sager and Harper 1964; Berendse 1983; Van Groenendael 1985 a). The species occurs in grasslands and

Table 5.23. Germination (% ± S.E.) of seeds of *P. lanceolata* in bags buried in September 1985 in the soil in chalk grassland with intact vegetation and at an adjacent site cleared of vegetation. (Pons 1989)

Vegetation	Date		
	March 27, 1986	Sept. 22, 1986	March 24, 1987
Intact	5 ± 2.9	11 ± 0.7	10 ± 2.3
Removed	14 ± 2.3	23 ± 6.4	35 ± 7.3

depends on sufficient open space for establishment from seed. The seedlings are better equipped for survival between grass than those of *P. major* (Grime 1966; Van der Toorn and Pons 1988).

The first seed shedding may already occur in early summer. Germination will be prevented to some extent by primary dormancy, although some seeds are able to germinate even under a grass leaf canopy due to the limited inhibition of low R/FR ratios in summertime with high temperatures. Emergence in summer is known (Sagar and Harper 1964), but most emergence occurs in spring (Sect. 4.1). Seedling emergence will then be due to the dormancy alleviating affect of ageing and low winter temperatures. Non-germinated seeds can enter the soil, although not as easily as in *P. major* with its smaller seeds, they are the part of a persistent seed bank (Thompson and Grime 1979; Van Groenendael 1985a).

Two factors have been identified that can break dormancy of seeds in the soil: light and nitrate, and they can function as a gap detection mechanism. Soil disturbance may expose seeds to light and thus stimulate germination (as in *P. major*), although less effectively. But emergence is also stimulated by destruction of the vegetation without disturbance of the soil (Sagar and Harper 1961). Stimulation of germination of buried seeds in the absence of vegetation was also found in a field experiment (Table 5.23). It has been shown that nitrate concentrations are too low for stimulation of germination under intact vegetation, but they can rise to levels of stimulation of germination when absorption by plants ceases, as in the case of destruction of the vegetation (Pons 1989). Hence, dormancy breaking by light as well as dormancy breaking by nitrate can function as gap detection mechanisms in *P. lanceolata.*

5.7 Phenotypic Plasticity in *Plantago* Species

5.7.1 Introduction

The role of phenotypic plasticity in the establishment of individual plants in closely related ecological habitats is quite different for *Plantago major* L. and for *Plantago lanceolata* L. Plasticity is not essential in the life history of *P. major:* in transplant experiments, germination and successful establishment of

P. major seeds were observed only in one single habitat, in which *P. major* did not originally occur (Sagar and Harper 1961; Groot and Boschhuizen 1970). Reciprocal transplant experiments of lawn and grass verge types of *P. major* showed a persistence of morphological features, indicating a lack of phenotypic plasticity (Warwick 1980; Warwick and Briggs 1979, 1980; Sect. 7.2). The observed large phenotypic variation within *P. major* is almost completely due to genetic differentiation (Mølgaard 1976), whereas phenotypic plasticity is more evident in *P. lanceolata* (Wolff 1988). This has also been shown in demographic studies (Boecher 1943; Barber et al. 1968; Sagar and Harper 1961; Van Groenendael 1985a) and in physiological experiments (Teramura and Strain 1979). The latter experiments demonstrated a remarkable phenotypic plasticity in photosynthetic characteristics of individual plants, originating from habitats with quite contrasting light conditions.

Little information is available about the significance of phenotypic plasticity in species with a narrow ecological amplitude, *P. media* L., *P. coronopus* L. and *P. maritima* L. Results from physiological studies (Kuiper and Kuiper 1979a,b; D. Kuiper 1984a; Lambers et al. 1981a) and from ecological studies (Blom 1983; Waite and Hutchings 1982) point to a rather high level of phenotypic plasticity in these genetically variable species (Dodds 1953; Gregor 1938; Jerling 1981; Sagar and Harper 1961).

Contrary to ecologists and geneticists, plant physiologists have neglected phenotypic plasticity. Controversy exists about the interpretation of the phenomenon "plasticity". In the ecological and genetical literature, plasticity has been described as a neutral process, whereas physiologists generally interpret it in adaptive terms. In this section "phenotypic plasticity" is defined as "all intra-genotypic variation" (Bradshaw 1965). Observation of plastic responses of plants and the evaluation of their adaptive value are two separate subjects of investigation. Plasticity and stability of plant characteristics may replace each other to various degrees; both factors may have adaptive value when responses contribute to the fitness of the plants. In this section we will provide examples of a plastic response of *Plantago* species and discuss possible adaptive consequences. Phenotypic plasticity is genetically determined and it may be involved in environmental specification between populations and within species. Plastic responses (Fig. 5.32) may reveal differences between *Plantago* species, which may explain their distribution in the field.

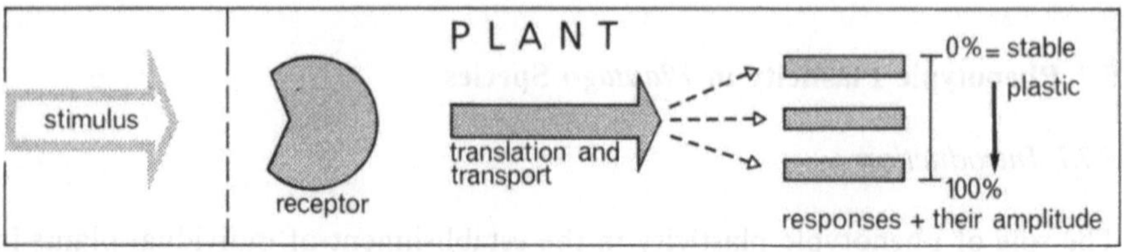

Fig. 5.32. Schematic presentation of the plastic response of plants to an environmental stimulus

5.7.2 The Response of Plantago Plants to Alterations in Mineral Supply

Genotypes of *P. major* L. and populations of *P. lanceolata* L. were exposed to high and low nutrient conditions. Part of the plants were transferred from one to the other nutrient condition in order to study the responses of physiological parameters to a change in nutrient supply. The results showed distinct differences between the two closely related species: shoot to root ratio, respiration and microsomal ATPase activity of the roots and the enzymes, involved in N-metabolism, remained stable (unaltered) in *P. major* ssp. *major*, while plants of *P. lanceolata* responded (Kuiper and Kuiper 1979a, b; Lambers et al. 1981b, c; Stulen et al. 1981a, b). Such differences are well known in literature: shade-tolerant *Sectaria* species showed a plastically responding shoot to root ratio when grown under three levels of irradiance, as opposed to shade-avoiding *Setaria* species (Lee and Cavers 1981). Drew (1982) observed a positive correlation between survival as affected by a range of temperatures and plasticity of the shoot to root ratio: a responsive shoot to root ratio was essential for individuals which functioned in a wide range of environments.

Four inbred lines (Tables 5.24) covering *P. major* ssp. *major* and ssp. *pleiosperma*, were exposed to nutrient switch experiments (Kuiper 1982, 1983a). The response of the shoot to root ratio differed much between the inbred lines: line 1, ssp. *major*, (Fig. 5.33) did not change its shoot to root ratio upon a change in mineral nutrient supply, whilst the other three lines, 2, 3 and 4, belonging to the ssp. *pleiosperma* (Fig. 5.33) were affected by the nutrient supply. The response of the shoot to root ratio of line 2 disappeared during the course of the test. Lines 3 and 4 did not stabilize throughout the experiment; they differed in the speed of response. Similar results for the shoot to root ratio of the four lines were obtained from measurements on the Mg^{2+}-dependent ATPase activity from a microsomal root fraction (Fig. 5.34): no response in line 1 and an extreme response in line 4. The results of measurements of root respiration, including the cytochrome and the alternative pathways, of nitrate reductase activity in shoot and roots and relative growth rate are summarized in Table 5.25; they also reveal genotypic differences in plasticity as mentioned above.

Table 5.24. Plant and habitat characteristics of inbred lines of *P. major*

Line	Seeds per capsule	Description of original habitat	Annual/perennial life cycle
1	11	Frequently mown lawn	Perennial
2	18	Temporarily disturbed grassland at the edge of the inner dunes. It is sometimes flooded by sea water	
3	19	More or less wet arable land	Annual
4	33	River bank and flooded in winter	Annual

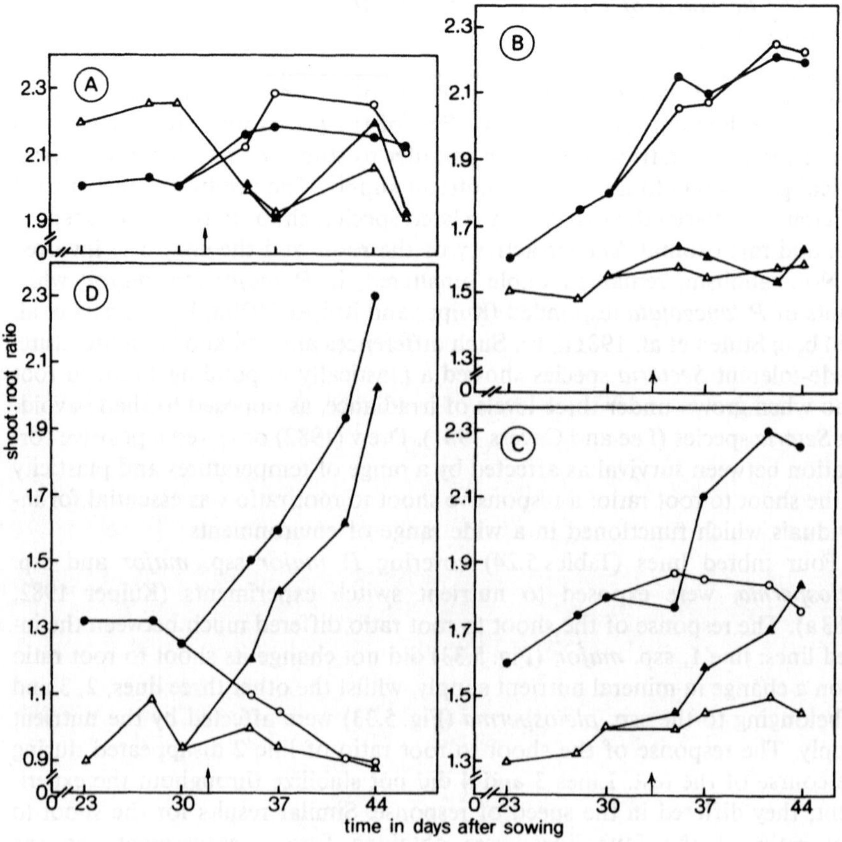

Fig. 5.33. Time course of the shoot to root ratio of four inbred lines of *Plantago major* during the experiment. **A** line 1; **B** line 2; **C** line 3; **D** line 4. Plants were grown in full (100%) or diluted nutrient solution (2%) and transferred to the other condition (100→2%; 2→100%): ●——●, 100%; ○——○, 100→2%; △——△, 2%; ▲——▲, 2→100%. *Arrows* indicate time of transfer. SD 10% or less

Table 5.25. Properties of inbred lines of *P. major*. RGR, relative growth rate; ATPase, adenosine triphosphatase

Parameter	ssp. *major*	ssp. *pleiosperma*		
	line 1	line 2	line 3	line 4
ATPase	Stable	Temporarily plastic	Plastic	Plastic
RGR	Slow response	Temporarily plastic	Plastic	Plastic
Root respiration	Stable	Temporarily plastic	Plastic	Plastic
Shoot to root ratio	Stable	Temporarily plastic	Plastic	Plastic
Nitrate reductase	Stable	Slow response	Rapid response	Rapid response

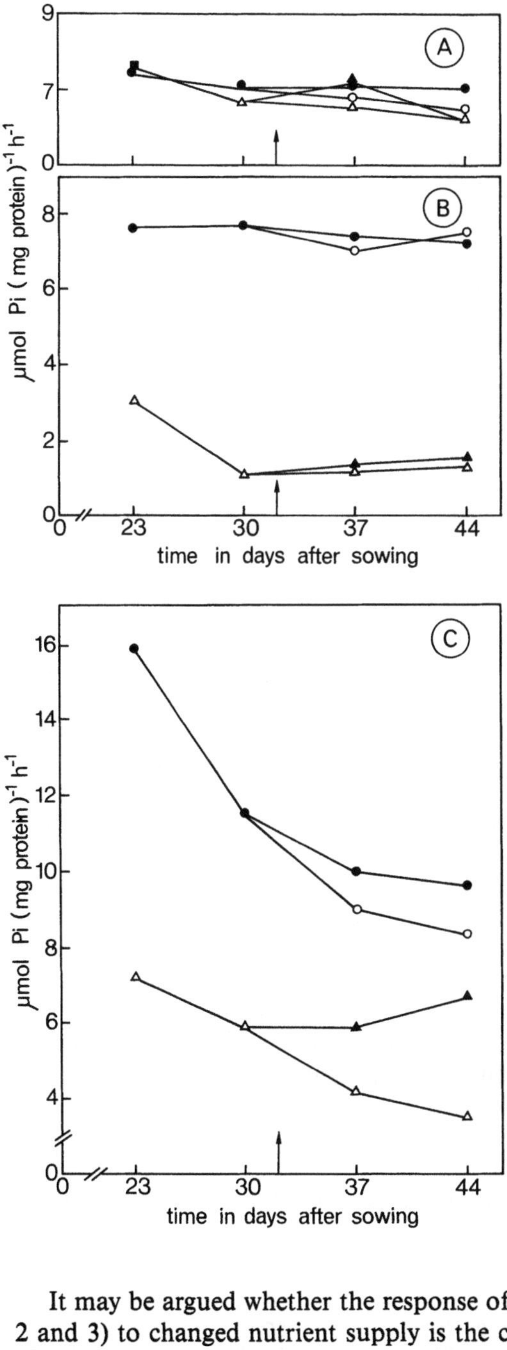

Fig. 5.34. The V_{max} values of the Mg^{2+}-stimulated ATPase activity from roots of four inbred lines during the experiment. **A** line 1; **B** line 1; **C** line 3; **D** line 4. Nutrient conditions: ●——●, 100%; ○——○, 100%→2%; △——△, 2%; ▲——▲, 2%→100%. *Arrows* indicate time of transfer. SD 10% or less

It may be argued whether the response of line 1 (and to a lesser extent line 2 and 3) to changed nutrient supply is the consequence of disturbed growth. When the growth rate may be modulated by changed enzyme properties/capacities as a response to a changed environment, it could well present adaptive plasticity. On the other hand, an altered environment may induce another

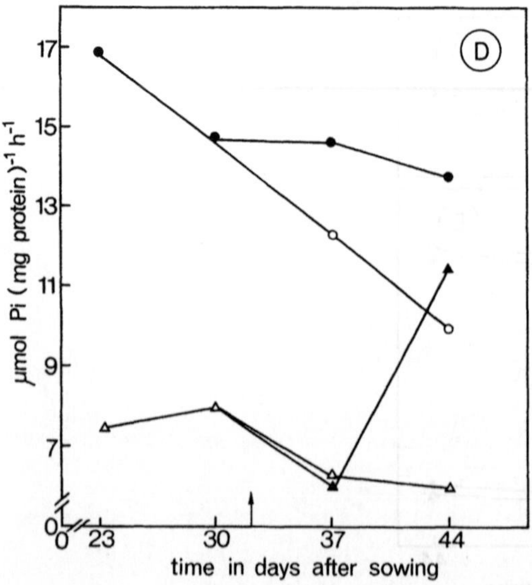

Fig. 5.34D

growth rate, and this in turn may affect enzyme activities; in this case the response probably is non-adaptive. Fitness experiments need to be done to provide conclusive answers to this dilemma. Adaptive and non-adaptive responses occur in *P. major* and there is also evidence for an age-dependent plastic response in line 2 (Figs. 5.33, 5.34). Warwick and Briggs (1979) and Corré (1984) observed stability in morphological characteristics in *P. major* ssp. *major*. It is obvious that there is no connection between heterozygosity and phenotypic plasticity (see also Sect. 7.2).

Competition experiments were set up to investigate the performance of plants of line 1 and 4 to differences in frequency of fluctuation in nutrient supply. Both genotypes responded in the same manner in monoculture. When grown in mixed culture, young plants of line 4 performed better with increasing frequency of fluctuation. Presumably a rapid response, such as observed in plants of line 4 of *P. major*, is of advantage in an environment in which the concentration of a limiting nutrient varies throughout the season (Poorter and Lambers 1986).

Table 5.26. Characteristics of the habitat of three populations of *P. lanceolata*. W, Westduinen; M. Merrevliet; H, Heteren

Population	W	M	H
Total organic matter (%)	2–20	86	33
Moisture content soil (%)	4–30	69–86	18–44
Soil	Sand	Peat	Clay
Height of vegetation (cm)	15	50	70

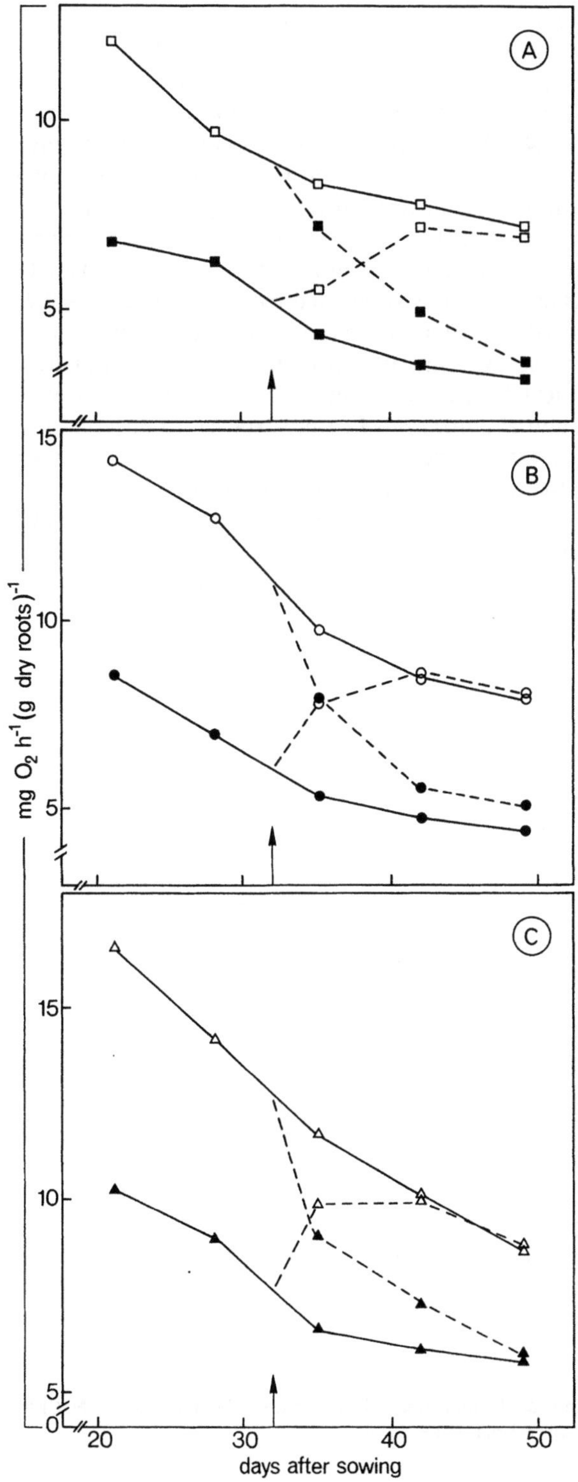

Fig. 5.35. Total root respiration in the plants of three populations at Westduinen (**A**), Merrevliet (**B**) and Heteren (**C**). Nutrient conditions were as follows: **A**: □——□, 100%; ■——■, 2%; ■– – –■, 100%→2%; □– – –□, 2%→100%. **B** ○——○, 100%; ●——●, 2%; ●– – –●, 100%→2%; ○– – –○, 2%→100%. **C** △——△, 100%; ▲——▲, 2%; ▲– – –▲, 100%→2%; △– – –△, 2%→100%. *Arrow* indicates time of transfer of plants from one nutrient condition to the other

Three populations of *P. lanceolata* were studied in the nutrient switch experiment (Table 5.26). Seeds from the populations were located in different vegetations (Mook et al. 1981), in different abiotic conditions (Noë and Blom 1980; Van der Aart 1985). Plants from the various populations differed in relative growth rate (D. Kuiper 1984). Phenotypic plasticity was found in all three populations: as an example the time course of total root respiration is shown (Fig. 5.35). Besides genetic differences in respiratory activity, growth parameters and ATPase activity (not shown), the plastic response was uniformly present in all measured parameters. The data suggest that phenotypic plasticity is species-characteristic. This conclusion is supported by studies of different *P. lanceolata* populations exposed to different irradiance (Teramura and Strain 1979; Teramura 1983). In these experiments, specialization to natural irradiance on population level next to a high degree of plasticity was observed. When plants of *P. media* L. and *P. coronopus* L. were exposed to different nutrient regimes, growth and microsomal root ATPase activities were responsive, as observed in *P. lanceolata* (Kuiper and Kuiper 1979b). Phenotypic plasticity in *P. coronopus* was evident from salinity experiments. Within 24 h after NaCl addition, plants of *P. coronopus* accumulated sorbitol in roots and shoots at the expense of a corresponding temporary fall in the activity of the alternative

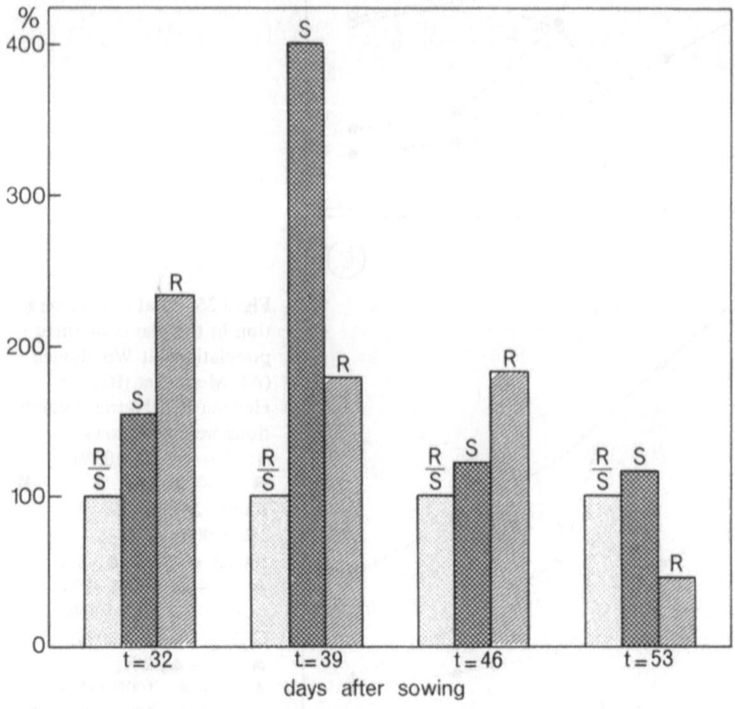

Fig. 5.36. Sorbitol accumulation in root (*R*) and shoot (*S*) in salt-treated plants (50 mM NaCl for 24 h) expressed as percentage of control plants (*R/S)*

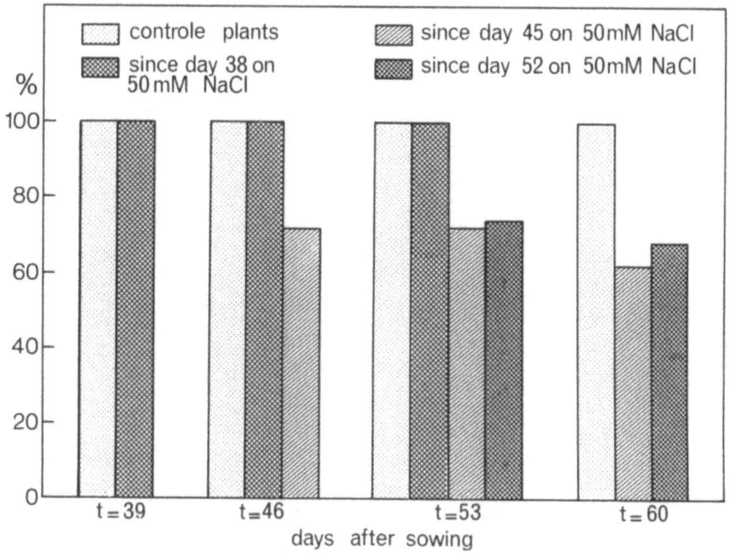

Fig. 5.37. Photosynthesis of salt-treated plants, expressed as percentage of control plants. Otherwise as for Fig. 5.36

pathway (Lambers et al. 1981c, Sect. 5.6). This response was age-dependent (Fig. 5.36) and the loss of ability to accumulate sorbitol, in 40-day-old plants, was reflected in a low photosynthetic activity (Fig. 5.37).

5.7.3 The Response of Plantago Plants to an Alteration in Irradiance

Seedlings of two genotypes of *P. major* ssp. *major* from an exposed and from a shaded habitat were exposed to two levels of irradiance. Several physiological parameters were followed during the experiment and responses of these parameters to a transfer of plants from one level of irradiance to the other were also studied. The aim was to obtain information on the degree of plasticity in field material compared with that of the inbred lines of *P. major.* Secondly, to gain insight in genetic differences to light intensity, since it is likely that such differentiation within *P. major* is important in exploiting its broad ecological amplitude (see also Sect. 7.2). The experimental design involved eight groups of plants (Table 5.27). Figure 5.38 shows the rate of dry matter accumulation (D. Kuiper 1984). Shoot and root growth of plants grown continuously at a high level of irradiance was equal in both genotypes. At a low level of irradiance shoot and root growth of plants, originating from the shaded habitat, exceeded that of the plants from the exposed habitat. Seedlings of both genotypes were able to respond adequately to growth light condition. Plants transferred to low light and control plants at low light showed a similar growth pattern. At the

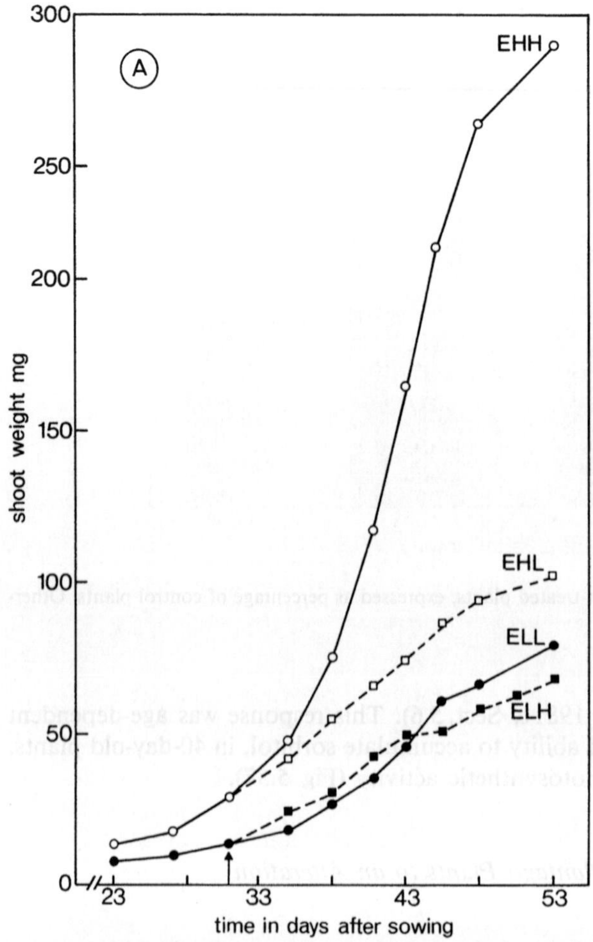

Fig. 5.38. Dry matter accumulation in shoots of two genotypes of *Plantago major* ssp. *major* of an exposed (*E* in **A**) and a shaded (*S* in **B**) habitat. ○———○, EHH; ●———●, ELL; △———△, SHH; ▲———▲, SLL; ■ – – – ■, ELH and SLH; □ – – – ■, EHL and SHL; **E** exposed habitat; *S* shaded habitat; *H* and *L* refer to high and low light intensity during the experiments and *HL* and *LH* to changes in light conditions during the experiment. For further details, see Table 5.26. SD 20% or less (n = 6); *Arrows* indicate time of transfer

end of the experiment the relative growth rate of the plants transferred to low irradiance was considerably lower than that of the other plants (D. Kuiper 1984), indicating a non-adaptive response: a lack of phenotypic plasticity. Consequently, the initially present plasticity had disappeared during the growth of the *P. major* plants. Results on chlorophyll content (Fig. 5.39) confirmed this assumption: a high chlorophyll content (per gram fresh weight) in low irradiance plants was evident in both genotypes, but the effect of shading did not take place in old and newly formed leaves of plants tranferred to low

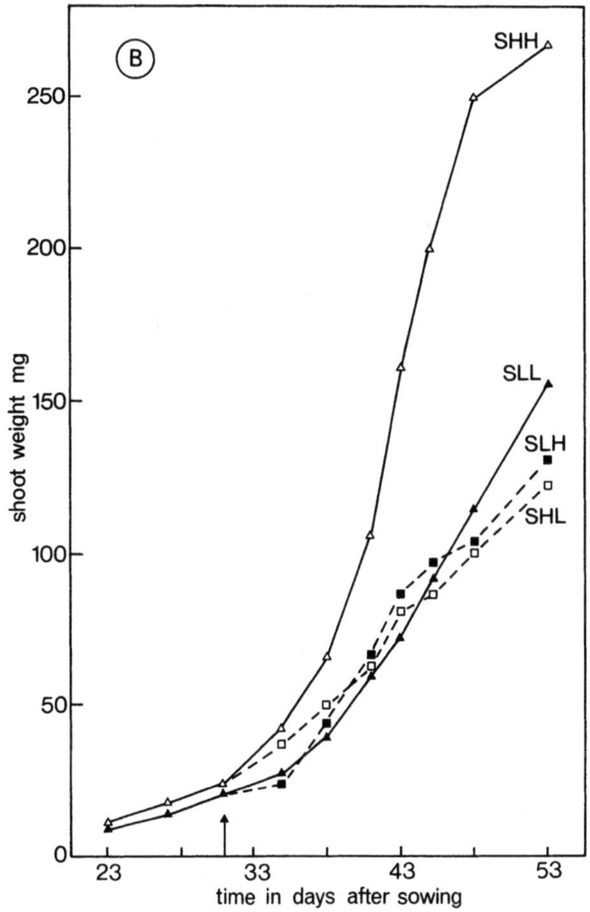

Fig. 5.38 B

Table 5.27. Information about eight groups of the tested plants and explanation of the abbreviations used throughout the chapter

Group	Plants from seeds collected on	Light condition during the two periods of the experiment
EHH	Exposed habitat	High – High
ELH	Exposed habitat	Low – High
EHL	Exposed habitat	High – Low
ELL	Exposed habitat	Low – Low
SHH	Shaded habitat	High – High
SLH	Shaded habitat	Low – High
SHL	Shaded habitat	High – Low
SLL	Shaded habitat	Low – Low

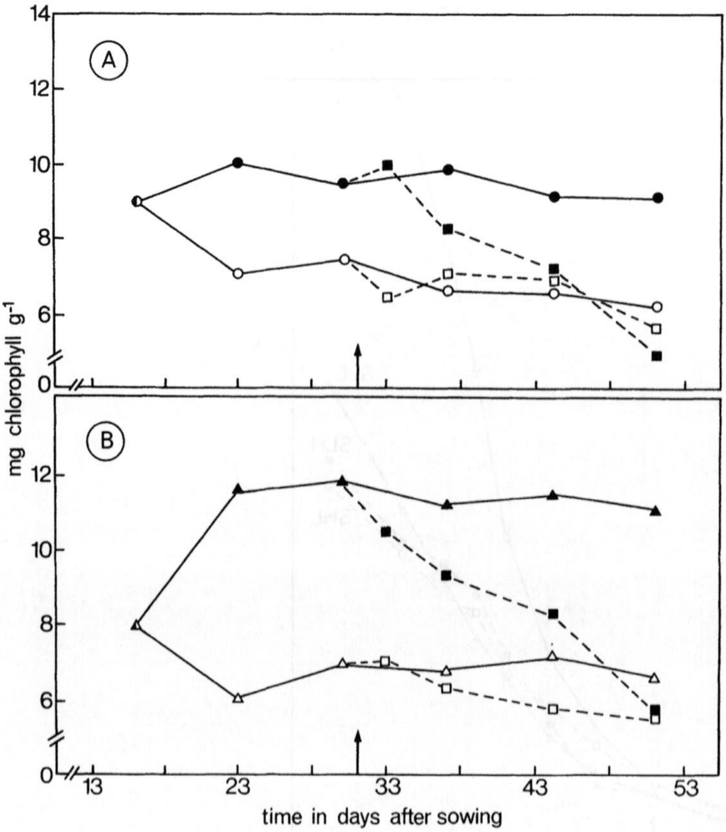

Fig. 5.39. Chlorophyll content of the two genotypes of *Plantago major* ssp. *major* from an exposed (**A**) and a shaded (**B**) habitat. For explanation of symbols see the legend of Fig. 5.38. SD 10% or less (n = 6)

Table 5.28. Photosynthesis [mg CO_2 h^{-1} (g fresh shoot weight)$^{-1}$] measured on day 37 and 51 at the level of irradiance at which the plants had been grown

Group	Plants from seeds collected on	Light condition during the two periods of the experiment	Photosynthesis	
			At day 37	At day 51
EHH	Exposed habitat	High – High	0.85	0.71
ELH	Exposed habitat	Low – High	0.48	0.23
EHL	Exposed habitat	High – Low	0.46	0.31
ELL	Exposed habitat	Low – Low	0.61	0.43
SHH	Shaded habitat	High – High	0.78	0.61
SLH	Shaded habitat	Low – High	0.55	0.29
SHL	Shaded habitat	High – Low	0.51	0.36
SLL	Shaded habitat	Low – Low	0.68	0.53

irradiance. The performance of the latter plants, transferred from a high to a low irradiance, was remarkable; no rise in chlorophyll content was observed and photosynthesis was low, also in plants transferred to high irradiance (Table 5.28). Plants of *Solanum dulcamara* L. from shaded habitats (Gauhl 1976), *Solidago virgaurea* L. (Bjoerkman and Holmgren 1966) and in *Viola* species (Curtis 1984), transferred to a level of high irradiance, also showed a low photosynthetic activity, combined with a low protein and chlorophyll content, which was probably due to lack of anti-photooxidantia (Bjoerkman 1981).

The response of *P. major* plants which were transferred to each of both light conditions is unlikely to be adaptive, since all physiological parameters showed lowered values compared with plants grown continuously at a low level of irradiance. The response of *P. major* ssp. *major* to irradiance resembled that in plants of line 2 to nutrient supply. Such temporary plasticity has also been ob-

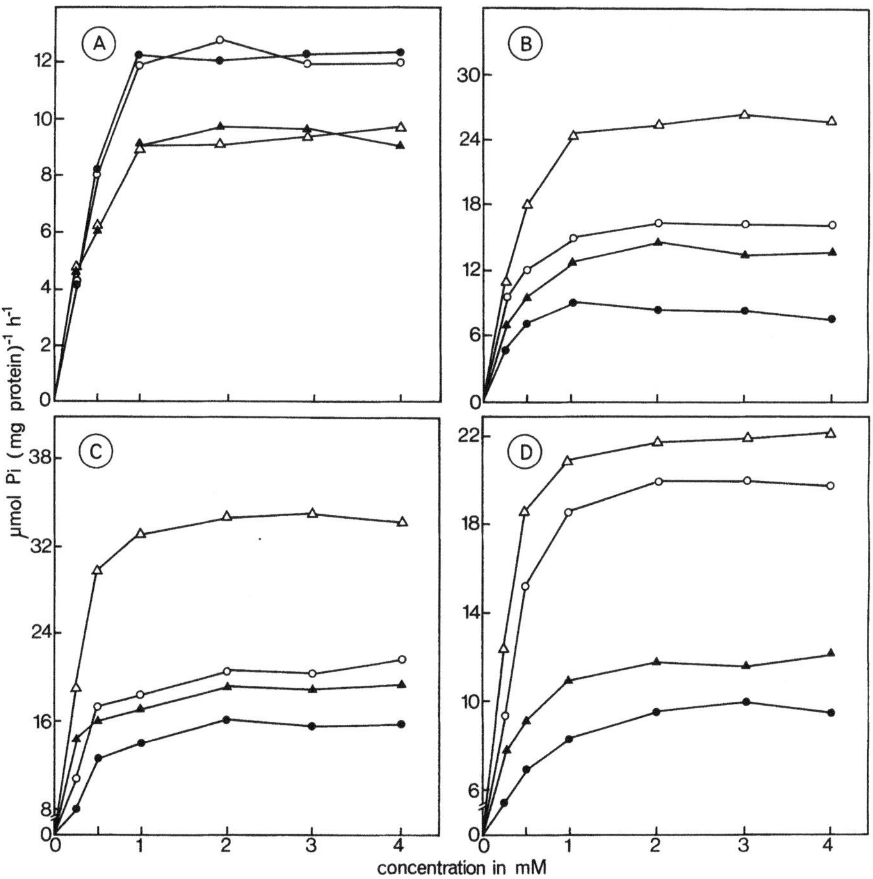

Fig. 5.40. Stimulation of ATPase from roots of *Plantago major* ssp. *major* (**A**), *P. lanceolata* (**B**), *P. media* (**C**) and *P. coronopus* (**D**) by Ca^{2-} (△, ▲) and by Mg^{2-} (●, ○). *Open* and *closed symbols* refer to plants grown at high and low levels of irradiance, respectively. SD 10% or less

served in *Trifolim repens* L. (Dennis and Woledge 1983), in *Betula pendula* L. (Öquist et al. 1982), *Hedera helix* L. (Hoflacher and Bauer 1982) and *P. coronopus* L. (Sect. 6.1).

Only a few experiments were carried out to test the response of *P. lanceolata, P. media* and *P. coronopus* to different levels of irradiance (Kuiper 1983b). Generally the three species responded in the same fashion: a higher shoot to root ratio (not shown) and a decreased root ATPase activity (Fig. 5.40) at a low level of irradiance, thus showing phenotypic plasticity to irradiance. Plants of two populations of *P. lanceolata,* separated by less than 8 m and exposed to different light conditions, an open and a sunflecked site, were capable of growing in full sunlight or in heavily shaded environments. Despite their close proximity, evidence was found for ecotypic differentiation. The population which naturally occurred in periodic shade was better adapted to growth at low irradiance. No evidence was shown for age-dependent plasticity (Teramura and Strain 1979; Teramura 1983). *P. lanceolata* had a phenotypically determined growth rate and it responded to nutrients (Kuiper 1985), to irradiance (Teramura and Strain 1979; Teramura 1983; Kuiper 1983b) and to temperature (Teramura et al. 1981).

5.7.4 A Physiological Model for Phenotypic Plasticity

A model for phenotypic plasticity using irradiance and mineral supply as stimuli for phenotypic responses is presented in Fig. 5.41 (for more details see

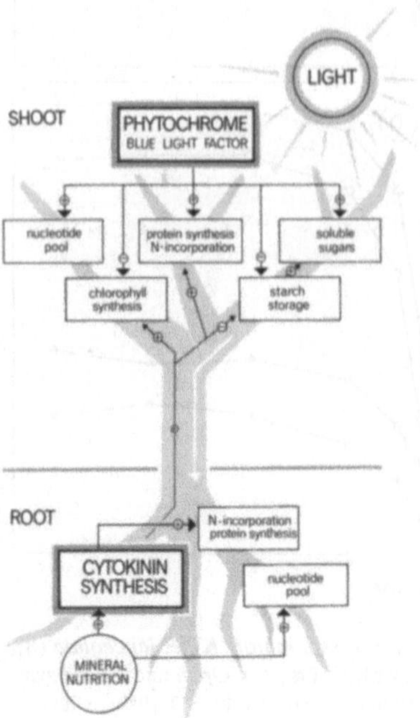

Fig. 5.41. A physiological model for phenotypic plasticity. + and − present positive and negative effects, respectively

Kuiper 1984a, Kuiper and Kuiper 1988). In mineral nutrition, NO_3^- is an ion likely to act as stimulus. Low NO_3^- availability will decrease cytokinin synthesis in the roots (Torrey 1976; Darral and Wareing 1981), also in *Plantago* (Kuiper and Kuiper 1988); synthesis of its couterpart, abscisic acid will be enhanced (Palmer 1981). A low cytokinin production may presumably result in decreased transport of this growth substance to the shoot, resulting in decreased protein syntheses. Consequently, enzyme levels of nitrate reductase and ATPase will decline, and growth, especially that of the shoot, will be retarded and starch accumulation will be promoted. Nucleotide levels will be lowered, due to the change in growth rate; possibly, there is also a direct effect of mineral nutrition on the nucleotide pool via the phosphate level. Further details on this model are presented in Section 7.2.

Recently it was found that gibberellins were involved in differences in RGR_{max} of two inbred lines, W_9 and A_4, of *P. major* (Dijkstra 1990, Dijkstra and Kuiper 1989, Dijkstra et al. 1990).

Light quality (and quantity) is perceived by phytochrome (Bennet 1983; Oelmuller and Mohr 1984). Low irradiance will favour a transition in phytochrome from the far-red to the red light-absorbing form. Such a shift in equilibrium between the two pigment forms will be reflected in a decreased rate of protein synthesis, of sugar production, a decrease in nucleotide pool, and an increased chlorophyll content. Growth reduction of roots is more pronounced than that of the shoot.

The lack of plasticity in lines 1 and 2 and in field material of *P. major* ssp. *major* may be due to a disturbance of the perception of the receptors or to a disturbance of the translation of the signal: differences in sensitivity and/or amplitude of plastic responses are probably caused by differences in properties of the receptor/translation system and/or protein synthesis. A major role for protein synthesis in phenotypic plasticity was demonstrated in experiments with wheat exposed to different temperature regimes (Weidner et al. 1982).

6 Genetic Variation Within *Plantago* Populations

6.1 Allozyme Variation and Genetic Structure in *Plantago* Species

6.1.1 Introduction

Genetic variation is one of the means which enable a species to occupy an extensive geographic distribution and to occur in large numbers in heterogeneous environments. Genetic differentiation within a species can be manifested on different levels. Geographic races may occur in geographically different parts of the species range. In that case only a limited exchange of genes between remote parts of the distribution will occur. Genetic differences between populations can be brought about by chance processes, by selection and by a combination of both. If selection is the main cause for differentiation both geographical distance and genetic isolation will be relatively less important factors in structuring the genetic variation. In that case different specific genotypes (ecotypes) each adapted to specific environmental conditions, will occur. In many cases populations will consist of a single, particular, ecotype though also several ecotypes within a population may occur, depending on the level of hybridization and the habitat heterogeneity.

The characters involved in ecological differentiation are generally polygenic in nature: many genes contribute to the trait. For the understanding of the genetic structure of a population or species it is therefore important to analyze the relation between genotypes and environment and to reveal the underlying selection mechanisms. Knowledge of the breeding system is also of uttermost importance for the understanding of the genetic structure of a population as it governs the way genes are passed on from generation to generation and influences gene flow to a certain extent. Allozymes can be used effectively as a tool to study such different topics as population structure, level of gene flow and the localization of genes involved in ecotypic differentiation. For these purposes they may be considered as selectively neutral and can be used as markers.

6.1.2 Allozyme Variation in Plantago Populations

Allozyme variation in relation to population structure has been studied extensively in *Plantago major, P. lanceolata* and *P. coronopus* (Van Dijk and Van

Delden 1981; Van Dijk 1985a; Van Dijk et al. 1988). The number of loci investigated in *P. major* and *P. coronopus* was 36, while in *P. lanceolata* 39 loci were identified. Table 6.1 summarizes the results for the loci which were variable in at least one species; the number of subunits of the enzymes is also given. The linkage relations between the loci are presented in Table 6.2 as far as they are known. The use of identical designation of the loci in different species, such as e.g. in *Est-4* does not necessarily mean that the loci are evolutionary homologous. The three species are probably too remote in an evolutionary sense to expect uniformity in linkage relationships in the three species between homologous loci. This is illustrated by the fact that the haploid chromosome number in *P. coronopus* is five, while in *P. major* and *P. lanceolata* the haploid number is six.

For each of the three *Plantago* species a number of populations from different parts of The Netherlands were screened for allozyme variation. The four *P. major* populations sampled belonged either to the subspecies *major* (two populations), to the subspecies *pleiosperma* (one population) or to a mixed population of *major* and *pleiosperma* plants (one population). For *P. lanceolata* and *P. coronopus* seven respectively four populations were sampled. The results are summarized in Table 6.3a, where the fraction of polymorphic loci (P), the mean effective number of alleles per locus (A_e) and the mean gene diversity per locus (H_e) are given. The two latter measures are based on Σp_i^2, the sum of squares of the allele frequencies per locus; the effective number of alleles equals $1/\Sigma p_i^2$; the gene diversity equals $1-\Sigma p_i^2$.

The *P. major* populations listed in Table 6.3a were large and existed for a prolonged period of time. For three small, relatively recently founded, populations not included in Table 6.3a the allozyme variation was much lower, compared with the large and stable populations. The small populations were usually fixed for the most frequent allele of the variable loci, but sometimes for a less frequent one. This is to be expected as a consequence of genetic drift, occurring after recent colonization by only a few individuals. In *P. lanceolata* the Uddel (Ud) population was exceptional in that it was isolated from surrounding populations by extensive heath fields. The isolation was reflected in a lower level of allozyme variation, though the decrease is much less drastic than in the small *P. major* populations, which may be a consequence of the differences in breeding system among both species. An isolated *P. coronopus* population from Uddel (not listed in Table 6.3a) was fixed for all 13 allozyme loci investigated. This inland population is probably completely isolated from the coastal populations (*P. coronopus* occurs almost only along the coast in The Netherlands) and was most likely founded by a single individual. These data demonstrate that local populations may deviate dramatically from the regular level of allozyme variation found in the species as a consequence of historical events, isolation or limited population size.

The two subspecies distinguished in *P. major* are almost equal in their level of allozyme variation. They possess, however, subspecies-specific alleles at the *Pgm*-1 and the *Got*-1 loci (Van Dijk and Van Delden 1981). The mixed population Hk is more variable than the other populations, consisting of only a single subspecies.

Table 6.1. Survey of the loci variable in at least one species with linkage relationships and description of alleles

Locus	*P. major*			*P. lanceolata*			*P. coronopus*[a]		
	L	S	Alleles	L	S	Alleles	L	S	Alleles
Acph-1	–		–	–		–	1	1	S(30), N(35), F(40)
Acph-2	5	2	N, F	–		–	1	2	S(46), N(50)
Adh	–		–	–		–	1	1	S(21 1/2), N(23), F(25)
Est-1	–		–	–		–	2	?	N(9), O
Est-2	–		–	–		–	3	?	N(18), O
Est-3	–		–	–		–	?	1	S(29), I(33), N(38), O
Est-4	4	1	N, F	3	1	A, B, C, D, E, F, G, O	≠1, ≠4	1	N(43), F(47), O
Est-5				2	1	A, B, C, D, E, F, O			
Got-1	2	2	S, I, F	2	2	S, I, F	–		–
Got-2	3	2	S, N, F	4	2	S, F	–		–
Gpi-1	?	2	N, F	1	2	S_1, S_2, I, F	?	2	N(22), F(26)
Idh	–		–	2	2	S, N, F	4	2	S(33), N(35 1/2)
Lap-1	–		–	1	1	S, I, F	–	–	
Lap-2	–		–	4	1	S_1, S_2, N, F	2	1	S(36), N(40)
Me-1	3	4	S, N	–		–	–		–
6Pgd-1	–		–	5	2	S, N, F	–		–
6Pgd-2	4		N, O	5	1	S, N, F	5	2	S(26), F(29 1/2)
6Pgd-3	–		–	4	2	S, N, F	–		–
Pgm-1	1	1	S_1, S_2, N	5	1	S, N, F	?	1	N(30), F(33)
Pgm-2	–		–	3	1	S_1, S_2, S_3, N, F_1, F_2	≠1, ≠2	1	S(38), I(41), N(44)
Shdh	3	1	S_1, S_2, N, F	–		–	–		–
To-2	–		–	3	2	N, F	–		–

[a] The electrophoretic mobilities of the *P. coronopus* alleles are given in brackets (bromophenol blue being 100). Invariable loci are indicated by –. Est-5 has not been detected in *P. major* and *P. coronopus*.
L = linkage group; ? = not determined; S = no. of subunits.

Table 6.2. Linkage relations in *P. major, P. lanceolata* and *P. coronopus*

P. major	
Linkage group	Loci
I	Pgm-1
II	Got-1
III	Me-1——Shdh
IV	6Pgd-2——Got-2——Est-4
V	Acph-1
P. lanceolata	
Linkage group	Loci
I	Gpi-1——Lap-1
II	Got-1——Idh——Est-5
III	mr2——To-2——Pgm-2——Est-4
IV	Mr3——6Pgd-3——Got-2——Lap-2
V	mr1——6Pgd-1/6Pgd-2——Pgm-1——SI
P. coronopus	
Linkage group	Loci
I	Acph-1——Adh——Acph-2 or Adh——Acph-1/Acph-2
II	Est-1——Lap-2
III	Est-2
IV	Idh
V	6Pgd-1

Mr and mr indicate male sterility loci; SI indicates self-incompatibility locus.

The data for polymorphism and heterozygosity obtained for the three plantain species are well within the range found for other plant species (see reviews in Brown 1979; Hamrick et al. 1979; Nevo et al. 1984).

6.1.3 Population Differentiation in Plantago Species

The three *Plantago* species investigated for allozyme variation can be compared for the degree of interpopulation differentiation. Table 6.3 b gives the data for the differentiation between populations and subpopulations. The measure for population differentiation used is the ratio D_{ST}/H_e (Brown 1979). This measure, with a scale ranging from zero to infinity, is biased by a small number of subpopulations and low sample sizes. When corrected for both k (the number of subpopulations) and N_h (the harmonic mean of the population sample sizes), the estimate of population differentiation becomes unbiased and comparisons for different situations are possible. The unbiased measure is:

$$\frac{D_{ST}}{H_e} \cdot \frac{k}{k-1} - \frac{1}{N_h} \text{ (Van Dijk et al. 1988).}$$

P. major has the highest population differentiation, but the genetic differentiation of the two subspecies is largely responsible for this phenomenon.

Table 6.3a. Levels of allozyme variation

Species	Population	% Polymorphic loci (P)	Mean no. of alleles per locus (A)	Mean effective no. of alleles per locus (A_e)	Mean gene diversity per locus (H_e)
P. major	Br	13.9	1.25	1.08	0.042
	He	11.1	1.14	1.07	0.035
	Hk	19.4	1.31	1.12	0.065
	An	16.7	1.28	1.08	0.048
	Mean	*15.3*	*1.24*	*1.09*	*0.047*
P. lanceolata	He	30.8	1.77	1.36	0.131
	Pa	35.9	1.82	1.36	0.134
	Ud	28.2	1.67	1.28	0.104
	Br	33.3	1.90	1.35	0.134
	Wd	35.9	1.79	1.37	0.124
	Me	33.3	1.82	1.41	0.133
	We	35.9	1.87	1.39	0.128
	Mean	*33.3*	*1.81*	*1.36*	*0.127*
P. coronopus	Kw	30.6	1.44	1.15	0.088
	Wd	30.6	1.44	1.16	0.093
	Vc	30.6	1.33	1.13	0.078
	Vr	30.6	1.36	1.15	0.093
	Mean	*30.6*	*1.40*	*1.15*	*0.088*

Table 6.3b. Differentiation between populations and subpopulations

Species or population	No. of (sub-)populations (k)[a]	Harmonic mean of sample sizes ($\bar{N}_h$)	$\frac{D_{ST}}{\bar{H}_e}$	Unbiased estimate of population differentiation[b]
P. major	4	67.7	0.379	0.491
P. major ssp. *m*	3	31.2	0.107	0.128
P. major ssp. *p*	2	50.3	0.077	0.134
P. lanceolata	7	87.6	0.045	0.041
P. coronopus	4	121.9	0.127	0.161
P. lanc. Wd	2 (90 m)	46.5	0.016	0.010
We	2 (25 m)	44.9	0.029	0.036
P. cor. Kw	2 (200 m)	67.1	0.015	0.015
Wd	2 (5 m)	65.9	0.100	0.185
Vr	2 (1300 m)	54.1	0.062	0.106

[a] The distances between the pairs of subpopulations are given in brackets.

[b] $\frac{D_{ST}}{\bar{H}_e} \cdot \frac{k}{k-1} - \frac{1}{\bar{N}_h}$.

When populations of the same subspecies are compared, the degree of differentiation is much lower and is about equal in both subspecies. The Hk population of *P. major*, which is a thorough mix of the two subspecies, allows an estimation of the differentiation between the two subspecies without interference of geographical distance. The D_{ST}/H_e value (corrected for N_h only) in this

Table 6.4. Geographic distance and genetic differentiation for some combinations of populations

Species	Populations	(Mean) distance in km	Unbiased estimate of population differentiation
P. major	Br-He	27	0.096
	Br-Hk-m	165	0.089
	He-Hk-m	146	0.193
	An-Hk-p	147	0.134
P. lanceolata	He-Pa-Br-We	18	0.026
	Wd-Me	15	0.008
	(He, Pa, Br, We)-(Wd, Me)	130	0.022
P. coronopus	Kw-Wd	10	0.112
	Vc-Vr	2.5	0.043
	(Kw, Wd)-(Vc, Vr)	175	0.161

case is 0.105. Differentiation within the *P. major* subspecies is about equal to that of *P. coronopus*. *P. lanceolata* shows a much lower degree of differentiation. The figures found for the *Plantago* species are lower than the mean $D_{ST}H_e$ values given by Brown (1979) for a sample of inbreeding and outbreeding plant species (1.06 and 0.20, respectively).

Population differentiation already occurs over distances less than 1 km, and the degree of differentiation over such distances can be equal to the differentiation found over The Netherlands, including populations separated by more than 100 km (Tables 6.3b and 6.4). Especially in *P. lanceolata*, the greatest part of the total allozyme variation is present within each population. Large-scale geographic patterns in allele frequencies appear to be absent. This holds also for *P. major* and *P. coronopus*, though to a lesser extent.

6.1.4 The Use of Allozyme Variation in the Determination of Gene Flow and Outcrossing Rates

The way in which allozyme variation is distributed over a population provides significant information about the genetic system:

1. how genes are passed on to next generations;
2. the spatial distance between parents and offspring;
3. the degree of outcrossing.

An ideal situation of refer to is the Hardy-Weinberg situation, which assumes a population with infinite population size, where alleles are randomly combined to diploid genotypes. Hardy-Weinberg conditions have two consequences: the homozygosity is equal to the sum of squares of the allele frequencies, and the allele frequencies are equal in all parts of the population. Gene flow is infinite under these conditions and the selfing rate is zero.

In some animal species Hardy-Weinberg conditions may occur when population sizes are very large, and dispersal is such that the individuals in each generation are thoroughly mixed. In plant species like *Plantago* it is unlikely that these conditions are fulfilled. Only neighbouring plants will pollinate each other with a reasonable probability, and seeds will mostly land near the female parent. If differences in allele frequencies between parts of the population exist, the restricted gene flow will not be able to equalize these frequencies completely; the remaining differences will depend on the effect of genetic drift, which effect will be larger with decreasing local effective population size (the "neighbourhood size"). Sampling from the total population will lead to an excess of homozygotes compared with the Hardy-Weinberg ratios (the "Wahlund effect"). Homozygosity is also higher than expected when selfing occurs. Clearly the homozygosity/heterozygosity ratio is an overall indicator of deviation from the Hardy-Weinberg equilibrium. The relative deviation of heterozygosity from the expected value is expressed in the fixation index F_{IT} of the population:

$$F_{IT} = \frac{H_{exp} - H_{obs}}{H_{exp}} \quad \text{or} \quad \frac{Hom_{obs} - \Sigma p_i^2}{1 - \Sigma p_i^2},$$

where H_{exp}, H_{obs} and Hom_{obs} stand for the expected heterozygosity, the observed heterozygosity and the observed homozygosity, respectively. F_{IT} can be averaged over loci because it is a measure independent of allele frequency. In the *Plantago* populations studied, F_{IT} is generally positive, indicating selfing of the plants and/or genetic substructuring of the population (see Table 6.5). Some populations are split up into two parts, because they do not occur in a continuous area, or have been sampled at two remote places. *P. major* showed very high homozygosity levels compared with the other two species which were not extremely far from Hardy-Weinberg equilibrium.

The next point of interest is the contribution of selfing and local differentiation to the F_{IT} value. Local differentiation is measurable like homozygosity; homozygosity is the probability that both alleles in one individual plant are the same, local differentiation can be defined as the probability that two alleles drawn from two different individuals at the same place are identical. This probability must be estimated with the aid of identity relationships between plants at larger distances, because more than one plant can never be selected at a single selected spot. The estimated probability should be converted into a F_{IT}-like value to make it independent from allele frequencies. Like homozygosity, its expectation under Hardy-Weinberg conditions is Σp_i^2. This F, called $F_r(0)$, is zero under Hardy-Weinberg conditions; it is also zero when gene flow is infinitely large, irrespective of the selfing level. An extensive treatment of the properties of $F_r(0)$ is given in van Dijk (1987). The maximum value of $F_r(0)$ is 1, in the case that only groups of individuals containing just a single allele do exist. The values of local differentiation in the *Plantago* populations studied are shown in Table 6.5.

The mean values of the species indicate that populations of *P. major* and *P. coronopus* usually show much more local differentiation than populations of *P. lanceolata*. The $F_r(0)$ values represent the maximal contribution of local

Table 6.5. Homozygosity and local differentiation within populations, and mean values for the loci with the highest gene diversities

Species	Population	No. of loci used	F_{IT}	F_r (0)
P. major	BR	3	0.749	0.541
	He	3	0.369	0.261
	An[a]	5	0.892	0.255
P. lanceolata	He	6	−0.011	0.001
	Pa	6	0.092	−0.071
	Ud	6	0.118	0.031
	Br	6	0.090	0.002
	Wd	6	0.084	0.025
	Me	6	0.081	0.014
	We-1	6	0.049	0.096
	We-2	6	0.045	0.292
	We-1+2	6	0.047	0.179
P. coronopus	Kw-1	6	0.122	0.275
	Kw-2	6	0.149	0.209
	Kw-1+2	6	0.140	0.254
	Wd-1	6	0.327	0.801
	Wd-2	6	0.055	0.247
	Vc	6	−0.050	0.072
	Vr-1	6	0.113	−0.059
	Vr-2	6	0.047	0.177
	Vr-1+2	6	0.064	0.182

[a] Mainly juveniles used instead of only adults.

differentiation to homozygosity: viz. the case that there is no gene flow at all. The real contribution to F_{IT} will lie between $F_r(0)$ and zero, depending on the gene flow level. It is clear that F_{IT} of *P. major* is considerably larger than $F_r(0)$; its selfing rate should be about 83% in Br, 50—60% in He and about 94% in An.

The Br population has been investigated in more detail than the other two populations: gene flow per generation could be estimated as a few decimeters: the neighbourhood size was about 2.5. Gene flow was largely due to seed transport, because pollen flow had only a limited distribution, due to the high selfing rate.

P. lanceolata also was on the average slightly more homozygous than could be explained by local differentiation. This species, however, is self-incompatible. The reason for this additional homozygosity remains unclear.

In *P. coronopus*, a contradictory situation was observed. On the one hand gene flow should be high in order to obtain the low homozygosity values found (this species is self-compatible, which gives reason to expect much higher levels of homozygosity), and on the other gene flow cannot be above the level necessary to cause the rather high degree of local differentiation. Which factors lead to this paradoxical situation is not clear at all. It can only be concluded that the selfing rate in *P. coronopus* could not be very substantial.

A more direct estimation of outcrossing rates in *P. major* and *P. coronopus* has been achieved by using a multilocus estimation procedure (Shaw et al. 1981). In this method the allozyme genotypes of adult plants and their natural progenies are determined and the outcrossing rates are calculated from the numbers of apparently outcrossed progeny and allele frequencies. It appeared that outcrossing rates, determined by this multilocus procedure (tm), in *P. major* populations ranged from 0 to 8%. This is in general somewhat lower than outcrossing rates of the same populations, determined by using Wright's fixation index (tF) (Wolff 1990b). *P. coronopus* populations appeared to have outcrossing rates (tm) ranging from 0.34 to 0.93, these different values are probably due to differences in plant densities (Wolff et al. 1988). For *P. coronopus* tm values are also lower than tF values from the same populations. The differences between tm and tF for both species indicate that adult plants, whose genotype frequencies are used for Wright's fixation index, have a higher heterozygosity than the seed population and that probably selection plays an important role. In separate experiments it is indeed found that more outcrossed *P. coronopus* plants have a higher seed output than more inbred plants.

Summarizing, it can be said that *P. lanceolata* has a high level of gene flow, and *P. major* and *P. coronopus* a low level. *P. major* is distinctly a predominantly selfing species, whereas *P. lanceolata* is an outcrosser and *P. coronopus* can, dependent on the circumstances, take all intermediate positions.

6.2 Polymorphism for Caffeic Acid Esters in Populations of *Plantago major* ssp. *pleiosperma*

6.2.1 Introduction

This section presents an investigation of the distribution of two caffeic acid esters in natural populations of *Plantago major* in Denmark and The Netherlands. It contributes to the knowledge of the distribution of a monofactorial character in plant populations within a restricted geographical region.

Although publications of single-gene analyses in plants are abundant, investigations of single genes as markers in natural plant populations are rare.

Apart from the now classical reports on cyanogenesis in *Trifolium repens* and *Lotus corniculatus* (Jones 1972, 1973, 1977), only few genetic investigations of secondary plant products in plant populations have been published (*Thymus vulgaris*, Vernet et al. 1977). Population genetics are frequently based on protein content (allozymes) separated by electrophoresis (Gottlieb 1981). The underlying genetics regarding number of genes, alleles or loci, dominance and co-function is rarely known (see also Sect. 6.1).

The distribution of secondary metabolites in plant populations has been related to possible antiherbivore effects (Rhoades and Cates 1976; Crawley 1983), and the different types of chemical defense have been evaluated in relation to the occurrence of the species (Feeny 1976).

Plantago major	ssp. *major*	ssp. *pleiosperma*
Morphology		
Sum of character states	10 – 30	31 – 50
Seeds per capsule	8 (5 – 13)	18 (14 – 28)
Growth type	Erect	Prostrate-ascending
Leaves	Broad ovate, entire margin	Ovate-lanceolate, dentate margin
Biology		
Reaction to mechanical stress	Resistant	Sensitive
Growth rate	Slow	Faster
Longevity	Perennial	Annual (-perennial)
Ecology		
Plant communities	Stabile: Roadsides Grass leys	Unstable: Sea shores Arable land
Phytochemistry		
Caffeic acid ester with	Glucose	Glucose or rhamnose

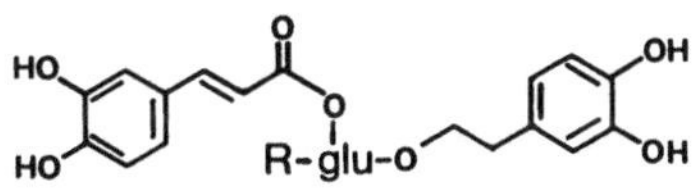

	Structure	Compound	Genotype
1)	R = glucose	Plantamajoside	AA, Aa
2)	R = rhamnose	Verbascoside	aa

Fig. 6.1. Features of the two subspecies of *Plantago major*

This investigation is based on a classification of *P. major* in two subspecies, ssp. *major* and ssp. *pleiosperma* Pilger (Mølgaard 1976). The classification was primarily based on morphological characters but has been supported by studies on genetic variation in enzyme loci (Van Dijk and Van Delden 1981), probably linked to ecological fitness of the subspecies. Differences in ecology and competitive ability are shown in Fig. 6.1. Part of the distribution pattern has recently been explained by a higher plasticity in physiological parameters as root respiration and activity of the alternative pathway in ssp. *pleiosperma* than in ssp. *major* (Kuiper 1983a).

Apart from morphological and ecological differences, caffeic acid, esterified with glucose and rhamnose, respectively, has been found in the leaves of the subspecies (Mølgaard 1986a). Recently, the structure of the two compounds has been elucidated as plantamajoside and verbascoside, respectively (Ravn and Brimer 1988), plant phenolics which are generally present in the genus *Plantago* (Andary et al. 1989), verbascoside also in related plant families (Mølgaard and Ravn 1988). In spite of their rather complex structure, the two compounds differ only in their sugar moiety (Fig. 6.1). They present exclusive characters which rarely occur together in the same plant. Plantamajoside was found in both subspecies, while verbascoside was found only in ssp. *pleiosperma.* The synthesis of esters is determined monogenetically with complete dominance for the caffeoyl glucose plantamajoside (Mølgaard et al. 1980); some population samples indicate that the genetics is not always that simple (Table 6.6).

This section deals with the geographical distribution of the two compounds in populations of *Plantago major* ssp. *pleiosperma* in Danish and Dutch plant communities. The distribution is considered in relation to the plant community structure, to longevity of the life span and to herbivore attack on the plants in the respective populations.

6.2.2 Material and Methods

Population samples were collected from several localities in Denmark (1981, 1983) and in The Netherlands (1983). The samples were composed of 30 plants from each locality; they were taken from a small and uniform area of less than 10 m^2; plants were dried in a plant press for morphological and chemical analysis. The damage to the plants due to herbivore attack was registered as number of holes per leaf and per plant.

Community samples were taken from the plant collections as used for the morphological analysis of the species complex (Mølgaard 1976). Samples were collected from localities throughout Denmark and represented arable land, grass leys, roadsides and sea shore communities. Three plants from each locality were transplanted and grown to maturity in a greenhouse; morphological and chemical screening was performed on herbarium material from this cultivation. The purpose was to describe the introgression between ssp. *major* and ssp. *pleiosperma* by means of a numerical system by giving scores from 1 to 5 for ten individual characters. The scale was arranged in such a way that a typical ssp. *major* character was given the value of 1 and a typical ssp. *pleiosperma* character the value of 5 (Mølgaard 1976). By adding the values for ten morphological characters, a typical ssp. *major* plant could theoretically obtain the sum of 10 and a typical ssp. *pleiosperma* the value of 50 (Fig. 6.4).

For a nutritional experiment plants were grown in a greenhouse from seeds derived from three different populations of *P. major* ssp. *pleiosperma*, i.e. Angeren, Oostvoornse Meer and Kwade Hoek, all in The Netherlands. Plants of each population were divided into three groups and supplied with three different levels of Hoagland solution (for further information, see Lotz and Blom 1986). Plants were harvested after 91 and 119 days (lowest nutrient level) and dried. One mature leaf from each plant was used for chemical analysis.

For phytochemical analysis 0.5 g dried leaves were extracted with 500 μl of methanol at 40 °C for 24 h; 10 μl of the extract was applied to a TLC sheet with microcrystalline cellulose (Merck) and eluted with n-butanol: acetic acid: water (4 : 1 : 5 by vol.). Two bright blue spots were detected in UV light, representing plantamajoside (Rf = 0.4 – 0.5) and verbascoside (Rf = 0.5 – 0.6; Mølgaard et al. 1980; Ravn and Brimer 1989). In the plants containing plantamajoside, traces of verbascoside were frequently found.

Quantitative determination of the caffeic acid esters was carried out by HPLC using a 125×4.6 mm column packed with Sphaerisorb RP – 18, 5 μm, using an isocratic eluent system with the solvent methanol: water: acetic acid (200 : 600 : 5 by vol.) and caffeic acid as an internal standard.

6.2.3 Results

6.2.3.1 Population Samples

The genotypical pattern of the relative distribution of caffeic acid esters in populations of *P. major* ssp. *pleiosperma* in Denmark (Fig. 6.2) and The Netherlands (Fig. 6.3) shows a remarkable similarity. The frequency of the recessive rhamnose ester verbascoside decreased towards the west. In The Netherlands, the decrease was associated with a difference in the plant communities from which the populations were derived; those from the west country had been sampled on saline soil, in shore meadows and in dune hollows, while the inland populations were collected from inundated banks along rivers and from ruderal sites. None of the Dutch samples were obtained from arable land.

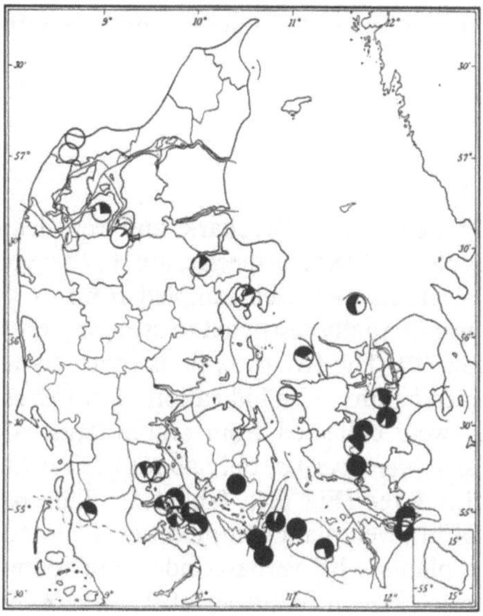

Fig. 6.2. Genotypical frequency distribution of verbascoside in populations of *Plantago major* ssp. *pleiosperma* in Denmark 1981 and 1983. Each sample is composed of 30 specimens. The *black area of the dots* indicates the proportion of plants with verbascoside

Fig. 6.3. Genotypical frequency distribution of verbascoside plants in populations of *Plantago major* ssp. *pleiosperma* in The Netherlands 1983. Each sample is composed of 30 specimens

The samples from Denmark were, to a large extent, obtained from arable soil. *Plantago major* is one of the most common weeds in Danish fields (Mikkelsen and Laursen 1966; Haas and Streibig 1981) and in annual crops ssp. *pleiosperma* is inevitably found in the lower areas all over the country. The frequency of the verbascoside type was highest in the southeastern part of the country, irrespective of which plant community the population sample was derived from.

6.2.3.2 Community Samples

Samples were obtained from nearly 300 sites, all over Denmark, and were dealt with according to taxonomy and morphology of the subspecies, and to the type of plant community and crop type the plants had been sampled from.

Figure 6.4 shows that verbascoside was virtually restricted to ssp. *pleiosperma* (sum of character states higher than 30). The five exceptions that may be referred to ssp. *major* all had high character sums as an indication of introgression of ssp. *pleiosperma* genes. They were collected from arable land. Although verbascoside was restricted to ssp. *pleiosperma*, only about half of the specimens of this taxon contained this ester (Fig. 6.4; Table 6.6). Table 6.6 shows the distribution of the subspecies in seven plant communities. On roadsides, ssp. *major* was present and no plants with verbascoside were found. From the agricultural areas there was a decrease in frequency of ssp. *pleiosper-*

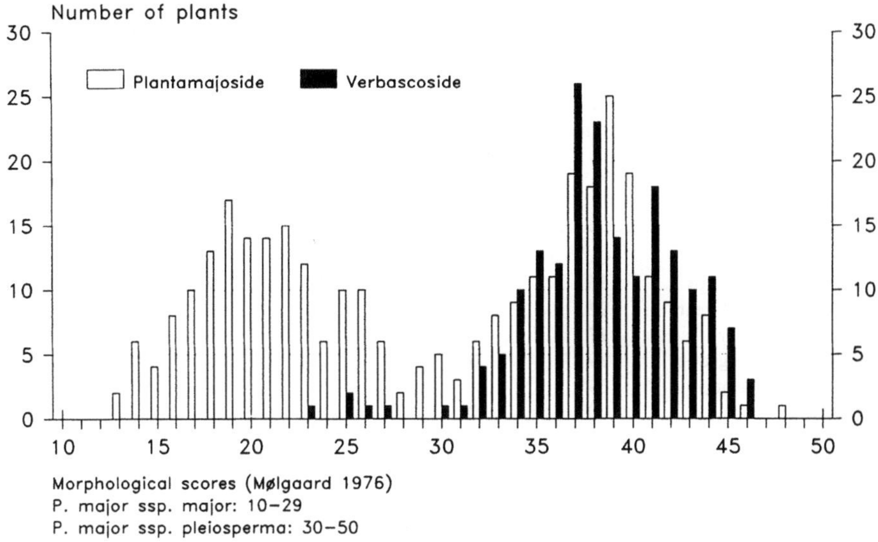

Fig. 6.4. Frequency distributions of *Plantago* major specimens from Denmark, containing plantamajoside and verbascoside according to a morphological index (on the *abscissa*)

Table 6.6. A total of 576 specimens of *Plantago major* from Denmark grouped according to original plant habitat, division into subspecies and content of plantamajoside (glu) and verbascoside (rha), respectively

	No. of plants	ssp. *major*		ssp. *pleiosperma*	
		glu	rha	glu	rha
Annual groups	107	5	1	23	78
Overwintering crops	42	5	1	11	25
Grass leys, 1 – 2 years old	139	43	4	53	39
Grass leys, older	72	55	0	8	9
Shore meadow group	61	22	0	28	11
Sandy shore group	91	28	0	44	19
Roadsides	64	64	0	0	0
Total	576	222	6	167	181

ma with longevity of the crop; at the same time the frequency of the verbascoside phenotype decreased in the order annual crops, overwintering crops and younger to older grasslands. The two communities shared in common a similar proportion of the subspecies; the same pattern of caffeoyl esters occurred, i.e. one third of ssp. *pleiosperma* contained verbascoside.

The geographical distribution of the verbascoside type in the community samples (three specimens from each locality) of ssp. *pleiosperma* (Table 6.7) was similar to that in the population samples (30 specimens from each locali-

Table 6.7. The occurrence of *P. major* ssp. *pleiosperma* in the two geographical regions of Denmark, Jutland (West) and the Islands (East). The samples have been divided according to original plant habitat and proportion of plantamajoside (glu) and verbascoside (rha)

	No. of plants	*Plantago major* ssp. *pleiosperma*					
		West			East		
		glu	rha	% rha	glu	rha	% rha
Annual crops	101	17	11	39	6	67	92
Overwintering crops	36	2	1	(33)[a]	9	24	73
Grass leys, 1–2 years	92	52	34	40	1	5	(83)
Grass leys, older	17	8	4	33	0	5	(100)
Shore meadow group	39	3	1	(25)	25	10	29
Sandy shore group	63	31	16	34	13	3	19
Roadsides	0	0	0	0	0	0	0
Total	348	113	67	37	54	114	68

[a] Figures in brackets are based on six or fewer specimens.

ty). Most plants from habitats with annual and overwintering crops were derived from the east country of Denmark (the Islands) and had a high frequency of verbascoside. The grassland samples were mainly from the west country of Denmark and had a low frequency of verbascoside.

Most of the samples from the shore meadow group had been collected in the east country, where this community type is most abundant (Table 6.7); the samples from the sandy shore group were mainly obtained from the west country; no plants had been found on the west coast of Jutland. However, the frequency of the verbascoside type was comparable in the two geographical regions (when plants from sea shore communities are considered).

Otherwise, Table 6.7 shows a most interesting distribution pattern. The frequency of the verbascoside type was almost the same in all communities in the west country, whereas on the Islands there was a pronounced variation, with a high frequency of verbascoside in the agricultural areas and a low one in the coastal plant communities. The data of this pattern was supported by the results on population samples (Fig. 6.2).

6.2.3.3 Nutritional Influence

There was a discrete variation in the concentration of the caffeic acid esters with (controlled) nutrient supply (Table 6.8): the three studied populations did not respond in the same way to the variation in nutrient level. Irrespective of the sugar moiety, the plants from Kwade Hoek attained their highest concentration of caffeic acid esters when the nutrient supply was high, whereas the plants from Angeren and Oostvoornse Meer obtained their highest concentration when no nutrients were added to the pots. In all populations, an intermediate nutrient level led to low concentration of caffeic acid esters.

Table 6.8. Concentration (mg per g dry weight) of caffeic acid esters in leaves of *Plantago major* ssp. *pleiosperma*. Plants from each of three populations were grown in quartz sand with three different levels of nutrition, six plants in each group (cf. Lotz and Blom 1986)

Population locality	Nutrition level			
	Full	Half	No	F
Angeren	12±7[a]	4±2	20±12	6.9**
Oostvoornse Meer	2±1	1±0.3	18±5	71.8***
Kwade Hoek	27±9	6±3	7±3	23.5**
F	22.3***	7.5**	6.1*	

[a] Mean of six replicates ± standard deviation. In the ANOVA, *, **, *** indicate $P<0.05$, 0.01, and 0.001, respectively.

Very likely phosphate deficiency was most pronounced in this experiment (Lotz and Blom 1986), a situation which may be rather common in nature. Therefore, the variation in soil parameters in nature may influence the concentration of the caffeic acid esters in the plants to a high degree.

As all plants with the same nutrient supply obtained almost the same total weight, the variation between the three populations in the concentration, and total amount of the caffeic acid esters may be genetically determined and of ecological importance. Kwade Hoek is a eutrophic site whereas Oostvoornse Meer is oligotrophic, and the populations from these localities seem to respond to the variation in available nutrients by a different concentration of the esters. The plants from Kwade Hoek had a high concentration of caffeoyl esters (verbascoside) when they were in good nutritional condition, whereas the plants from Oostvoornse Meer had a high concentration of caffeoyl esters (plantamajoside) only when the nutrient supply was low.

6.2.3.4 Herbivore Damage to Plant Populations

The effect of herbivores on mature plants of three selected population samples from Denmark is shown in Table 6.9 and Fig. 6.5. Two of the samples were unusual because of the high number of plants with presence of both esters in almost the same concentration.

In samples 83–32 there was a fairly equal representation of all three phenotypes. Nevertheless, there was a significant difference in the attack to the plants as seen from the number of holes after nerbivore feeding on the leaves. The plants with only plantamajoside were attacked to a much higher degree than those with a mixture of the two esters; those with verbascoside were not touched at all.

Samples no. 83–31 and no. 83–33 may be regarded as mirror images of each other: plantamajoside was rare in 83–31 (5 plants of 30), and verbascoside in 83–33 (5 plants of 31). In both populations, the least abundant

Table 6.9. Herbivore damage to three populations of *P. major* ssp. *pleiosperma* from Denmark 1983. 83-31 and 32 are plants from the isle of Møn in SE-Denmark, 83-33 is from Jutland

Ester with[a]	No. of plants	Damaged plants	Damaged plants %	No. of leaves	Damaged leaves	Damaged leaves %	No. of holes	No. of holes per dam. leaf
83-31								
Glu	5	1	20	37	1	3	1	1.00
Rha	26	14	54	211	23	11	30	1.30
Total	31	15	48	248	24	10	31	1.29
		0.99		0.27	2.19		3.34	
83-32								
Glu	11	4	36	94	10	11	10	1.00
Glu/Rha	12	2	17	107	2	2	3	1.50
Rha	7	0	0	69	0	0	0	0.00
Total	30	6	20	270	12	4	13	1.08
		2.94		0.83	12.77**		10.82**	
83-33								
Glu	14	8	57	76	16	21	24	1.50
Glu/Rha	11	5	45	44	12	27	16	1.33
Rha	5	1	20	25	4	16	6	1.50
Total	30	14	47	145	32	22	46	1.44
		1.09		2.64	0.99		0.77	

[a] Ester with glucose (Glu) = plantamajoside. Ester with rhamnose (Rha) = verbascoside.
*, ** $P<0.05$ and 0.01, respectively.

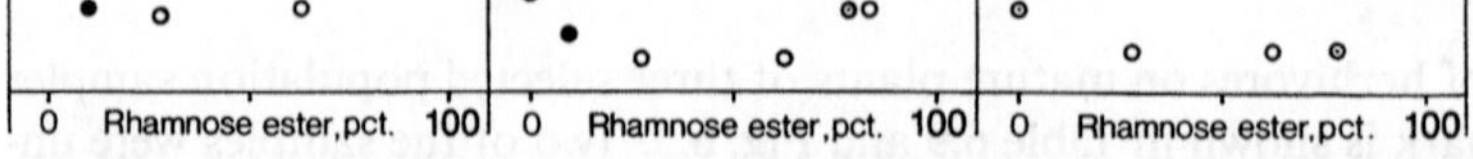

Fig. 6.5. Herbivore damage to plants of *Plantago major* ssp. *pleiosperma* from Denmark 1983 in relation to the frequency of verbascoside plants in the population. ○, barley fields; ●, lake shores and river banks; ⊙, two sites in the same field with dry and wet soil, respectively

type had the lowest degree of attack to the leaves, although this pattern did not significantly deviate from a random (neutral) distribution.

The relation between ester type frequency and herbivore attack is shown in Fig. 6.5: the highest damage is seen in populations which were most homogeneous, i.e. in populations with a low degree of polymorphism. When the population had a high degree of polymorphism (i.e. equal or almost equal representation of the two phenotypes), the damage to the plants was highly reduced.

This observation was valid, irrespective of the measurement of the damage (number of damaged plants, leaves and holes per attacked leaf).

6.2.4 Discussion

The geographical distribution of the frequency of verbascoside plants was similar in Denmark and The Netherlands and may be associated with the distance to the ocean. In The Netherlands, however, there was also a change in plant community from the inland samples to those from the Islands; the latter were all from shore meadows, whereas the former were from river banks or waste grounds. In Denmark, the situation was not that simple because the geographical variation occurred within the same type of plant community, arable land, and hence, the change was not merely a function of community selection.

It is tempting to see the observed distribution in the light of herbivore pressure, and in this connection, snails and slugs may be the most important predators. Feeding experiments with land mollusks (Mølgaard 1986a) showed that plants with caffeic acid and condensed tannins were least attractive to these herbivores (Table 6.10).

Next to the qualitative variation in caffeic acid ester content in the plants, there was a pronounced variation also in the total content of the esters. This variation was a function of the available nutrients and showed the highest content of caffeic acid esters when the plants were supplied with nutrients to a level similar to that in the original habitat: plants from an oligotrophic habitat (Oostvoornse Meer) had the highest content when the nutrient level was low and a population from a eutrophic habitat (Kwade Hoek) had the highest content when nutrients were plenty. Finally, the plants from a river bank with changing water levels (Angeren), and subsequently wide variations in the ion concentration of the soil water (Sect. 3.2.), had high levels of the esters at low as well as at high nutrient supply.

In North American growth experiments with *Zostera* a high concentration of caffeoyl esters was seen in plants at low nutrient levels (Buchsbaum et al. 1989). A limited supply of nitrogen leads to surplus carbon in the plant, which may be accumulated as caffeic acid derivatives. This explanation may also be valid to the results from the oligotrophic Oostvoornse Meer.

The present results suggest that the presence of caffeoyl esters are important for the performance of the population. The chemical structure is determined monogenetically with the two sugar esters, plantamajoside and verbascoside, as the phenotypical expression, and the concentration of the esters may be expressed genetically in relation to the abiotic conditions of the habitat.

That a suitable high concentration of active compounds may be important for protection against herbivores, is known from monoterpenoids in *Satureja douglasii*, where the content was highest in areas with high herbivore pressure (Lincoln and Langenheim 1979).

Verbascoside was found only in the annual ssp. *pleiosperma*. This ester possibly was most efficient as a protector against feeding pressure on seedlings

Table 6.10. Relative food plant preference by land mollusks. Plant families ranked according to decreasing palatability

Plant species	*Arion ater*	*Helix pomatia*	Major plant constituents
Fabaceae			
Phaseolus vulgaris	130		Amino acids, flavonoids
Trifolium repens	117		Amino acids, flavonoids, HCN
Solanaceae			
Solanum nigrum	123		Alkaloids
S. tuberosum	125		Alkaloids
Convolvulaceae			
Calystegium sepium	107		Alkaloids
Sambucaceae			
Sambucus nigra	100^3	59	HCN
Asteraceae			
Taraxacum officinale	76^4	110^2	Bitter principles, terpenes
Tussilago farfara	98	92^2	Bitter principles, terpenes, alkaloids
Polygonaceae			
Rumex sanguineus	95		Flavonoids
Brassicaceae			
Brassica napus	98	103	Sinapic acid, glucosinolates
Sinapis alba		79	Sinapic acid, glucosinolates
Alliaria petiolata	4		Sinapic acid, glucosinolates, saponin
Lunaria annua	8		Sinapic acid, glucosinolates, alkaloid
Apiaceae			
Aegopodium podagraria	50	39	Coumarins, essential oils
Plantaginaceae			
Plantago major	44	59^6	Caffeic acid, aucubin
P. media	16		Caffeic acid, aucubin
P. lanceolata		29	Caffeic acid, aucubin
Lamiaceae			
Lycopus europaeus	48		Caffeic acid, essential oils
Stachys sylvatica	17	0	Caffeic acid, essential oils
Rosaceae s.l.			
Rubus idaeus	14	21	Condensed tannins
Alchemilla vulgaris		12	Condensed tannins
Geum urbanum	0		Condensed tannins
Prunus cerasus	10		Condensed tannins, HCN, acids
Fagaceae			
Fagus sylvatica	0		Condensed tannins

The results are based on experiments where the mollusks had the choice between various combinations of food plants, two by two at the same time. Indices indicate the number of replicates that compose the mean. For further explanation see Mølgaard (1986b).

Table 6.11. Preference values for the two subspecies of *Plantago major* to four species of mollusks (Mølgaard 1986b).

	Arion ater		*Arion rufus*	*Helix pomatia*	*Deroceras reticulatum*
	Spring	Summer			
P. major ssp. *major*	107 ± 18[2]	44 ± 32[3]	–	55 ± 8[5]	56[1]
P. major ssp. *major*	73	44	43	62	79[a]
P. m. ssp. *pleiosperma*	34	25	78	82	26[a]

[a] Only one slug in this experiment. Other results are means of results with ten individuals. The indices indicate the number of replicates with ten individuals.

during the establishment of plants. Snails and slugs are not frequent herbivores in salt-exposed plant communities near the coast (Ellis et al. 1977a, b), which could explain a low frequency of verbascoside plants in populations from coastal areas.

That verbascoside was a more efficient repellent than plantamajoside was supported experimentally (Table 6.11) with the slugs *Arion ater* and *Deroceras reticulatum*; however, *A. rufus* and the snail *Helix pomatia* showed preference for the verbascoside type. It is not known which predator is most important for the survival of seedlings in the spring, but the palatability of *P. major* was high in spring and decreased during the season (Table 6.11). Only *D. reticulatum* may be important in arable land, where the results in Fig. 6.5 and Table 6.9 have been obtained.

The hypothesis was inconsistent with results in which both esters had the same ability to reduce feeding (Table 6.9), the least common of the phenotypes was shown to be the least attractive and less attacked. The herbivores could be conditioned to one of the chemotypes, and when they cannot recognize the type, when moving on to a new plant of the same species, they may be confused and will probably prefer to go to another taxon with predictable phytochemical composition.

The mechanism of frequency-dependent selection will lead to a stable situation of polymorphism and homozygosity of the population (Clarke and O'Donald 1964). A restricted gene flow, as shown in *P. lanceolata* (Bos et al. 1986), and which is likely to be similar in *P. major*, is not even needed to explain a stable polymorphism with homozygotes.

A similar case may be deduced from the results of Ennos (1981); cultivars of *Trifolium repens*, polymorphic for cyanogenesis, showed a positive selection in favour of plants with linamarase genotype (acLi). This was the least common phenotype, as almost 90% of the seedlings contained the cyanogenic glycoside (AcLi or Acli). Hence, selection for linamarase genotype was simply a way to maintain polymorphism in the population. It may be concluded that polymorphism in itself is vital for protection of the plant population against uncontrolled predation and hence present part of the driving force for evolutionary selection.

6.3 Selection for the Maintenance of Sex Polymorphism in *Plantago*

6.3.1 Introduction

An important topic in plant ecology is the adaptation of plants to their environment. In hayfields, for instance, leaves of *Plantago lanceolata* grow upright, whereas in meadows they lie flat on the ground (Sect. 7.3). The difference in leaf angle can be interpreted in terms of adaptation: in hayfields an upright position of leaves favours the interception of light better than a flat one, while the latter prevents grazing more efficiently. In order to justify such an interpretation it is necessary to show the existence of genetic variation for leaf angle. Secondly, selection pressure should induce the proper reactions to grazing and light interception. From a genetic point of view, a population will move towards a paricular genetic composition, which is determined by all environmental factors relevant to the fitness of the individual plants. The contribution of population genetics to the understanding of ecological processes is to analyze the impact of the environment on the genetic structure of a population by selection.

There is another interface between genetics and ecology. Instead of the question which leaf angle is optimal in which environment, one may ask why leaf angle presents a plant character at all. In other words: why do plantain plants have leaves? – since not all species exhibit leaves, or stems (as most plantain species, for example), or flowers concentrated on a spike, or self-incompatibility, etc. The question then is: what determines the plan of morphogenesis of a species? The answer obviously does not lie at the level of the population, but at the level of species, genus or even higher. The question concerns an interface between population genetics, developmental biology and ecology. There is a relation between population genetics and ecology because evolutionary processes take place at the population level and changes in the environment may induce evolutionary changes. Therefore, the study of present-day differences, for instance at the species level, should include what might be called "the ecology of the past".

In the above examples, characters, determined in each plant separately, could be measured. Characters like sex ratio in dioecious and gynodioecious species can only be assessed for group of plants or for a population. Although such a distinction is not relevant for the genetic and evolutionary approaches to ecology, the study of sex ratio is different from that of leaf angle or spike length: there is no a priori expectation of what leaf angle or spike length should be realized in a species, and characters are only noteworthy if their expression differs between (sub)populations. In contrast, the sex ratio in a dioecious species should be unity, according to Fisher's argument (1958); whenever this expectation is not satisfied for a population, there is a problem relating to the action of natural selection. In a gynodioecious species, an a priori expectation of sex ratio also exists, which (as explained below) probably is never satisfied in nature. Thus, the study of sex polymorphism within a population offers an excellent opportunity to analyze the action of natural selection. The a priori expectation does not immediately relate to ecology. Nevertheless, when sex ratios differ between populations, the ecological consequences are important, as was found for dioecious species (Barret and Helenium 1981) and gynodioecious species (Dommée et al. 1978; Brockmann and Bocquet 1978; Van Damme and Van Delden 1982), and for considerations on the evolution of breeding systems (Bawa 1980; Beach 1981 and others).

In this section the breeding system gynodioecy (see Sect. 2.6) will be considered. Both population genetic and evolutionary aspects will be discussed, with some emphasis on ecology. Particularly, the question will be addressed which factors determine the distribution of male steriles within a single population.

6.3.2 The Problem of Maintaining Gynodioecy

A male sterile mutant, introduced into a hermaphrodite population, is at a disadvantage because it can only reproduce as a female; hermaphrodites derive their fitness from male and female sexes. This asymmetry in sexuality among individuals leads to the a priori expectation that gynodioecy should not exist since there is strong selection against male steriles (Lewis 1941). Nevertheless, many gynodioecious species are known, scattered over many taxonomic groups (Delannay 1978); this suggests that the systems have evolved repeatedly and independently of each other. This conclusion has led to theoretical studies of how male steriles could compensate for their disadvantage in order to be maintained. It follows intuitively from the extreme sexual asymmetry that male steriles should reproduce better as females than hermaphrodites. This can be achieved in many ways. Assuming single gene inheritance of male sterility, Lewis (1941) showed that male steriles needed at least a doubled seed production for compensation. Differences in characters like seedling survival and pollen tube growth (Lloyd 1974), adult suvival (Van Damme and Van Damme 1986) and vegetative reproduction (Stevens and Van Damme 1988), could, when associated with sex, contribute to a better production of offspring and thus to maintenance of gynodioecy.

The conditions for maintenance of gynodioecy depend on the mode of inheritance (Lewis 1941; Lloyd 1974, 1975; Charlesworth 1981; Delannay et al. 1981; Ross and Gregorius 1985). If male sterility only inherited through nuclear genes, the number of loci, alleles, and their interaction should be unimportant (Van Damme and Van Damme 1986). If nuclear inheritance, cytoplasmic inheritance or the combination of both are involved in male sterility, different requirements for the maintenance of the system are present: opposed to the doubled seed production, required for nuclear inheritance as the only factor, the slightest difference in favour of male steriles will be sufficient when the inheritance is cytoplasmic (Lewis 1941; Lloyd 1974). With a combined nuclear-cytoplasmic inheritance of male sterility, fitness differences between genotypes within sex phenotypes will be required; because of the complexity of the latter system no precise relations between male sterile frequency and such fitness differences are known at the moment.

A second way for analysis of the maintenance of gynodioecy is to ask why the system exists al all. Many authors have proposed that the adaptive value of gynodioecy is the enhancement of outcrossing (Charlesworth and Charlesworth 1979; Lloyd 1982; Willson 1982; Sun and Ganders 1986). The advantage of obligately outcrossing male sterile plants over partially selfing hermaphrodites is either expressed via heterosis (or a lack of inbreeding depression) in their progeny or via the production of a more varied progeny. However, gynodioecy does not necessarily present adaptive value. If the advantage of male steriles over hermaphrodites is not due to the avoidance of inbreeding depression, it should originate from pleiotropic effects of the male sterility genes (Van Damme 1985). Since conditions for maintenance of gynodioecy are rather stringent, most of the male sterility mutations that occur in nature will be lost because the conditions for fitness are not met. The occurrence of gynodioecy will depend on the developmental programme of the plant, which should permit appropriate pleiotropic effects of the male sterility genes. Differences between species with and without gynodioecy thus would involve differences in developmental constraints.

6.3.3 Gynodioecy in Plantago

Strictly speaking, gynodioecy is not the correct term for a description of the sexual system in *Plantago* species. In all three gynodioecious species, *P. coronopus, P. lanceolata* and *P. maritima,* partially male steriles, IN for short, were found in considerable frequencies (Sect. 2.6). In *P. coronopus* and *P. maritima* frequencies above 50% were observed. The male fertility of IN plants ranged in all three species from almost zero to almost one, so that male fertility in fact was a continuous variable. Although gynodioecy was originally defined as the joint occurrence of completely male steriles (MS) and male fertiles (hermaphrodites), the term gynodioecy will be used throughout this paper. The frequency distributions of the sexual types in the gynodioecious *Plantago* species were already presented in Section 2.6. Both the male steriles and intermediates occurred in sufficiently high frequencies to speak of a sex ratio problem since the a priori expectation that MS (and IN) plants do not occur in nature was not fulfilled.

In view of the crucial importance for the sex ratio of the mode of inheritance of male sterility, the phenotypic and genetic variation for sexual type will be considered. The situation in *P. lanceolata* is best known (Van Damme and Van Delden 1982; Van Damme 1983 a, b; Van Dijk 1985 b; Wolff 1987). Three MS types were found in this species which could be distinguished by the morphology of the male reproductive organs: the anthers of the first type; MS 1, were relatively small, brown and without any pollen and their filaments were short. In the second type, MS 2, there were no anthers present but eight instead of four coralla lobes. Finally, in the third type, MS 3, the anthers contained numerous bright yellow but completely sterile pollen grains. The moment of breakdown of pollen development differed for the three MS types: the earliest

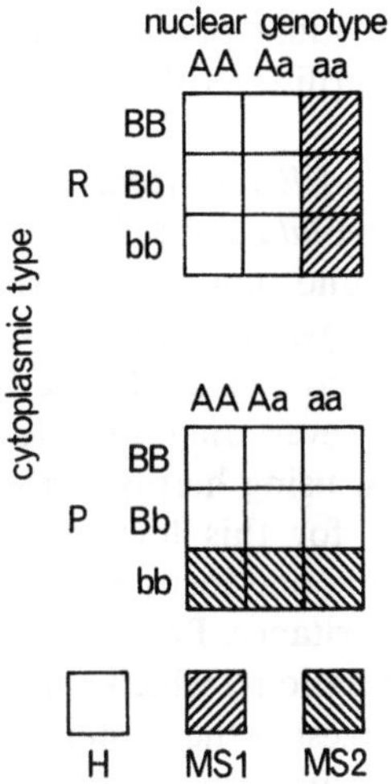

Fig. 6.6. A scheme of the mode of inheritance of male sterility types MS1 and MS2. Each *square* represents one genotype. The nuclear genes are simple designated *A/a* and *B/b* for convenience. For further explanation of the scheme see text. (Van Damme 1985)

derailment was in MS 2, followed by MS 1 and MS 3, respectively. There were two partial male sterility types, IN 1 and IN 2, with morphological differences corresponding to those of MS 1 and MS 2, respectively, (Van Damme and Van Delden 1982).

Inheritance of two MS types, MS 1 and MS 2, was nuclear-cytoplasmic. A simple scheme of this mode of inheritance is shown in Fig. 6.6. The two MS-types were expressed in plants with a different cytoplasmic type, *R* and *P*, in combination with a particular nuclear genotype. In contrast, the third male sterility type, MS 3, was only determined by nuclear genes and independent of cytoplasmic type. The number of loci for each MS type was larger than one (Table 6.12). The genes for each MS type interacted multiplicatively, i.e. all loci, together with the cytoplasmic type, had to be in the "sterile" condition in order to result in an MS phenotype. Alternatively, each MS type can be considered to be determined by a set of independently acting restorer genes. The partial male sterility types, IN 1 and IN 2, probably resulted from incomplete dominance of the restorer genes (Van Damme 1983). The phenotypic expressions of all sex types were, to a limited extent, environmentally determined, which

Table 6.12. Minimum numbers of male sterility genes for the three male sterility types in *Plantago lanceolata*. (Van Damme 1983 b)

Male sterility type	Male sterility genes				
	Recessive	Minimum number	Dominant	Minimum number	With expression in cytoplasm
MS1	mr_i	2	Mr_i	3	R
MS2	mp_i	3	–	–	P
MS3	my_i	2	–	–	R, P

sometimes made nuclear-genotypic identification of individual plants difficult. Most genotypes covered a small range on the male sterility-fertility scale and genotypic ranges could overlap considerably.

In the other two gynodioecious species, *P. coronopus* and *P. maritima*, the situation was in some respects comparable to that in *P. lanceolata* and different in others. In *P. coronopus* two male sterility types were found that resembled MS1 and MS3 of *P. lanceolata* with respect to morphology. A preliminary study of the genetics of the "brown-anther" MS-type showed a single MS-locus, male sterility being recessive (Van Damme 1991). However, only a single cytoplasmic source was involved in this study; test crosses, using hermaphrodites that are heterozygous and homozygous dominant for this locus (as males), yielded sex ratios that clearly did not fit the simple one-locus model (Table 6.13). The results suggested nuclear-cytoplasmic inheritance. Recently, a more extensive genetic study was undertaken in three populations, which confirmed a nuclear-cytoplasmic mode of inheritance, and suggested the existence of two cytoplasmic types (Koelewijn and Van Damme, unpubl. results). There is a difference with the situation in *P. lanceolata*: partial male fertility in *P. coronopus* seems to be largely a plastic reaction of hermaphrodite genotypes. Preliminary greenhouse studies showed that male fertility may depend on the flowering stage of the plants, nutrient conditions and temperature. Partial male sterility could also be induced in hermaphrodites by hormone treatment (auxine; unpubl. results). Little is known of male sterility in *P. maritima*, but the extremely high frequencies of IN plants in the field (Table 2.3 in Sect. 2.6),

Table 6.13. Crosses between MS plants collected from two wild populations on the isle of Schiermonnikoog (The Netherlands) and H plants that are heterozygous (*MS*+) and homozygous (++) at the *MS1* locus. (Van Damme 1991)

Male parent: Expectation: Cross	Heterozygous 1:1 Offspring MS	H	Homozygous 0:1 Cross	Offspring MS	H
Population Beach plain (BP)					
BP-11×40−20	0	29***[b]	BP-11×36−15	0	24
BP-14×36−13	14	13	BP-14×40−28	0	24
BP-23×36−8	0	27***	BP-23×36−10	0	25
BP-29×36−13	12	14	BP-69×36−10	0	31
Population Waddendyke (WA)					
WA-1×36−13	15	15	WA-1×36−15	1	38
WA-5×36−2	31	8***	WA-5×36−10, 15[a]	38	0***
WA-16×36−2, 13[a]	17	22	WA-16×40−11	0	38
WA-29×36−13	15	7	WA-29×36−15	34	0***
WA-31×36−13	22	20			

[a] The second H plant was used for additional pollination because of a lack of pollen of the first one.

[b] This figure includes four male steriles, morphologically and genetically distinct from the MS type studied in the present section (unpubl. result).

***, crosses deviating from expectation ($P<0.001$).

and personal observations of the author of perfect hermaphrodite flowering of plants transplanted to the greenhouse, suggest that this species closely resembles *P. coronopus* in this respect.

Summarizing, the mode of inheritance of the gynodioecious breeding system in *Plantago* species is nuclear-cytoplasmic inheritance. This implies, as was discussed earlier, that the conditions for the maintenance of a particular sex ratio are complex.

6.3.4 How Are Male Steriles Maintained in P. lanceolata?

Male steriles should compensate their disadvantage through a higher seed production. Since seed production over the whole life cycle is a complex character, the first question that will be addressed concerns differences in components of sexual reproduction between the sexes. The second question concerns the frequencies at which male steriles are maintained, and finally a computer model will test whether the observed differences in components of fitness can explain both the frequencies and the spatial distribution of male steriles in nature.

6.3.4.1 Differences between Sexes in Life Cycle Characters

Components of seed production have been studied for the male sterility types MS1 and MS2 and the intermediates IN1 and IN2 in several hayfields. Individual plants were tagged at the beginning of the flowering season and all spikes were collected shortly before mowing (Van Damme 1984). A summary of the results is given in Table 6.14. The MS types produced more ovules per plant than H plants. MS2 plants had less seeds per ovule than H plants, which was caused by partial female sterility of MS2 plants (Van Damme 1984). The net result was that seed production per plant did not differ between MS2 and H. In addition, the seeds produced by MS2, weighed less than seeds of H plants.

The situation for MS1 was different: this sex type was equally female fertile as H and seeds of MS1 plants were heavier than those of H plants (Table 6.14). The IN types also differed from each other: the IN1 plants were, depending on the component of seed production, either not different from MS1 and H plants or intermediate between them (Van Damme 1984 and Table 6.14). In contrast, the IN2 plants performed much better than MS2 and H in one of the studied populations; they produced far more ovules and they were fully female fertile. The weight of seeds, however, was as low as that of MS2 (Table 6.14).

The sex types MS1 and IN1, present in an old dune grassland, were studied in more detail. The seed production per season for MS1 (Table 6.15) agreed with those of the hayfield (Table 6.14). A large difference between MS1 and H was found for adult mortality, which factor strongly affects the number of

Table 6.14. Frequencies and components of sexual reproduction of sex types in *Plantago lanceolata*. (Van Damme 1984, and unpubl.)

Population	He			Gr	
Sex type	*H*	*MS1*	*IN1*	*H*	*MS1*
Frequency	0.70	0.22	0.07	0.83	0.09
Relative male fertility	1	*0*[a]	*0.39*	1	*0*
Ovules/plant	577	*1.86*	*1*	–	–
Seeds/ovule	0.80	*1*	*1*	0.58	*1*
Weight/seed (mg)	1.44	*1.12*	*1*	1.40	*1.17*

Population	Me			An			Gr	
Sex type	*H*	*MS2*	*IN2*	*H*	*MS2*	*IN2*	*H*	*MS2*
Frequency	0.90	0.04	0.06	0.91	0.08	0.01	0.83	0.06
Male fertility	1	*1*	*0.32*	1	*0*	*0.38*	1	*0*
Ovules/plant	297	*1.67*	*1*	60[b]	*1.49*	*3.15*	–	–
Seeds/ovule	0.47	*0.80*	*1*	0.39[b]	*0.75*	*1*	0.51	*0.54*
Weight/seed	1.95	*0.82*	*0.88*	1.49	*0.88*	*0.82*	1.53	*1*

[a] Figures in italics represent *ratios* of the value of that character of the concerning sex type and that of the corresponding hermaphrodite.
[b] Three flowers per mm spike were assumed.

Table 6.15. Frequencies and components of sexual reproduction of sex types in a dune grassland population (Westduinen) at two spots (cf. Fig. 6.8). (Van Damme and Van Delden 1984; Van Damme 1986, unpubl.)

Spot number	P1[a]			P2[a]		
Sex	*H*	*MS1*	*IN1*	*H*	*MS1*	*IN1*
Frequency	0.66	0.29	0.05	0.73	0.25	0.02
Relative male fertility	*1*[b]	*0*	*0.63*	1	*0*	*0.53*
Seeds/plant	23.9	*1.38*	*1.38*	64.7	*1.20*	*1.37*
Weight/seed (mg)	0.72	*1*	*1*	0.91	*1.24*	*1*
Adult mortality	0.83	*1*	*1*	0.45	*0.27*	*0.27*

[a] The results for spot *P1* are based on 1 year of observation, those of spot *P2* on 2 years.
[b] Figures in italics represent *ratios* of the value of a character of the concerning sex type relative to that of the corresponding hermaphrodite.

seasons available for seed production (Table 6.15). The IN1 plants differed in seed production (per season) and adult mortality from H plants in the same way as MS1. The sexes were also compared in the field for germination and vegetative growth rate, but no differences were found (Van Damme and Van Delden 1984). Apparently, fitness differences between MS1 and H were confined to the adult life stage.

Variation in pollen production between MS and non-MS plants is also important for maintenance of sex polymorphism. The relative male fertility values of IN1 and IN2 sex types, shown in Tables 6.14 and 6.15, varied around 0.5. Furthermore, in H1 and H2 plants with cytoplasmic type *R* and *P*, there appeared to be a significant difference in anther volume, and probably also in pollen production; 24% difference in favour of H1 was observed in a garden culture of a series of reciprocal crosses (Van Damme 1984).

It is clear that many components of sexual reproduction are involved in the dynamics of *Plantago* sex polymorphism. None of the differences was very large, except for the difference in survival rate in one of the plots of the dune grassland. Since the expression of this difference depended on environmental stress, viz. catastrophic droughts in the dune grassland, Van Damme and Van Delden (1984) suggested that the impact of the difference in adult survival and seed weight (Table 6.15) depended on the pattern of temporal and spatial variation in environmental circumstances in this population.

6.3.4.2 The Spatial Distribution of Sex Types and Sex Genes

As shown in Section 2.6, the MS1 type occurs on an average in higher frequencies than MS2 (Fig. 6.7). This is also clear from the comparison of the populations studied in the previous section (Tables 6.14 and 6.15). In a survey, the frequencies of MS1 and MS2 appeared to be negatively correlated (Van Damme and Van Delden 1982) (Fig. 6.7). In most populations both types were found together; in five, MS1 was absent. These five populations all belonged to the same vegetation type, *Calthion palustris* and *Junco-Molinion*, wet hayfields with a high groundwater table (−10 cm). In contrast, other populations in this vegetation type had very high MS1 frequencies (Van Damme and Van Delden 1982). A difficulty in the study of these populations was the extreme vulnerability of the wet hayfield vegetation; only a relatively small area of several hundred m^2 could be examined. Later investigations on the five populations lacking MS1, over much larger areas, showed the presence of two small patches

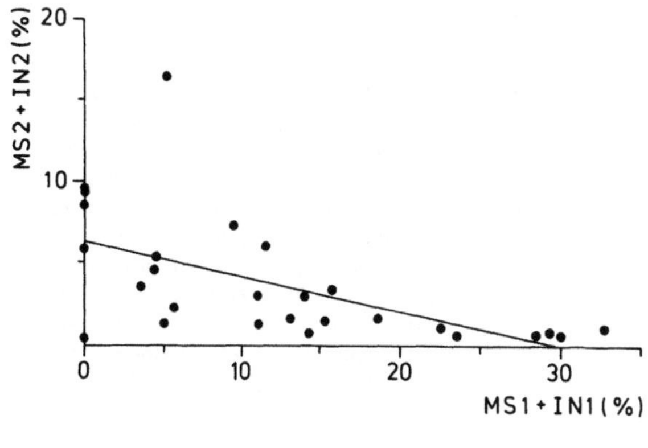

Fig. 6.7. Relationship between frequency of type 1 steriles (MS1+IN1) and frequency of type 2 steriles (MS2+IN2) in natural populations of *Plantago lanceolata*. (Van Damme and Van Delden 1982)

of about 10 m^2 with high MS 1 frequencies up to 50% in one of them (Van Damme and Graveland 1985). A patchy distribution of MS 1 could have escaped of earlier frequency recordings and focusses attention on the question at which surface scale frequencies of male steriles should be evaluated.

A single population, the dune grassland Westduinen, was examined in detail. Figure 6.8 shows a highly clustered distribution of MS 1 in a 7-ha area. This area was sampled by collecting plants at regular distances and using them as males in test crosses with MS 1 plants of known genotypes (Van Damme 1986). Analysis of the crossing results permitted estimates of the nuclear and cytoplasmic genes for MS 1, averaged over the whole field, and locally in the four largest MS 1 spots (Fig. 6.8). The largest part of the area contained only plants with cytoplasm *P*. The other cytoplasm, *R*, seemed to be restricted to the spots with MS 1 and IN 1 plants and constituted only 6% of the total area. Averaged over the field, the sterility alleles at the MS 1 loci had high frequencies: 0.98 for the recessive and 0.92 for the dominant alleles (cf. Table 6.12). This implies that most plants in the field had a "male sterile" nuclear genotype combined with a "male fertile" cytoplasm, resulting in hermaphroditism. In the four largest MS 1 spots (Fig. 6.8) the frequency of cytoplasm *R* was far higher than average, up to about 40%, and the frequencies of the nuclear sterility alleles were significantly lower (Fig. 6.10). Spatial differentiation at the phenotypic level was apparently paralleled by similar genetic differences. Furthermore, these results suggest that the area scale, at which population structure with respect to male sterility occurs, is of the order of several m^2.

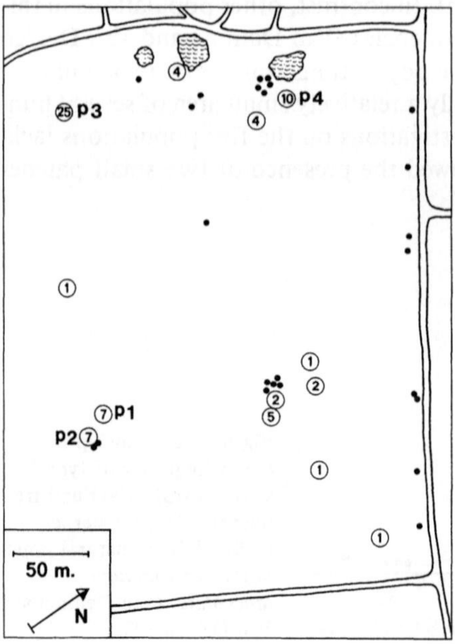

Fig. 6.8. Distribution of spots containing MS1 and IN1 plants of *Plantago lanceolata* in the Westduinen. Single plants are represented by *dots* and groups of plants by *circles*. *Within each circle* the spot area is given in square meters. The *four largest spots* are designated *p1* to *p4*. The *three shaded areas* are dune pools used as watering places by cattle. The spots *p1* and *p2* correspond with plots 1 and 2 in Table 6.15. (Van Damme 1986)

A question arises whether the MS 1 spots are distributed at random or not. This was investigated in two wet hayfield populations that showed a limited and patchy distribution of MS 1 plants (as in the Westduinen population). Vegetation relevés were made at five subsites in each population and at two additional experimental sites in one of the populations (see below). Per subsite, varying from 25 to 250 m^2, up to five plots of 1 m^2 were sampled. The plots were selected in such a way that coverage of *P. lanceolata* was at least 2%; the percentage cover of all higher plant species was then estimated. The plot relevés were ordinated according to standard techniques (Hill 1973, 1974). The distributions of individual plots along the first two axes, as shown in Fig. 6.9, accounted together for 42% of the observed variation. In addition, the lengths of the axes, expressed in standard units, were relatively small (Hill 1974), implying small differences in vegetation composition among subsites. Plots within subsites lay close together as did subsites within populations, in particular along the first axis (Fig. 6.9). While differences among populations were apparent, there was no differentiation between sites with low and high MS 1 fre-

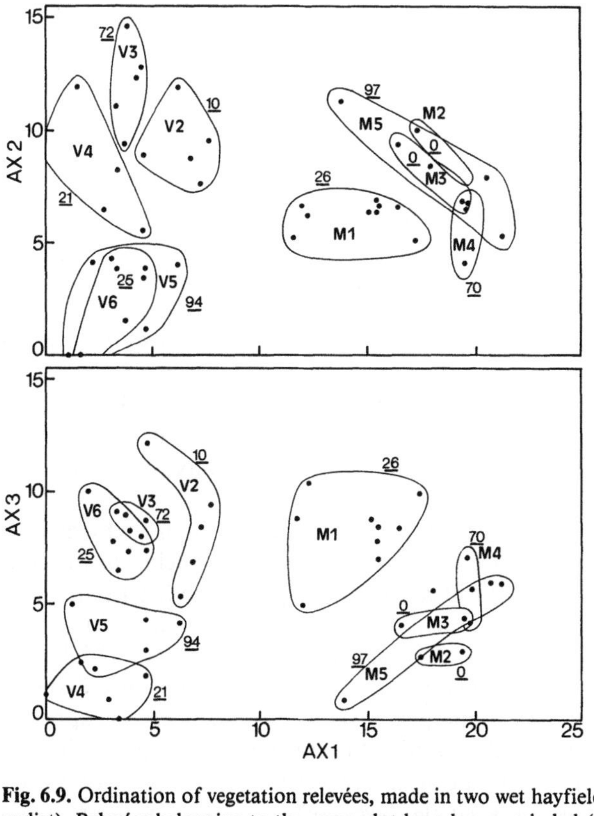

Fig. 6.9. Ordination of vegetation relevées, made in two wet hayfields (*V* Veerslootlanden; *M* Merrevliet). Relevées belonging to the same plot have been encircled (*V1, V2* etc.). For each plot the local MS frequency is *underlined.* (Van Damme and Graveland 1985)

quencies. None of the correlations of the MS 1 frequency in a site and the position of the site along one of the axes was significant. Since the vegetation composition probably strongly depended on environmental conditions, the findings support the view that spots with high MS 1 frequencies were distributed at random with respect to environmental variables.

A field experiment was performed in a wet hayfield population (Merrevliet), in which seedlings with cytoplasm *R* or *P* were transplanted into a non-MS 1 area. Two nuclear backgrounds were used (from the population *Wd* and *Gr*; see Tables 6.14 and 6.15), and as controls seedlings with unknown cytoplasmic type were used from the Merrevliet population itself. Differences between nuclear background were shown for several characteristics (Table 6.16); no differences between cytoplasmic types were present. Seedlings with cytoplasmic type *R* performed equally well as seedlings with the *P* as far as growth and mortality were concerned. The exact proportions of sex types among the seedlings used was not known, except that the majority of the *R* seedlings with *Wd* nuclear background was MS 1. Apparently, MS 1 plants were quite able to grow at places where only H plants of *P. lanceolata* were found.

In summary, populations clearly differed in the frequencies of male sterility, but the distribution patterns of male steriles, MS 1 in particular, within populations suggest that the relevant scale to consider selection processes for sex polymorphism was not the population in the widely used sense; i.e. a site, but an area of only about 10 m^2. There was no evidence that environmental factors were important to explain the distribution of MS 1 within the study sites.

6.3.4.3 Modelling the Spatial Distribution of Male Sterility, Type MS 1

Theoretical models on the maintenance of gynodioecy start with assumptions concerning the mode of inheritance of male sterility and with assigning values of components of fitness to sex genotypes. They predict gene and male sterile spatial frequencies. The purpose of the present section is to model the spatial dynamics of a gynodioecious population of *P. lanceolata*. Since all three components – genetics, fitness values, sex type frequencies and gene frequencies – had been studied in the dune grassland population, this population was chosen for a model.

As pointed out elsewhere (Van Damme 1986), the very high frequencies of sterility alleles (i.e. the very low frequencies of the restorer alleles) at the MS 1 loci in cytoplasm *P* area (Fig. 6.8) will cause the progeny of an MS 1 individual, that migrated into such an area, to be almost entirely male sterile. In spite of chance effects, a MS 1 spot may arise because the MS 1 plants had a higher fitness than hermaphrodites (see Table 6.15). Due to the low rate of seed dispersal in the Westduinen (Chap. 6.4), a compact structure of such an MS spot was expected and found. A simplified representation of the genetic composition of the four largest spots (Fig. 6.8) is shown in Fig. 6.10, in which the five nuclear MS 1 genes are represented by a single "average" gene. The position of the spots in Fig. 6.10 suggests that after an initial increase of cytoplasm *R*, when

Table 6.16. Transplant experiment of seedlings of different cytoplasmic types. Averages per group and significance levels of contrasts among groups, based on analysis of variance, are given. Measurements were made 3.5 months after transplantation

Nuclear background	Control	Wd		Gr		Significance levels		
Characteristic cytoplasmic type	–	R	P	R	P	Control (Wd + Gr)	Wd-Gr	R-P
Relative growth rate (% per day)	3.80	3.41	3.61	3.80	3.80	NS	*	NS
Estimated biomass[a]	12631	5558	5792	7671	8892	***	***	NS
Turnover rate (%)	20.4	23.2	21.0	20.9	20.4	NS	NS	NS
Percentage dry weight	12.8	15.7	15.2	13.4	12.6	*	***	NS
Mortality	11.3	13.8	11.3	18.8	12.5	NS	NS	NS

[a] Sum of leaf length × width of widest leaf.
***, $P<0.001$; *, $P<0.005$; NS = not significant.

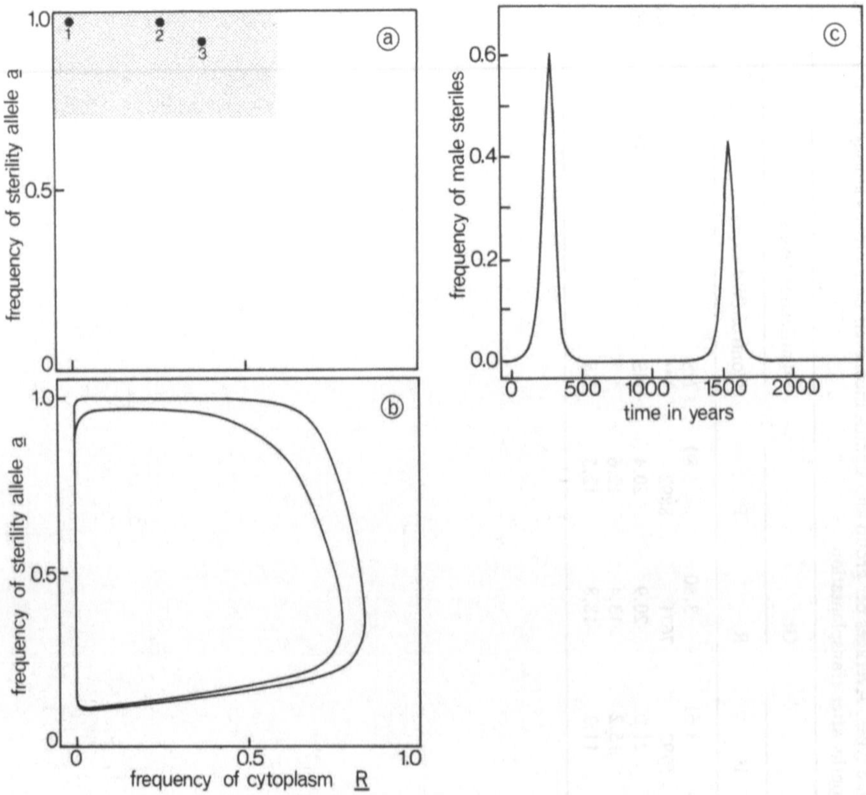

Fig. 6.10. a The frequency of the sterility allele *a* plotted against the frequency of cytoplasm *R* for the Westduinen: outside the MS1 spots (*1*) and in the four largest MS1 spots (*2*: p4 and *3*: p1, p2 and p3). The *shaded rectangle* indicates the criteria according to which computer runs have been selected (see text). **b** The frequencies of *a* and *R* during an example computer run resulting in a limit cycle. Parameter values for seed production for genotypes RAA, RAa, Raa, PAa and Paa were 0.6, 1, 1, 0.8, 0.8, 0.9, respectively, and for survival per year 0.72 for RAA, and 0.76 for the other genotypes. The first 2300 computer years are shown. **c** The MS frequency plotted against time of the run in **b**

the MS 1 spot is increasing in size, the restorer allele frequency (which was very low in the beginning) would increase with time.

The actual development of a spot after foundation is unknown. The spots *pl*, *p2* and *p3* indicated that indeed they might be in equilibrium since their genetic compositions were very alike (Van Damme 1986). Secondly, the small size of the MS 1 spots suggested that there might be limits of their size. If the spot compostion is stable, which mechanism will then limit spot growth? There seems to be no obvious answer to this question; if the superiority of MS 1 plants enables them to successfully intrude into the cytoplasm *P* area and if the fitnesses of all genotypes are so that an equilibrium can be attained, then seed dispersal would ensure continuous spreading of MS 1 seeds into the surrounding cytoplasm *P* area.

In order to answer the question about the stability of the genetic composition, a computer model was constructed (Charlesworth 1981; Delannay et al. 1981; Ross and Gregorius 1985; Gouyon et al. 1991). The genetics of MS 1 was assumed to be determined by the cytoplasm (types *R* and *P*) and one nuclear locus (alleles *a* and *A*) so that *Raa* was MS 1 and *RAa* IN 1 and the other four genotypes, H. This corresponds to incomplete dominance of the nuclear gene in cytoplasm *R*. The assumption is fairly close to the inheritance of MS 1 and IN 1; it is simplified for the number of nuclear MS 1 loci and for possible interactions between genes for MS2 and MS 1 (Van Damme 1983 a, b). Because of a very low frequency at the Westduinen and a random distribution over the area, the sex type MS2 is ignored (Van Damme 1986). The model was deterministic, with infinite population size and with overlapping generations. The fraction of the population that died was replaced by seeds produced the year before. Plants were supposed to flower from the first year and no pre-reproductive period was considered.

Each genotype was assigned a value for relative pollen production, for relative seed production per year and for the fraction surviving the flowering season until next season. The values of these parameters were chosen to be close to the observed values in the Westduinen population (Tables 6.15 and 6.17; and Sect. 6.3.4.1). Pollen production values were kept constant; seed production values of hermaphrodite genotypes varied between runs with a step size increment of 0.1, and survival values of all genotypes with step 0.02. For each run a parameter set was tested with a starting frequency of the sterility allele, *a*, of 0.98 and of cytoplasm, *R*, of 0.001; the situation corresponds to a cytoplasm *R*-individual, migrating into a cytoplasm *P* area (with a genetic composition as in Westduinen). A run was assumed to have reached equilibrium when gene and MS 1 frequencies between 2 years differed less than 10^{-4}. Some runs did not meet this criterion within 5000 simulation years and were stopped.

More than five million parameter combinations were tested. About 23% of the runs reached a stable nuclear-cytoplasmic equilibrium, i.e. with both cytoplasmic types and both nuclear alleles present in frequencies higher than 0.001. In a so-called *R-a* plot (Fig. 6.10) in which the frequency of the sterility allele *a* is plotted against the frequency of cytoplasm *R*, these runs had their equilibrium represented as a point in the interior of the graph. If only those runs are selected that at equilibrium had a frequency of *a* higher than 0.70 and a *R* frequency between 0.02 and 0.60 (see the rectangle in Fig. 6.10a) *none* was found.

Table 6.17. Values and ranges of fitness components of sex genotypes in the model (see text). Step size was 0.1 for relative seed production and 0.02 for survival per year

Genotype Sex type	RAA H	RAa IN1	Raa MS;	PAA H	PAa H	Paa H
Relative pollen production	1.25	0.5	0	1	1	1
Relative seed production	0.4 – 1	1	1	0.4 – 1	0.4 – 1	0.4 – 1
Survival/year	0.70 – x	x	0.70 – 0.86(= x)	0.70 – x	0.70 – x	0.70 – x

The MS 1 spots *p1* to *p4* at the Westduinen all fall within the last-mentioned frequency ranges. Thus, the attempt to find a combination of pollen and seed production values and survival rates that approximated the observed values of these parameters in the MS 1 spots of the Westduinen, and on the other hand yielded gene and sex type frequencies similar to those found in the same spots, had all failed.

This result suggests that (some of) the MS 1 spots at Westduinen had a non-stable genetic composition. The earlier-mentioned apparent contradiction between stability of genetic composition and limits to the size of the MS 1 spots is absent and the question remains how to explain the observed spatial distribution of MS plants. An indication of what might actually be the case in the field comes from computer runs that did not meet the stability criterion within 5000 years. An example of such a run is show in Fig. 6.10b. The population reached a state of equilibrium, but it did not end in a point in a *R-a* plot, i.e. with stable gene and cytoplasmic frequencies. Instead, the population kept running around in a cyclic way. This type of equilibrium, a limit cycle, is known from predator-prey systems (Roughgarden 1979; Gilbert 1984). It presents a stable equilibrium, since there is a stable trajectory along which the population continuously moves, gene and sex type frequencies varying through time in a cyclic fashion (Fig. 6.10c). The conditions for a limit cycle in the present model were apparently stringent since only several tens of them were found out of five million attempts. In more complex genetic models limit cycles appear to be more common (Gouyon et al. 1991).

Limit cycles as in Fig. 6.10b may help explain the observed spatial MS 1 distribution at the Westduinen. At the start of the run there is a situation with only cytoplasm *P* and allele *a* and very few MS 1 individuals that have migrated into the area. Initially, the cytoplasm *R* frequency rises and the spot size grows due to seed migration. However, after a number of years the frequency of *a* drops to such a low value that the MS 1 plants disappear although cytoplasm *R* remains present for a while. This means that the MS 1 spot fades away, inplying that its growth is automatically stopped. Then cytoplasm *R* frequency returns to almost zero and after that the frequency of *a* slowly increases again until the next cycle starts. For several of the observed cycles the first cycle differed only little from the equilibrium cycle; the process of spot growth, fading away, being absent for some time and starting again may go on forever.

The situation described above presents an ideal one; it differs from the situation in the Westduinen in several respects. Firstly, the number of years required to reach an equilibrium cycle was usually very large. Some of the cycles needed 1000 years to complete one turn whereas the age of the Westduinen is about 300 years. Secondly, after completing 3/4 of the first cycle the population, due to the finite population size, will most probably collapse on the y-axis (Fig. 6.10), i.e. cytoplasm *R* will be completely lost. However, as was argued by Gouyon et al. (1991), a cycle line phenomenon will then occur, entailing the rise of *a* to unity, so that the population will reach the starting situation again. Then the emergence of a new MS 1 spot will depend on the migration of cytoplasm *R* seeds (or on mutation).

As stated previously, the correct spatial scale to consider selection processes is of the order of $10\,m^2$. The part of the Westduinen studied for the distribution of MS 1 plants enclosed 7 ha. The cycle process might occur in any of $10\,m^2$ areas, and if the stages they are in are distributed at random over the cycle trajectory, a spatial distribution pattern as in Fig. 6.8 would be expected: at any moment in time only a small number of areas are in the MS 1-spot phase.

Finally, the cycle phenomenon might theoretically also be responsible for the differences *between* populations in MS 1 and MS 2 frequencies (Fig. 6.7.). This was suggested by Gouyon et al. (1991), who used a model similar to the one presented in this section, except that two MS types were considered, one corresponding to each cytoplasmic type. Further, an asymmetry between cytoplasmic types in the components of fitness was used. Figure 6.11 represents a temporal pattern obtained with this model that can explain the variation in MS 1 and MS 2 frequencies between populations by assuming that all populations are in a different stage of the developing limit cycle (Fig. 6.11).

In conclusion, there is no evidence from the computer runs that the MS 1 spots at the Westduinen represent a stable situation: the spatial distribution of MS plants may be explained as a series of time points in a trajectory non-synchronized cyclic of plants within small areas in the population.

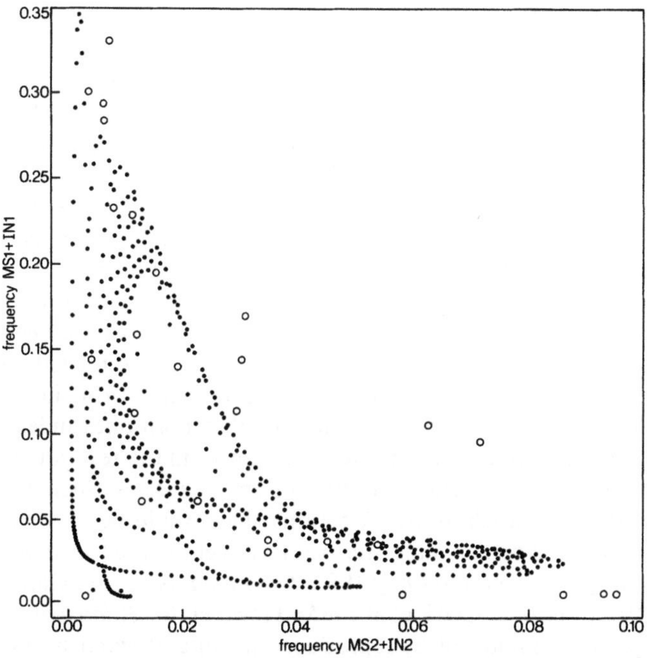

Fig. 6.11. The frequency of MS1 (+IN1) plotted against the frequency of MS2 (+IN2). *Circles* indicate observations of Van Damme and Van Delden (1982) (see Fig. 6.7) and the *points* indicate frequency combinations that occurred in a particular run of the computer model of Gouyon et al. (1991)

6.3.5 Why Does Gynodioecy exist?

In previous sections the proximate causes for the existence of gynodioecy have been discussed. Now attention will be focussed on the ultimate causes for the existence of gynodioecy. The fitness differences between male steriles and hermaphrodites may result either from inbreeding depression, i.e. an overall effect of many genes, or from pleiotropic effects of the male sterility genes (that is apparent pleiotropy due to closely linked genes; this distinction is irrelevant for the present purpose; Van Damme 1985). The former effect relates to a difference in outcrossing rate between obligatorily outcrossing male steriles and potentially selfing hermaphrodites. Evidence for the occurrence of inbreeding depression has been shown for several self-compatible gynodioecious species (Assouad et al; 1978; Perrot et al. 1982; Sun and Ganders 1986; Wolff et al. 1988). There is little doubt that such differences may contribute to the maintenance of gynodioecy in other self-compatible species as well. In *Plantago*, several of the gynodioecious species are self-incompatible (Sect. 2.6) such as, e.g., *P. lanceolata* (Ross 1973). The differences in life cycle characters between MS and H in this species, which were observed in the field (Tables 6.14 and 6.15), are probably due to pleiotropy (Van Damme 1984): both MS and H plants depend on pollen of other plants for seed set so that the amount of inbreeding depression will not differ noticeably between the sexes. In addition, the species is wind-pollinated in the study populations, implying that the differences in the number of seeds per ovule between MS2 and H (Table 6.14) are not due to pollination conditions but to the genotypes themselves, i.e. to pleiotropy.

Thus, sex genotypes in *P. lanceolata* did not differ only in male fertility but in a series of morphological and life history characters, a syndrome. One should expect to find differences between sexes in physiological characters that are developmentally linked to this set of traits. The genetics of male sterility might, for example, prove to be the genetics of a hormone level, acting as a switch in plant development, allowing differences in a "hermaphrodite" and a "male sterile" developmental pathway. Evidence for the involvement of hormones exists in *P. lanceolata*, where differences in cytokinin levels have been found (Fig. 6.12; Olff et al. 1989). Differences in hormone levels related to male sterility have been found for other species as well (Sawney 1974; Louis and Durand 1978; Ahokas 1982), supporting the idea that hormones indeed may be involved. Musgrave et al. (1986) reported differences in the respiratory pathway between male steriles and hermaphrodites in several species. Such differences would be related to mitochondrial functions. This would be of interest, since the cytoplasmic factor for male sterility has been located in mitochondrial DNA (*Zea mays*, Leaver and Gray 1982; *Nicotiana*, Avid and Galun 1980; Belliard et al. 1979; *Petunia*, Clark et al 1985). However, for *P. lanceolata* the difference in respiration could not be reproduced in spite of extensive experiments under various conditions (Van Dijk and Kuiper 1989), making the involvement of the respiratory pathway in male sterility doubtful.

To maintain nuclear-cytoplasmic polymorphism, fitness differences are required between male steriles and hermaphrodites, but also between the her-

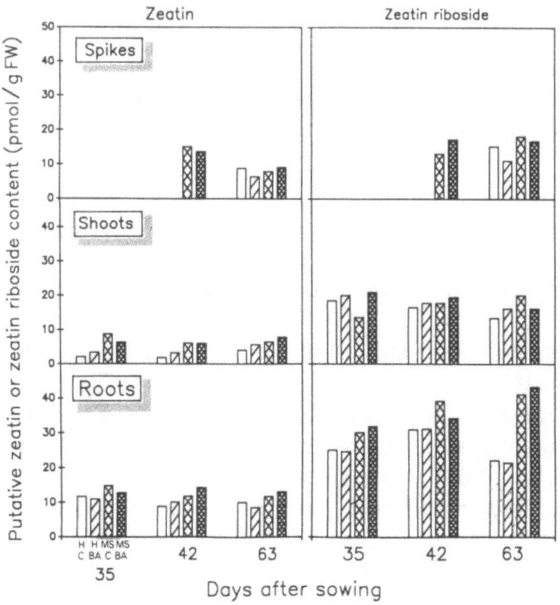

Fig. 6.12. The concentrations of putative zeatin and zeatin riboside in the spikes, shoots, and roots of control H1 plants (▮), H1 plants sprayed with benzyladenine BA (▨), control MS1 plants (□), and MS1 plants sprayed with BA (▩) at 35, 42, and 63 days aftter sowing. The concentrations were determined by enzyme-linked immunoassay (ELISA). *FW* fresh weight. (Olff et al. 1989)

maphrodite genotypes (Charlesworth 1981; Delannay et al. 1981; Ross and Gregorius 1985; Gouyon et al. 1990). Conditions are rather stringent: cytoplasmic and restorer mutations that do not meet these conditions are lost, or nuclear male sterility will be established, which will subsequently evolve towards dioecy (Ross 1978). Establishment and maintenance of gynodioecy thus depends on the possibility that cytoplasmic and restorer mutations occur with specific associated effects on fitness. The reason why relative to other genera so many *Plantago* species have gynodioecy (Sect. 2.6) may be a developmental pattern allowing frequent occurrence of such mutations (Van Damme 1985).

The question why gynodioecy exists at all is more general. It is really striking how often male sterility has been found to be inherited through both nuclear and cytoplasmic genes (Frankel and Galun 1977; Van Damme 1983b; Kaul 1988). Genetic information from both the cytoplasm and the nucleus generally seems to be necessary for stamen development. An explanation can be found in the existence of an "intragenomic conflict" (Cosmides and Tooby 1981): in animal and most plant species cytoplasmic genes are transmitted to the next generation only through ovules, whereas nuclear genes are transmitted through both ovules and sperm/pollen. This evolutionary conflict of interests should be regarded as the primary cause for the existence of gynodioecy (Gouyon and Couvet 1985); it leads to an everlasting disagreement between information from cytoplasm and nucleus to develop male fertility/male sterility. The conflict is the basis of the spiral behaviour of a population in *R-a* plots (Fig. 6.10) when it approaches equilibrium and the limit cycles present the ultimate expression of the conflict: frequencies of cytoplasmic and nuclear genes rise and fall in succession forever. More detailed studies on the dynamics of this system with the help of molecular markers (Rouwendal et al. 1987) are now underway.

6.4 Gene Flow Characters and Population Structure in *Plantago lanceolata*

6.4.1 The Gene-Dispersing Apparatus in Plantago lanceolata

The ribwort plantain is a perennial hermaphrodite species with obligate outbreeding, resulting from its self-incompatibility system (Sect. 2.4). Therefore, inter-individual pollen transport is necessary for sexual reprodution. The flowers are in general wind-pollinated, although biotic pollination has been observed (Clifford 1962; Stelleman 1984b), and are arranged along a compact, cylindrical inflorescence. The female and male parts of the reproductive apparatus (Van Damme and Van Delden 1982; Bos et al. 1985) are extremely well exposed to wind and insects (Faegri and Van der Pijl 1979), and allow an excellent air-mediated exchange. The breeding system is described in Section 2.6.

The species possesses a number of properties which may influence the distribution and composition of the pollen cloud around the spikes and thus affect the levels of inbreeding or gene flow in the population. One of these properties is protogyny. Protogyny provides the species with a system which separates the female and the male function of the spike and its flowers *in time*: in a comparison of eight Dutch populations (Bos et. al. 1985 and Table 6.18), female flowers appeared 5.3 to 6.2 days before male flowers in hermaphrodite spikes. In the same survey we found that female flowering spikes were also separated from the hermaphroditic flowering spikes *in space*: the combination of protogyny in the spike with continued elongation of the scape resulted in the

Table 6.18. Protogyny and scape length parameters in eight *P. lanceolata* populations (after Bos et al. 1985) in greenhouse conditions (20 plants/population; three scapes/plant)

Population	Management	Protogyny (days ± S.D.)[a]	Scape length (cm ± S.E.)	
			♀ stage	♂ stage
Junner Koeland (JK)	Pasture; cows	5.3 ± 0.9	20.1 ± 1.5	30.6 ± 1.8
Westduinen (Wd)	Pasture; cows, horses	5.3 ± 0.8	15.3 ± 1.3	26.2 ± 1.8
Loenermark (L)	Pasture; sheep	5.6 ± 1.0	19.3 ± 1.3	34.4 ± 1.6
Anloo (An)	Wet hayfield	6.1 ± 1.1	14.8 ± 1.3	31.7 ± 1.7
Zeegse (Ze)	Wet hayfield	5.8 ± 0.8	16.5 ± 1.3	32.5 ± 1.5
Bruuk (B)	Hayfield	5.7 ± 0.8	23.4 ± 1.7	42.1 ± 1.9
Heteren (He)	Hayfield	6.2 ± 1.1	22.7 ± 1.5	44.9 ± 1.4
Merrevliet (Me)	Wet hayfield	5.8 ± 0.7	21.5 ± 1.0	39.6 ± 2.0

[a] n = 60.

stigmas being presented 10 to 22 cm lower in the vegetation than the pollen were released: *protogyny-growth syndrome*. Pollen release is synchronized in the early hours of the day (Hyde and Williams 1946; Stelleman 1984b). Furthermore, gynodioecy is present in the species: within populations a frequency range of 0 to 23% of male sterile individuals was found (Sect. 2.3). The above properties enable the species to vary gene flow patterns. This conclusion seems to contradict the statement of Waddington (1983): "I agree with Price and Waser (1979) that it is difficult to see how plants influence pollen flow patterns".

Seeds of *P. lanceolata* have no special adaptations for wind dispersal. There is a distinction between primary and secondary seed dispersal. Primarily, most of the seeds fall at the circumference of a circle with a radius of scape length around the plant: during ripening, scapes bend more and more to the earth. Primary dispersal is short (Bos et al. 1986; Van Damme 1986). Secondary dispersal provides for long distance gene flow. Some incidental observations are available about dispersal by animals and scientists (pers. observ.): wet seeds stick to fur or shoes because of the presence of mucilage (Sagar and Harper 1964); viable seeds have been found in the droppings of cattle, pigeons and sparrows (Salisbury 1961).

Pollen grains and seeds are the vehicles for nuclear genes. Cytoplasmic (mitochondrial and chloroplast) genes are exclusively transported in the species by seeds. The main aim here is to describe flow of nuclear genes. From recent

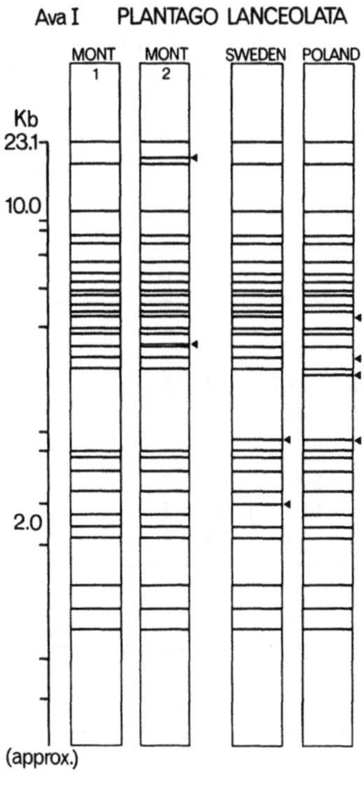

Fig. 6.13. Restriction endonuclease Ava I pattern in three *P. lanceolata* populations (schematic)

restriction enzyme analysis (Bos et al. in prep.), we know that within species variation – (even within population variation (Montpellier population): Fig. 6.13) exists. Therefore flow of cytoplasmic genes can be studied (for mt-DNA, see Rouwendal et al. 1987).

6.4.2 Gene Flow Measurement in P. lanceolata

Recently, four publications have appeared presenting estimates of gene flow in the anemophilous species *P. lanceolata* (Tonsor 1985a, b; Van Dijk 1985; Bos et al. 1985). Some other aspects of wind pollination (Primack 1978) or insect pollination (Stelleman 1984b; Clifford 1962; Hammer 1978) were published earlier. The relevant aspects of the biology of the species are summarized by Sagar and Harper (1964) and Cavers et al. (1980) and this volume (Sect. 2.1). Levin and Kerster (1974) define potential gene flow as "the deposition of pollen and seeds from a source as a function of distance". In contrast, they define actual gene flow, as "the incidence of fertilization – in the case of pollen – and establishment of reproductive individuals – in the case of seeds – as a function of the distance from a source".

Table 6.19 summarizes the gene flow data of both types. Applicable data are presented for an open dune pasture, extensively grazed by cattle (Westduinen) and for a hayfield situation (Haren). Tonsor's (T) estimate of gene flow was based on pollen flow in a wind tunnel and on an outdoor experimental field with male sterile plants as mothers. Bos et al. (B) derived their estimates independently in the same period from a wind tunnel experiment, from an exper-

Table 6.19. a) Gene flow in a pasture (WD) and a hayfield (Haren) situation. Density (d) pasture = 64, hayfield = 70 plants m^{-2}. Windspeed in wind tunnel 1.5 m sec^{-1}. (Bos et al. 1986). b) Gene flow data from other sources

Method	Source	σ_p (m)	σ_s (m)	N_e	N_a (m^2)	M (m)	
a)							
Pasture							
1. Disp. allozyme var.	Bos et al. (1986)	0.23	0.08	13.7	0.4	0.2	
2. Wind tunnel[a]	Bos et al. (1986)	0.76	0.08	20.0	3.7	0.7	
Hayfield							
1. Disp. allozyme var.	Spaak (pers. commun.)	0.40	0.51	290	4.3	0.7	
2. Wind tunnel	Bos et al. (1986)	1.50	0.51	619	17.4	1.5	
3. Field	This Section	1.24	0.42	70	4.4	1.2	
b)							
1. Outdoor exp.	Tonsor (1985b)	1.33	–	111	11.1	1.2	(d = 10 m^{-2})
2. Wind tunnel[a]	Tonsor (1985b)	0.58	–	21	2.1	0.5	(d = 10 m^{-2})
Disp. allozyme var.	Van Dijk et al. (1985)[b]	–	0.4	50	120.0	–	

[a] Values from comparable wind speed: Tonsor 1.24 m s^{-1}, Bos et al. 1.50 m s^{-1}.
[b] Westduinen population.

imental garden with male steriles and from a (Westduinen) field population. The table shows that data collected at comparable wind speeds (see subscript of Table 6.19) agreed.

Wright's neighbourhood model indicated that the area containing a panmictic unit within a continuous and uniform array of organisms can be estimated by $4\pi\sigma^2$, where σ^2 was the parent-offspring dispersal variance measured around a zero mean and relative to a single reference axis through a population. Crawford (1984) and Van Dijk (1985a) independently argued that the two components of dispersal variance (σ_p^2, pollen dispersal variance, and σ_s^2, seeds dispersal variance) should be combined as

$$\sigma^2 = 1/2\,\sigma_p^2 + \sigma_p^2 \;.$$

In 1987 Gliddon et al. extended for the formulae with the variance of vegetative dispersal.

Both studies T and B give roughly similar standard deviations for pollen flow (σ_p) in spite of the different experimental conditions: about 0.6 and 0.8 m, respectively, in the wind tunnel test and 1.3 and 1.5, respectively, in the outdoor experiments. The outdoor experiments probably gave an overestimation, since they were done in the absence of local pollen: the dispersal of pollen from a known source to male sterile (MS) mother plants. The values of σ_p of the pollen flow in Tonsor's paper were compared to those in Bos et al. – probably reduced by a much shorter wind tunnel (2.4 and 20 m, respectively) producing a restricted tail in the distribution. In addition, the flow parameters were affected by the high stickiness of the pollen in T's population and the smaller plant density in the experimental population. Other aberrant conditions were the pollen source height (T: at plant height; B: at 0.15 m) and the pollen collection height (T: plant height; B: 0.10 m).

Van Dijk (D) (1985a) developed a method for the calculation of gene flow parameters from allozyme distributions in field populations of *P. lanceolata*. The method was used by Bos et al. and Van Dijk (Table 6.19) and gave smaller values for gene flow. The method assumes that the estimates are done in populations in genetic equilibrium, which implies that the estimates are an average effect of years, individuals and conditions. A straightforward measure of isolation by distance is the parameter M from Van Dijk (1985a) ("the mean gene transport per generation"). This parameter estimates the actual gene flow in populations.

$$M = \sqrt{1/2\pi\,(\sigma_s^2 + 1/2\sigma_p^2)} = \sqrt{1/8\,N_a} \;.$$

In the two "outdoor experiments" (papers T and B), M was found to be 1.2 and 1.3 m, respectively, while the estimates from the genetic structure of the Wd population wer "low" and about 0.2 m. From Bos et al. (1986) we know that estimates vary with wind speed, pollen release, collection height, etc., so significant differences are no surprise. Van Dijk et al.'s Table 13 (1988) shows extremely low gene flow in all pasture and roadside populations and a higher flow in the hayfield (He). Yet, although *P. lanceolata* has in some situations

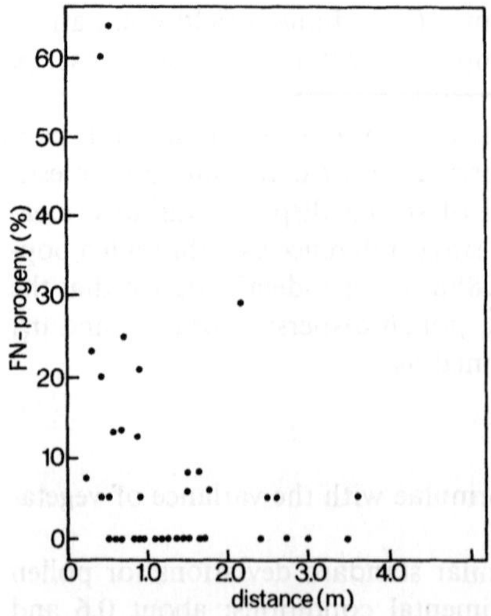

Fig. 6.14. Pollen flow from an Idh-FF genotype in the Haren hayfield: frequency of FN individuals in the progeny of neighbours at given distances from the FF plant

a relatively high gene flow – compared with the selfing species *P. major* and *P. coronopus* – (gene flow), expressed as M, probably does not exceed 2 m, which is in fair agreement with our Table 6.19.

In a final experiment in *P. lanceolata*, gene flow was measured directly in a hayfield (Haren) with the aid of an allozyme marker. In this population the *F* allele of the *Idh* (isocitrate dehydrogenase) locus did not occur. A unique *Idh-FF* homozygote was introduced in this population in the spring of 1987. In the autumn ripe seeds were collected from 40 neighbouring plants (*Idh-NN*). The percentage of *FN* descendants in a sample of 20 plants is plotted in Fig. 6.14 against distance from the *FF* homozygote. Frequencies decreased with distance and the average σ_p of *F* pollen was 1.24 m, which was in the range of the experimental data (Table 6.19). Seed flow was estimated in the same population by electrophoresis of leaf sample of 51 neighbouring plants in November 1988 which were scored for *Idh* genotype: 3*SN*, 5*FN* and 43*NN*. Thus, 1 year after introduction of the *FF* homozygote at least five descendants (*FN*) were established, which comprise about 10% of the neighbours. The distance from the mother was 0.42 m on the average, which is σ_s under the condition that the *FN* plants were descendants. Gene flow parameters derived from these field data are compared with previous experimental data in Table 6.19.

6.4.3 The Role of Pollen- and Seed-Mediated Gene Flow in P. lanceolata

Pollen grains and seeds are the transporters of single or double sets of nuclear genes, respectively, and are produced in numbers and forms adapted to their abiotic and biotic environments. Even between ecotypes seed numbers differ greatly. Van Groenendael (1985 a) describes the "dry" ecotype with a flat, small rosette which produces many small seeds with innate dormancy, resulting in a seed bank, while the taller "wet" ecotype makes fewer and bigger seeds without dormancy, so that no seed bank is formed. Transplantation experiments showed that both types are unable to survive in alien habitat.

The inflorescence of *Plantago lanceolata* is a typical cylindric spike with wind-adapted characteristics. Although wind often acts as a long-distance dispersal medium, the reviewed papers make clear that the average pollen-mediated gene flow in *P. lanceolata* by wind is for the greater part short-distance gene flow. Bos et al. (1985) and Van Damme (1986) give evidence that the majority of seeds also land near the mother plants.

Two questions are of particular interest with respect to gene flow. Firstly, does genetic variation in gene flow distances exist, and which characters influence the average distance? Secondly, how does gene flow influence population structure? An increasing number of publications is devoted to the study of characters in the context of gene flow instead of the characters per se (Price and Waser 1979; Waser and Price 1983; Waddington 1983; Tonsor 1985a): they include topics such as assortative mating, variation in the timing of reproduction, variation in floral characters relating to the proportion of selfed vs. outcrossed offspring. These phenomena influence the proportion of zero-dispersers and therefore the average distance of gene flow.

In the following, two additional characters which influence the pattern of gene flow in *P. lanceolata* are described in some detail. The first character is the variation in the amount of pollenkit on the pollen grain surfaces and consequently the variation in the tendency of the pollen grains to cluster (Stelleman 1982; Tonsor 1985a, b). Stelleman noted in *P. lanceolata* a positive correlation between the tendency to cluster and the degree of shelter against wind of the studied populations (and therefore probably the importance of insect pollination). Tonsor measured the dispersal distributions of classes of pollen grain clusters in a wind tunnel. From these graphs it is clear that populations without clustering of pollen grains have a completely different pattern of pollen flow than populations with clusters of large size (Fig. 6.15a).

The second character is the so-called protogyny-growth syndrome (Bos et al. 1985): female flowering scapes (including spikes) are on the average lower in height than when they are male flowering (in hermaphrodites). Populations grown under identical greenhouse conditions were found to differ in this character. When we compare the pollen distribution on the level of the pollen sources (type 1, leptocurtic) and on the lower level of the female flowering (type 2, vulcano-form), we find distributions (Fig. 6.15b) which are very similar to those in Fig. 6.15a. Obviously, populations can vary in the height differences between pollen release and collection levels, and in this way the pollen distribution patterns will vary. To analyze such effects, pollen flow simulation experiments were performed in which the pollen distribution patterns type 1 and 2 from Fig. 6.15b were used (for more details: see Bos and Van der Haring 1988). Pollen distribution type 1 is the leptocurtic distribution generally accepted for pollen dispersal. Distribution pattern type 2 – considered three-dimensional – is a kind of a "vulcano" form pattern found for the wind-pollinated species *P. lanceolata* (see above). Assuming a situation with a population

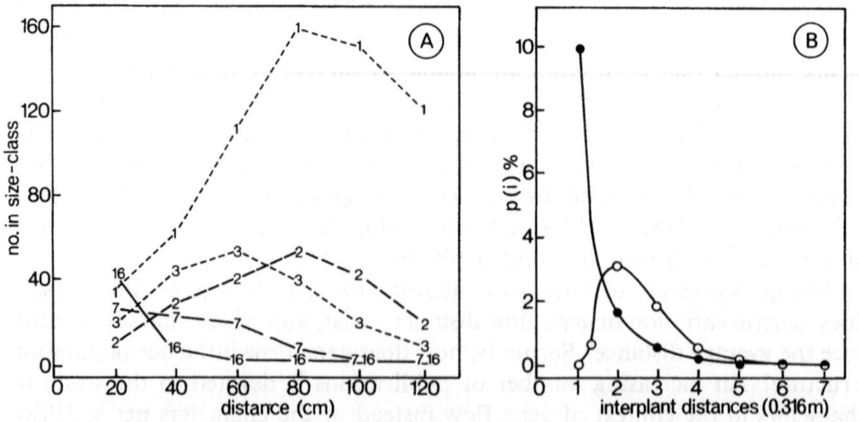

Fig. 6.15. A (After Tonsor 1985a) The dispersal distributions of five size classes of pollen grain clusters in the wind tunnel. The *numbers* in the body of the graph indicate the lower limit for the number of grains in the size class to which that line pertains (*1* single grains; 2 paired grains; *3* three to six grains; *7* seven to 15 grains; *16* 16 or more grains). The *vertical axis* indicates the number of clusters in each size class, averaged over the six runs of the wind tunnel. **B** The relative pollen distribution on two levels in the vegetation: *1* a leptocurtic distribution on the level of the (○) pollen sources; *2* platycurtic distribution on the level of female flowering. (Bos and Van der Haring 1988; Fig. 1)

of 2500 hermaphrodite, self-incompatible (140 *SI* alleles) individuals, with low seed flow ($\sigma_s = 1/2$ interplant distance and normal seed distribution), the simulation showed that after about 50 generations the population reached a lower level of fixation (F_{IT}) in the situation of pollen distribution of type 2 than with distributions type 1 (see Fig. 6.16). Distribution type 2 is the distribution on the natural level of female flowering in *P. lanceolata* (see above). This study also visualized that the rate of evolution of population structure is dependent on the pollination mechanism (wind, insect-pollinated) and on gene flow characters of the organism itself, like the protogyny-growth syndrome (see also Turner et al. 1982). The evolution of substructuring in *P. lanceolata* populations is strongly retarded by the syndrome (Fig. 6.17b). The relation between the syndrome and some population characteristics is brought together in Table 6.20.

So far we only have theoretical studies and correlations between gene flow distances and populations characteristics. Experiments are needed to verify the simulations. Table 6.20 presents a comparison of various parameters in three *Plantageo* species. Positive relations exist between genetic variability, gene flow, outcrossing rate, protogyny-growth syndrome and amount of pollenkit; a negative relation is found for population differentiation. At the moment it is impossible to unravel the relative and quantitative importance of the mechanisms (e.g. high outcrossing rate or high expression of the syndrome) for the increased genetic variation of the decreased population differentiation in the self-incompatible species *P. lanceolata*, compared to the two selfing species

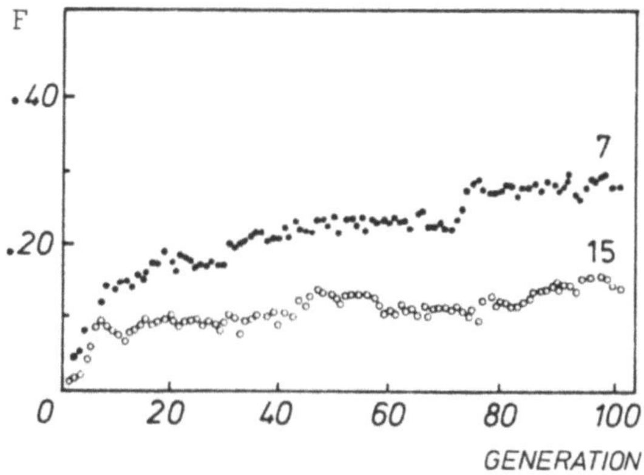

Fig. 6.16. The course of F_{IT} (fixation index, total population) during 100 generations of simulations, affected by pollen pattern: *7* with leptocurtic pollen distribution 1; *15* with pollen distribution 2 (protogyny-growth syndrome) (see Fig. 6.15b). Conditions were: low seed flow and seed distribution normal; 140 SI alleles. Plant height 0.5 m; female flowering at 0.3 m (hayfield situation). (Bos and Van der Haring 1988; Fig. 3)

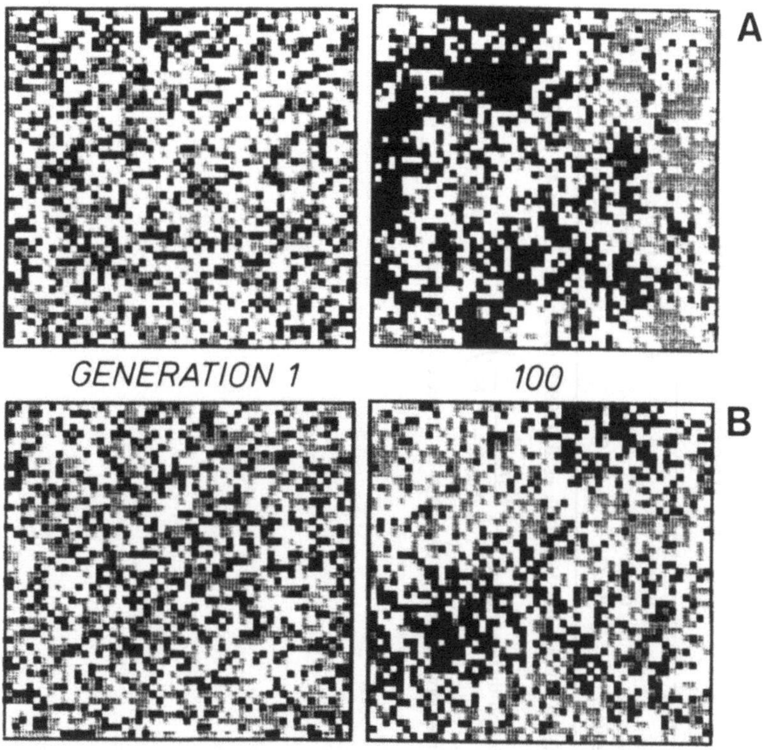

Fig. 6.17. Population structure after 100 generations of simulation. **A** (*above*) in a situation without, **B** (*below*) in a situation with the protogyny-growth syndrome. (Bos and Van der Haring 1988)

Table 6.20. Population characteristics [(1)–(4)] and the protogyny-growth syndrome (5) in three Dutch *Plantago* species

Species	(1) Gen. var. H	(2) Pop. diff.[a]	(3) Gene flow M (m)	(4) Outcr. rate $\bar{t}$	(5) Prot.-growth syndrome		
					stigma l. (mm)	[height ♂ – height ♀] Δ (cm)	T^b (days)
P. major	0.047	0.491	0.1–0.4 (0.22)	0.22	2.10	2	2.7
P. coronopus	0.088	0.161	0.05–0.5 (0.22)	1.0	3.53	4	4.8
P. lanceolata	0.127	0.041	0.2–2.0 (0.32)	1.0	4.57	10–20	5.6

Columns (1) (2) (4) after Van Dijk et al. (1988); (3) and (5) Bos et al. (1985, 1986).

[a] Population differentiation $= \left(\frac{D_{ST}}{\bar{H}_e} \cdot \frac{k}{k-1} - \frac{1}{\bar{N}_h}\right)$.

[b] T = time between first ♀ and first ♂ flowering of the spike).

P. major and *P. coronopus*. Experimental evidence is needed from selection experiments for reduced or increased gene flow distance and the study of the concomitant response in the above-described characters.

6.5 Ploidy Polymorphism in *Plantago media* and in Some Other *Plantago* Species

6.5.1 Introduction

In sexual organisms each parent transmits half of its chromosomes to the next generation by the processes of meiosis and fertilization. However, occasionally meiotic abberations produce non-reduced gametes that give rise to polyploid individuals (DeWet 1980). Because the reversal process (polyhaploidy) is very rare, polyploidy is predominantly an irreversible trend from lower to higher levels of chromosome number. While most animals are diploids, a remarkably high proportion of higher plants are polyploids (Lewis 1980; Jerling 1985b). Grant (1981) estimated that about 47% of all angiosperms are polyploids. Polyploidy is even more frequent in the genus *Plantago*. In more than two-thirds of the 122 *Plantago* species listed in Dietrich (1975b) and Dietrich (1980) polyploid chromosome numbers have been reported. In several *Plantago* species different ploidy levels (cytotypes) have been described: e.g. *P. media* (including *P. stepposa*) (2x, 4x, Rahn 1954, 1957), *P. coronopus* (2x, 4x, 6x, Gorenflot 1960), *P. major* (2x, 4x, 6x, Favarger and Vasudevan 1972; Vasudevan 1976), *P. alpina* (2x, 4x, Cartier 1971) and *P. maritima* (= *P. serpentina*) (2x, 4x, Cartier 1971).

In contrast to animals, plants physiologically tolerate the multiplication of entire chromosome sets. The high frequency of polyploidy in plant species asks for an explanation of the evolutionary success of these macromutants. Paradoxically, chromosome doubling by itself seems to be maladaptive (Stebbins 1971, but for an opposing view see Levin 1983). A major problem of chromosome doubling is that during meiosis homologous chromosomes form multivalents instead of bivalents. This often results in unbalanced gametes and loss of fertility (Gottschalk 1976). In 15 grass species, artificially produced tetraploids were consistently inferior to their diploid progenitors under more or less natural conditions (Stebbins 1985). Once formed, the probabilities for establishment of new polyploids in diploid populations are limited, because they will experience a minority cytotype disadvantage when inter-cytotype hybrids have reduced fitness (Levin 1975). In spite of the fact that many crop species are natural polyploids, artificial polyploids have been of little value in plant breeding (Dewey 1980).

By contrast to genome multiplication, polyploid gene segregation might be of adaptive value, since it tends to preserve favourable gene combinations by increased heterozygozity and epistatic gene interactions (Stebbins 1985). Heterozygous artificial polyploids often show hybrid vigour (Bingham 1980). Therefore Stebbins (1971, 1985) suggested that polyploid success would depend on increased heterozygosity due to either inter-racial hybridization (autopolyploids) or inter-specific hybridization (allopolyploids), either preceding or following genome doubling. However, there is very little direct evidence for this hypothesis in nature. (Here the terms autopolyploid and allopolyploid are used as intra- and inter-specific polyploids, respectively (Lewis 1980).) Many polyploids are allopolyploids (Grant 1981). In such species hybrids, polyploidization balances meiosis and restores fertility (Stebbins 1971). The importance of autopolyploidy in plant evolution is still a matter of much dispute (Stebbins 1985; Soltis and Rieseberg 1986). In spite of more than 60 years of polyploid research, the reasons for the polyploid success still remain largely obscure (Stebbins and Dawe 1987).

Ploidy polymorphism and distribution was studied in *Plantago media*. This species was previously investigated by Rahn (1954, 1957), who described two cytotypes: diploids ($2n = 2x = 12$) and tetraploids ($2n = 4x = 24$). Diploids were found in the Pyrenees and in Eastern Europe, tetraploids in Western and Central Europe.

6.5.2 Differentiating Characters Between Cytotypes

Rahn (1954, 1957) described two differentiating characters between diploid and tetraploid *Plantago media* when grown in a common garden. Firstly, the number of ovules per capsule, which ranged in diploids per population from 2.0 to 4.0 and in tetraploids from 3.1 to 8.0. Secondly, diploids had a shorter scape compared to the spike length than tetraploids. The spike/spike+scape ratio ranged in diploids from 0.31 to 0.42 per population and in tetraploids from 0.21 to 0.30 per population.

However, these data were based on a limited number of diploid populations. To investigate whether these differences were consistent in Pyrenean material, ten plants from four diploid and four tetraploid populations were grown in a glasshouse, as well as ten diploids and ten tetraploids from a mixed cytotype population (Formigal). Figure 6.18 shows that relative scape length is not a consistently differentiating character, although four of the six tetraploid populations have clearly longer scapes. By contrast, the number of ovules per capsule separates all diploids from all tetraploid populations: diploid range: 2.16–3.20, tetraploid range: 4.00–4.81 (Fig. 6.18). However, variation between individual plants within cytotypes does not permit ploidy determination by number of ovules per capsule: range in diploids: 2.00–3.85, in tetraploids: 3.10–5.50. In natural populations variation will be larger because of environmental and plant size variation.

A diploid population from the USSR that was measured for comparison had a considerably higher mean than the Pyrenean diploid material: 3.91 ± 0.20, ($\bar{x} \pm$S.E.; $N = 7$). Rahn (1957) also found in both diploids and in tetraploids the lowest number of ovules per capsule in the Pyrenean material. Obviously, this character varies geographically within the cytotypes. In both

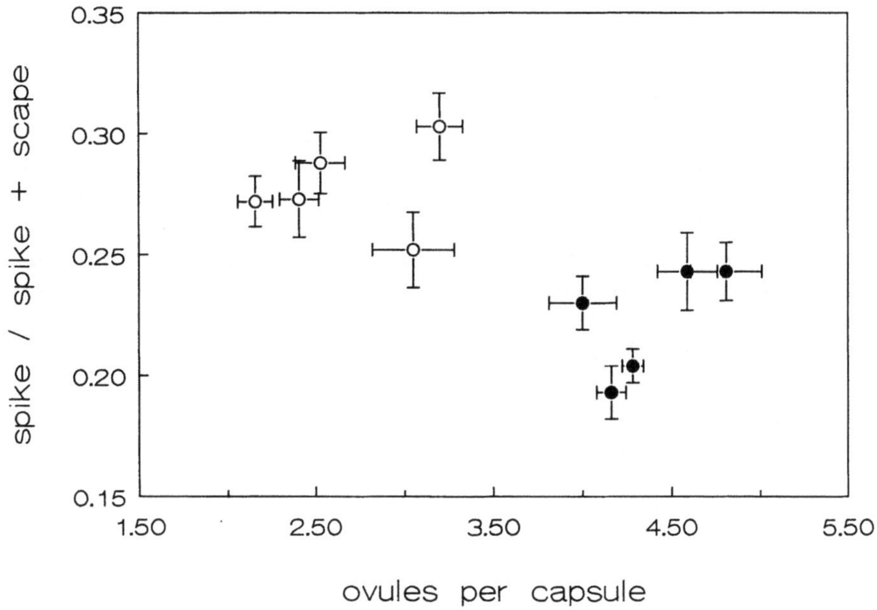

Fig. 6.18. The spike/spike + scape ratio and the number of ovules per capsule in 5 diploid and 5 tetraploid Pyrenean populations (diploids: *open symbols*; tetraploids: *closed symbols*; ten plants per populations). *Bars* standard error of the mean

diploids and tetraploids the number of ovules per capsule increases to the East in the Pyrenees. This trend may indicate parallel environmental adaptation or bidirectional introgression.

6.5.3 Evidence for Autopolyploidy in Plantago media

Several lines of evidence indicate autotetraploidy in *P. media* (Van Dijk and Van Delden 1990). Diploid and tetraploid *Plantago media* are morphologically very similar and distinct from other *Plantago* species. Strong reproductive barriers exist between the two ploidy levels, but occasionally 3x and 4x hybrids were produced in crosses with higher vigour. In contrast to diploids, colchicine tetraploids could easily be crossed with natural tetraploids but no longer with diploids. Reproductive isolation between ploidy levels was due to endosperm collapse in triploid hybrids. No prezygotic reproductive isolation between cytotypes was found. Furthermore, four diploid populations and four tetraploid populations shared the same allozymes for nine studied enzyme loci.

Detailed cytological investigations have revealed that quadrivalents are formed during pachytene of which some can be found in metaphasis I (pers. commun. Dr. J. S. Parker). In viable triploid hybrids, up to six trivalents per pollen mother cell may be observed, indicating homology between the 2x and 4x genomes. Altogether, the evidence strongly suggests autotetraploidy in *Plantago media*. Also in other plant genera autopolyploidy appears to be more common than once thought (Van Dijk and Van Delden 1990).

The fact that colchicine tetraploids were very vigourous indicates that *Plantago media* physiologically tolerates genome doubling. However, this tolerance is limited since artificial octaploids had leathery deformed leaves, flowered very poorly, and died early. In contrast, no sublethal effects were found in octaploid *Plantago coronopus* (Gorenflot 1963). Colchicine tetraploids of *Plantago media* had high seed and pollen fertility: 81 and 97% of diploids, respectively. These values are high in comparison to colchicine tetraploids in other species (Gottschalk 1976).

Stebbins (1971) considers that multivalent formation at metaphasis I by itself is unreliable for polyploid classification. The reason is that homologous pairing takes place before metaphasis, during pachytene. As has been shown in *P. media*, these quadrivalents can only be seen in metaphasis when fixed by chiasmata. Therefore the classifications of polyploid *P. major* and *P. coronopus* as allopolyploids seems doubtful (Favarger and Vasudevan 1972; Gorenflot 1960). In fact, in *P. coronopus* rare 3x hybrids were produced in crosses between 2x and 4x (Gorenflot 1961).

Because the vast majority of triploid hybrids are inviable, autotetraploidization by itself causes almost complete reproductive isolation between the cytotypes. Some gene exchange could be possible through rare but viable 3x and 4x hybrids, especially from 2x to 4x (unidirectional gene flow). Selectively advantageous alleles can rapidly pass even strong gene flow barriers (Barton and Hewitt 1985). Although the introgression of genes is possible, the gene pools cannot fuse and will ecologically behave as if they belong to two species. Clearly this situation challenges the biological species concept.

Artificial and natural polyploid *Plantago media* could easily be crossed, but some interesting morphological and cytological differences existed. First, the two cytotype-differentiating characters, the number of ovules per capsule and the length of the spike were not affected by colchicine treatment (Fig. 6.20). These differences are obviously not a direct consequence of polyploidization. Colchicine tetraploids showed increased cell and organ sizes (gigas effects), which generally accompany polyploidization (Stebbins 1971). In contrast, however, natural tetraploids of *Plantago media* showed no gigas effects at all (Figs. 6.19 and 6.20). A similar difference between natural and artificial tetraploids was found in *P. coronopus* by Gorenflot (1960). Natural hexaploids in *P. major* showed no marked gigas characteristics compared to diploids (Favarger and Vasudevan 1972). Bassett and Crompton (1968) also found no relationship between ploidy and pollen grain size in North American chasmogamous *Plantago* species. *In Plantago media* multivalent and chiasmata fre-

→

Fig. 6.19. Photographs of pollen grains of different cytotypes in *Plantago media* (**A** 2x; **B** natural 4x; **C** colchicine 4x). *Bar* 25 μ

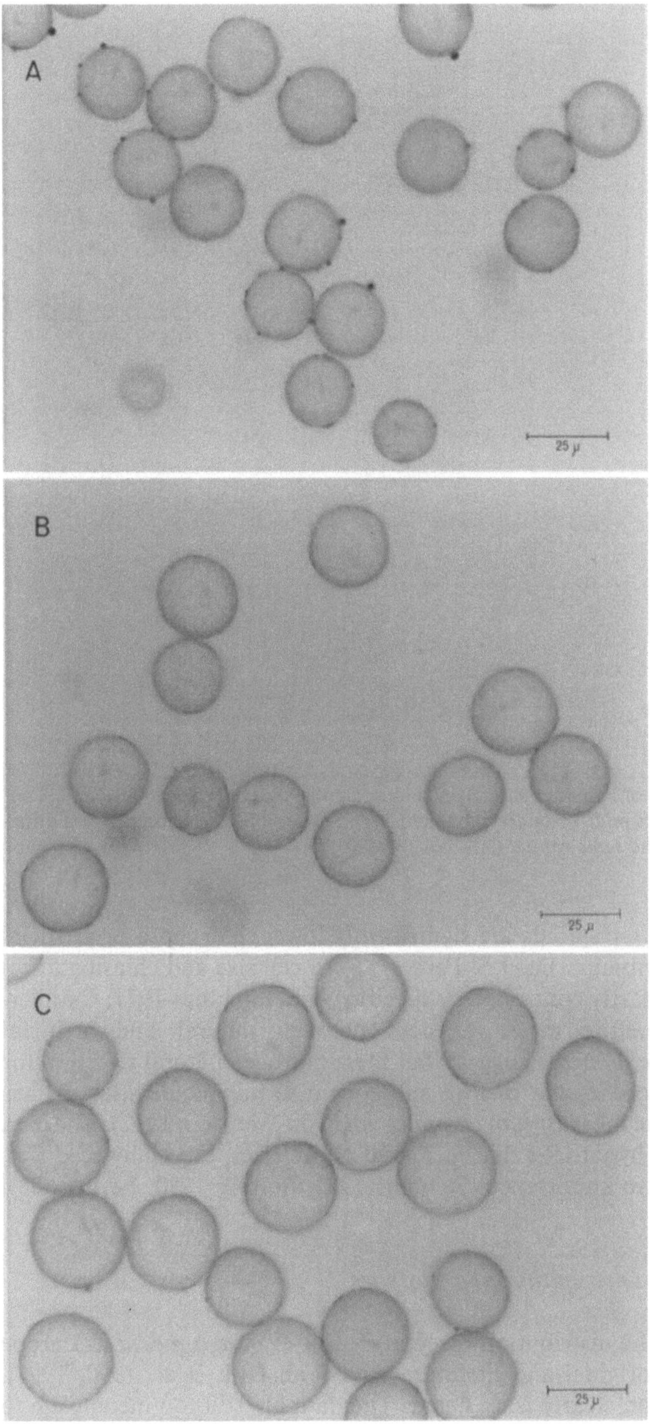
A
25 μ
B
25 μ
C
25 μ

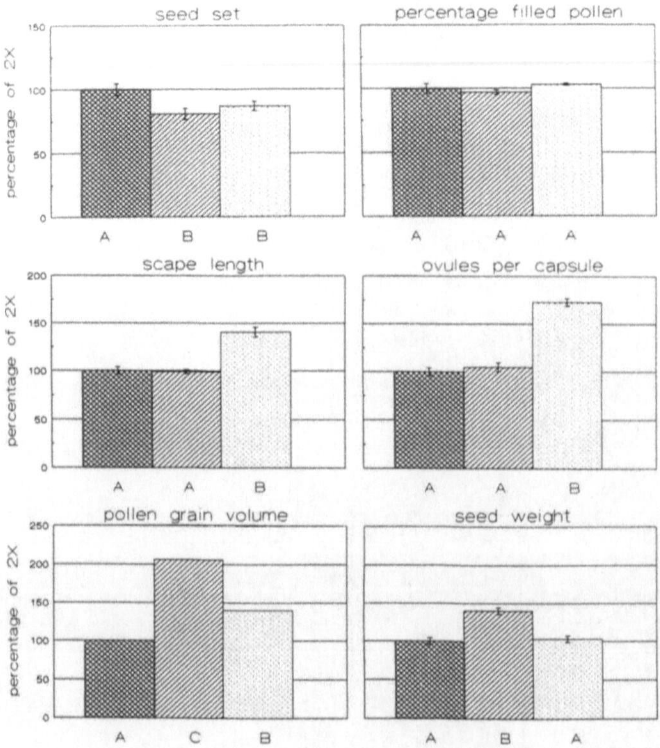

Fig. 6.20. Characteristics of Pyrenean natural and induced polyploids in *Plantago media* (diploids *left*; colchicine tetraploids *middle*; natural tetraploids *right*; means ± S.E.). Non-significant differences are indicated by the *same capital* (ANOVA, $p<0.05$)

quencies of natural tetraploids are significantly lower than those of artificial tetraploids (pers. commun. Dr. J. S. Parker). Both cell size and chiasmata frequency are at least partly controlled by specific genes (Stebbins 1971; Cavalier-Smith 1985). Therefore the differences between natural and artificial tetraploids in *Plantago media* suggest that there has been natural selection for smaller cells and more regular meiosis subsequent to polyploidization, as has been described in the moss *Bryum caespitosum* (Von Wettstein 1937). The data do not support the hypothesis that increased organ sizes, especially seed size, would be of adaptive importance (Stebbins 1971; Solbrig 1980; Levin 1983).

6.5.4 Geographical Distribution of Cytotypes

Figure 6.21 shows the distribution of cytotypes in *Plantago media* in Europe as compiled from original and literature data (Van Dijk et al. 1990, 1991). Cytotype distribution clearly is clustered. The data are still too incomplete to give a detailed pattern. Nevertheless, it is obvious that the geographic range of

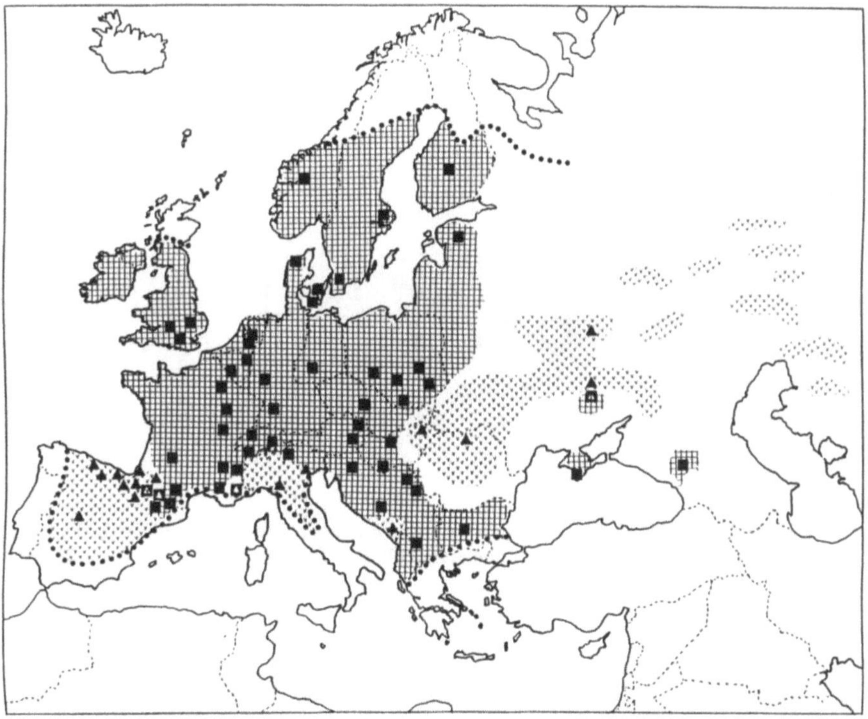

Fig. 6.21. The geographical distribution of 2x and 4x cytotypes of *Plantago media* in Europe, based on original data and data from literature (Van Dijk in prep.) (2x populations, ▲; 4x populations, ■; sympatric 2x–4x populations, □)

the ancestral 2x cytotype of *Plantago* media in Europe is more restricted than that of the derived 4x cytotype. A similar situation has been found in other polyploid complexes (Stebbins 1971; Stebbins and Dawe 1987). The distribution in the Pyrenees was studied more extensively. Forty six populations were sampled throughout the Central Pyrenees. Per population 5 to 20 seed parents were used. Figure 6.22 shows that the distribution of cytotypes is essentially parapatric: diploids occur in the Western Pyrenees, tetraploids in the Eastern Pyrenees. How can such large areas with single cytotypes be explained?

In contrast to *Dactylis glomerata* (Lumaret et al. 1987) and *Deschampsia cespitosa* (Rothera and Davy 1986), there is no indication for habitat differences between the cytotypes in *Plantago media.* Both 2x and 4x occur predominantly in dry basic grasslands between 0 and 2000 m altitude. There is a large overlap in ecological range between 2x and 4x *Plantago media.* The contact zone is not associated with an obvious environmental change. The parapatric distribution cannot be explained by differences in ecology between the cytotypes.

Alternatively to environmental selection, the pattern might well be explained in terms of historical events. In the Pyrenees the distribution of allozymes in the snail *Cepaea nemoralis* (Ochman et al. 1983) and of subspe-

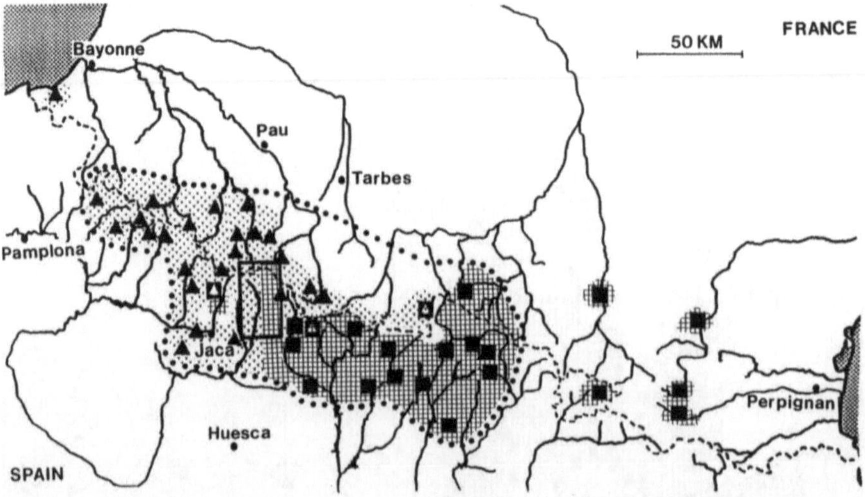

Fig. 6.22. The geographical distribution of 2x and 4x cytotypes of *Plantago media* in the Pyrenees, based on original data and data from Rahn (1954) (2x populations, ▲, 4x populations, ■; sympatric 2x–4x populations, ◘)

cies in the grasshopper *Chorthippus parallelus* (Hewitt 1989), has been strongly influenced by colonization patterns at the end of the Pleistocene, when new territory became available. The parapatry of cytotypes of *Plantago media* in the Pyrenees may best be explained by colonization from different refuges at the end of the Pleistocene, resulting in secondary contact, followed by exclusion, either ecologically or genetically.

There may be strong competition between cytotypes because of relatedness. Colonization of an empty territory may be fast, but may be retarded in presence of an ecologically equivalent species. A superior competitor may gradually replace the inferior cytotype after contact, thus producing parapatry.

When hybrids have reduced fitness, as is the case in *Plantago media*, mutual cytotype exclusion is expected because of genetic interactions (Levin 1975). In theory, a minority cytotype will be rapidly excluded, because it produces a higher proportion of unfit hybrids than the majority cytotype. Since migrants will be in the minority, they cannot intrude and therefore this process will result in parapatry. Even an adaptively superior cytotype may not be able to invade, since high fitness values are necessary to overcome the minority cytotype disadvantage. This situation is comparable to a hybrid zone that is maintained by hybrid unfitness (tension zone) (Barton and Hewitt 1985). Many of these hybrid zones in temperate regions are readily explained by secondary contact after the Pleistocene (Hewitt 1989).

Minority cytotype exclusion may well explain the macroscale parapatry in *Plantago media.* However, along the range's interface some sympatric populations have been found. This sympatry can be due to factors like prezygotic reproductive isolation (e. g. cytotype-specific pollen-pistil interactions, flowering time differences, selfing etc.). long-distance seed dispersal and long life span. As a matter of fact, selection for increased prezygotic reproductive isolation is expected within a tension zone (Barton and Hewitt 1985). At the moment these factors are being investigated in some sympatric cytotype populations. Sympatric cytotype populations are particularly important because here the demographic performance of the cytotypes and their interactions can be studied under common natural conditions (Solbrig 1980).

The relationship between polyploidy and geographical distribution has been studied for a long time. Various earlier hypotheses relating polyploidy in general to distribution (e.g. extreme temperature, drought, altitude, range size) have been rejected (Stebbins 1971, 1980: Stebbins and Dawe 1987). Stebbins and Dawe (1987) conclude that while many polyploid species are widespread, this is not an universal tendency. In many species diploids and polyploids differ in geographical distribution (Stebbins 1971, 1985; Ehrendorfer 1980). Polyploidy is often associated with the invasion of newly available territory due to climatic changes or human activities (Stebbins 1971, 1985; Ehrendorfer 1980). It is tempting to interpret wider distributions in terms of superior adaptations. However, such differences in distribution may also arise between adaptively equal cytotypes. Hewitt (1989) has argued that the edges of expanding ranges are very favourable for the establishment of new negatively heterotic mutants over large areas. Such marginal mutants may rapidly colonize the newly available territory, at the same time being protected against the invasion of central mutants by hybrid zones. This might well explain parapatric ranges in sexually reproducing polyploid complexes. Cytotype hybrids are likely to have low fitness because of endosperm collapse (Levin 1978) or because of reduced fertility as a consequence of uneven genome number (anorthoploids). This theory only predicts parapatric distributions between cytotypes and not a universal parameter association. Indeed the data on cytotype distribution in *Plantago major* in the Western Himalayas and of *P. alpina* and *P. maritima* in the French Alps suggest cytotype parapatry (Favarger and Vasudevan 1972; Cartier 1971). Other striking examples of cytotype parapatry have been found in *Lathyrus pratensisis* (Brunsberg 1971) and *Tolmiea menziesii* (Soltis 1984).

Because of its universal effects on plant functioning and its evolutionary success, polyploidization is of interest for physiologists and ecologists as well as for geneticists. Some of the many questions raised by polyploidy are very puzzling because they seem to be paradoxical. For instance, how is it possible that plants physiologically tolerate such increases in gene copies, whereas most animals do not? For what reason do natural polyploids not exhibit gigas characteristics whereas artificial tetraploids do? How can negatively heterotic mutants establish in a population?

Autopolyploidy is especially valuable for polyploid research since it uncouples the processes of polyploidization and species hybridization. As more becomes known and understood about the reasons why polyploidization has been of such importance in nature, we also may understand why induced polyploids in plant breeding have consistently fallen short of expectations. This may even lead to a revival of chromosome doubling in plant breeding (Stebbins 1980).

7 Genotypic and Phenotypic Variation Between *Plantago* Populations and the Relation with the Habitat

7.1 Phenotypic Plasticity in *Plantago maritima*

7.1.1 Introduction

Plantago maritima L. is a halophytic species occurring on beach plains, mud flats, salt marshes, sea dikes, coastal grasslands and sea cliffs. A great variety in morphological characteristics can be found; based on differences in leaf characters, Heukels (1910) distinguished three types.

Changes in morphological characters can be environmentally or genetically induced, referring to phenotypic plasticity and genetic differentiation, respectively (Bradshaw 1965). The nature of adaptive variation between individuals of plant populations was already discussed by Turesson (1922), who introduced the term ecotype as the genotypic response of a species to a particular habitat.

In this section the problem is discussed wheter differences in the morphological properties *P. maritima* plants are caused by phenotypic plasticity or by genetic differentiation.

7.1.2 Morphological Differences Between Populations in Relation to Distance

Gregor (1939) found differences in morphological characteristics of 52 populations of *Plantago maritima*, collected in eight regions in North America and Europe. The American populations differed from the European ones in bract, sepal and seed indices, and in anther, scape and spike lengths. Differences were also found between the American and European populations for spike density, number of seeds per capsule, leaf indices, leaf tooth grade, spots on leaves and ciliation of the floral parts. The American and Greenland sea plantains were self-fertile, and those of Europe self-sterile. Based on significant differences in morphological characteristics, Gregor (1939) distinguished these populations taxonomically into five groups of sea plantain: some of these were allied to *P. maritima*.

In a further study, Gregor and Land (1950) compared the range of morphological differences between populations from different areas in Britain (inland, island and coastal areas) with those between *P. maritima* populations within a single area, a coastal area in Britain. Between the populations of the three areas no significant difference was observed in the scape spread/height ratios. Surprisingly, great differences were found in scape and leaf length and in the ratios of scape spread and height between the populations of a mud, a salt marsh, a drained mud and a coastal meadow, all situated within the same coastal area. The latter results showed that in a particular coastal area the distribution of the sea plantain was continuous only in a relative sense; probably not all suitable habitats were occupied by the species.

7.1.3 Experiments on the Nature of Morphological Differences Between Populations

Experimental studies under controlled conditions are necessary to determine whether morphological differences between plants of different habitats are due to phenotypic plasticity or to genetic differentiation. Turesson (1930) showed in transplant experiments with *P. maritima* plants from salt marshes and sea cliffs that differences in height were reproducible every year with only very small fluctuations. These results indicate the existence of ecotypes. Other experiments are described by Gregor (1956); in an exposed area, occupied by a population of dwarf plants, shallow depressions occurred. The leaves and spikes of plants of *P. maritima* in the depression were taller than those of the higher parts of the area. Under experimental conditions, however, the differences disappeared in five out of six samples examined, indicating that changes in characteristics were mainly due to phenotypic plasticity.

This chapter describes the nature of relatively large differences in life characteristics between two adjacent populations of *P. maritima* (Blom 1983). The study was carried out at Kwade Hoek, a salt marsh on the island of Goeree, The Netherlands. The vegetation of the tidal area ranged from halophytic communities on heavy clay soils, to communities growing on inland dunes with a sandy soil (Westhoff et al. 1961). The boundaries between the communities of some sites were quite distinct, due to the presence of tidal creeks; on the more interior sites the transition was more gradual, as described for other salt marshes (Ranwell 1972, Beeftink 1975, 1977; Mahall and Park 1976). Part of Kwade Hoek is grazed by cattle from May to October.

Population I of *Plantago maritima* occurred mainly in the Armerion maritimae community with taxa like *Glaux maritima* L., *Juncus gerardii* Loisl., *Festuca rubra* L., *Aster tripolium* L. and *Artemisia maritima* L. The community was found on sandy clay soils, containing approximately 1200 mg Cl^- per 100 g dry soil.

Population II of *P. maritima* occurred in low numbers at transitions between the above community and the Saginetum maritimae community with *Sagina maritima* G. Don, *Plantago coronopus* L., *Cochlearia danica* L., *Honckenya peploides* (L.) Ehrh., *Centaurium littorale* (Turner) Gilmour, *Sagina nodosa* (L.) Fenzl. and *Festuca rubra* L. This community was found on a sandy soil with 4–5 mg Cl^- per 100 g dry soil.

Population I occurred in low-lying zones with a layer of sandy clay in the upper soil; individuals had an estimated age of more than 10 years. The plants mainly propagated vegetatively and only few seedlings were found; mortality of seedlings was very high. There was intensive grazing by cattle and regular flooding in this zone. The plants showed a prostrate growth form (Jerling and Andersson 1982).

Population II was located at the edge of the inner dunes. The zone was only flooded when the tide was extremely high. The vegetation was rarely grazed by cattle, and in winter the grazing pressure by rabbits was relatively high. In this upper zone, the individual plants were short-lived, about 3 years old. The reproduction of the plants was mainly by seed, which germinated rapidly, and seedlings were easily established. Most of the plants had an erect growth form.

The distance between population I and population II was about 300 m. In 1970 a dam was built in a neighbouring estuary and the high-water mark increased. It is probable that population II became established after this event.

Studies on allozyme variation and morphological characteristics suggested that the differences between the two populations did not have a genetic basis. Furthermore, there was a continuous mixing of seeds between both populations, due to spring-tide and the activities of wild and domestic animals. Therefore, we investigated if the differences in life-history characteristics could be explained as phenotypic plasticity. Since flooding with seawater and grazing were regular events at this site and in other sites with *P. maritima*, we investigated the effects of flooding and grazing on life-history characteristics.

The results of a sowing experiment (mixed seed sample) on sites of the habitats of both populations showed that emergence of *P. maritima* seedlings was significantly higher at the edge of the inner dunes (population II) than along the creek (population I); no difference was observed in seedling survival (Table 7.1). After 3 years, however, significantly more plants from sown seeds survived in the sites along the creek. Furthermore, plants from sown seeds flowered in significantly greater numbers at the edge of the inner dunes than along

Table 7.1. Emergence, survival and reproduction of *P. maritima* in sowing experiments in the field (Kwade Hoek, The Netherlands). In brackets: results obtained with naturally occurring plants. A mixed sample of seeds from both sites was used

Sowing sites at Kwade Hoek	Emergence (%)	Seedling survival (%)	Adult survival after 3 yrs[a] (%)	Reproduction after 3 yrs[a]	
				Generative (%)	Vegetative (%)
Along a creek (Population I)	18.3	60.0 (30.3)	72.5 (81.1)	12.9 (5.3)	48.9 (85.9)
Edge of inner dunes (Population II)	33.7	66.3 (74.8)	39.2 (32.7)	60.7 (72.4)	27.5 (16.1)

[a] The age of the naturally occurring plants along the creek is estimated to be 10 yrs.

the creek, where more daughter rosettes were found. Differences in reproduction between adjacent populations of *P. maritima* have also been described by Jefferies (1977): the start of the flowering period of a population on a beach plain was about 2 weeks earlier than that in an upper marsh. Even such relatively small differences in flowering time reduced gene flow between populations (Jefferies 1977).

The results of the sowing experiments on seedling survival at the edge of the inner dunes were in accordance with those of demographic studies of *P. maritima* plants at Kwade Hoek. At the grazed sites fewer (naturally occurring) seedlings were found than in the sown plots. No significant differences between survival of the naturally occurring adults and adults from sown seeds were observed. Between both habitats, the differences in form of reproduction were more pronounced than those observed in the sowing experiments. The differences in emergence were not due to the substrate (sand vs. sandy clay) but possibly to the salinity of the soil (Blom 1983). Germination of *P. maritima* was inhibited at relatively high NaCl concentrations, which can be found in the upper soil layer of the sites of population I (along the creek), whilst 25% sea water caused a significant decrease in emergence. In the field, seedling emergence increased after periods of rainfall, undoubtedly due to a decrease in the salinity of the soil. Jerling (1981) showed that the survival of *P. maritima* seedlings strongly depended on the microtopography of the soil. Mortality due to inundation has been ascribed to light deficiency and to toxic effects of H_2S. Seedling establishment and survival of *P. maritima* were also strongly reduced by trampling (Blom 1983).

The effects of increased salinity on growth of *P. maritima* was studied by Cooper (1982) and Blom (1983). Cooper concluded from an investigation of eight species occurring on salt marshes, that the growth of species from the upper marsh was strongly limited by salinity and by waterlogging (Table 7.2), in contrast with species of the lower marsh. The position of *P. maritima* in the salt marsh studied by Cooper (1982) was just between the upper and lower marsh. Shoot and root weights of *P. maritima* were reduced when grown under saline conditions for 2 months. This species was more tolerant to waterlogging and to salinity than *Festuca rubra*, *Juncus gerardii* and *Armeria maritima*, species from the upper salt marsh. In contrast, the growth of *Salicornia europeae* was stimulated by a high level of NaCl in the substrate (Figs. 7.1 and 7.2). Blom (1983) observed that plants grown for over 2 months under saline conditions started to produce more biomass. After about 4 months no differences were observed between the biomass of plants grown under saline and non-saline circumstances. At Kwade Hoek *P. maritima* occurred in saline and in less saline habitats. Floristic surveys of salt marshes in Britain also showed that the species occurs in both lower and higher zones of salt marshes (e.g. Gimingham 1964; Boorman 1971; Beeftink 1975; Adam et al. 1977). Under saline conditions *P. maritima* plants showed significantly more daughter rosettes than under non-saline conditions (Fig. 7.3). In both treatments the plants were grown from the same seed pool which indicates that the formation of daughter rosettes depended on the environmental circumstances.

Table 7.2a. The effects of waterlogging and salinity on the mean whole plant dry wt. yield (mg) of species from a salt marsh near Portaferry, Britain. (Results from Cooper 1982)

	Drained non-saline	Waterlogged non-saline	Drained saline	Waterlogged saline
Festuca rubra[a]	345.5	138.9	150.8	80.5
Juncus gerardii	150.5	77.3	59.4	34.8
Armeria maritima	278.5	158.0	189.9	81.1
Plantago maritima	723.6	712.8	521.6	499.7
Aster tripolium	904.2	908.3	740.3	489.0
Triglochin maritima	154.9	130.5	158.6	92.4
Puccinellia maritima	369.0	424.1	243.8	282.6
Salicornia europaea	31.6	19.4	40.7	35.4

[a] Sequence of species represents a seaward progression.

Table 7.2b. Summary of analysis of variance of the data in a

	Waterlogging	Salinity	Interaction
Festuca rubra	**	**	NS
Juncus gerardii	*	**	NS
Armeria maritima	**	*	NS
Plantago maritima	NS	**	NS
Aster tripolium	NS	**	NS
Triglochin maritima	**	NS	NS
Puccinellia maritima	NS	**	NS
Salicornia europaea	*	**	NS

*, ** Significance of F values at the $P = 0.05$ and 0.01 levels respectively; NS, not significant.

Next to salinity, grazing was an important environmental factor in differentiation between the habitats of the two populations. The separate effects of trampling and clipping on biomass production and propagation of *P. maritima* have also been studied (Blom 1983). Trampling caused a significantly higher biomass production than non-trampling or clipping. Furthermore, significantly more daughter rosettes were formed by trampling (Fig. 7.4).

In treatments of simulated inundation and of grazing, the treatment determined to a large extent the morphological characters of the individual plants. The differences found in the experiments were in accordance with the differences in morphological characteristics and demographic features of the populations in the field between the non grazed and rarely inundated habitat at the edge of the inner dunes and the grazed and inundated habitat along the creek. These results suggest a plastic reaction in growth and reproduction characteristics of *P. maritima*.

Most of the differences which had been recognized between *P. maritima* populations related to morphological characteristics such as growth forms and reproductive features. There were some examples of direct evidence of genetic differentiation. Jefferies (1977) and Jefferies and Perkins (1977) stated that in

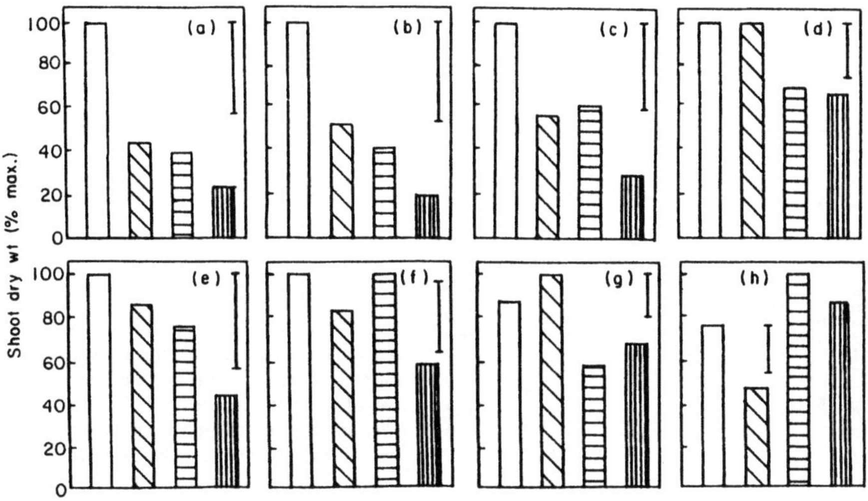

Fig. 7.1. The effects of waterlogging and salinity on the shoot dry weight yield of **a** *Festuca rubra*, **b** *Juncus gerardii*, **c** *Armeria maritima*, **d** *Plantago maritima*, **e** *Aster tripolium*, **f** *Triglochin maritima*, **g** *Puccinellia maritima* and **h** *Salicornia europaea*: ☐, drained non-saline; ▧, waterlogged non-saline; ▤, drained saline; ▥, waterlogged saline. *Vertical bars* indicate confidence intervals above which any two treatment means are significantly different. Results obtained after 2 months. (Cooper 1982)

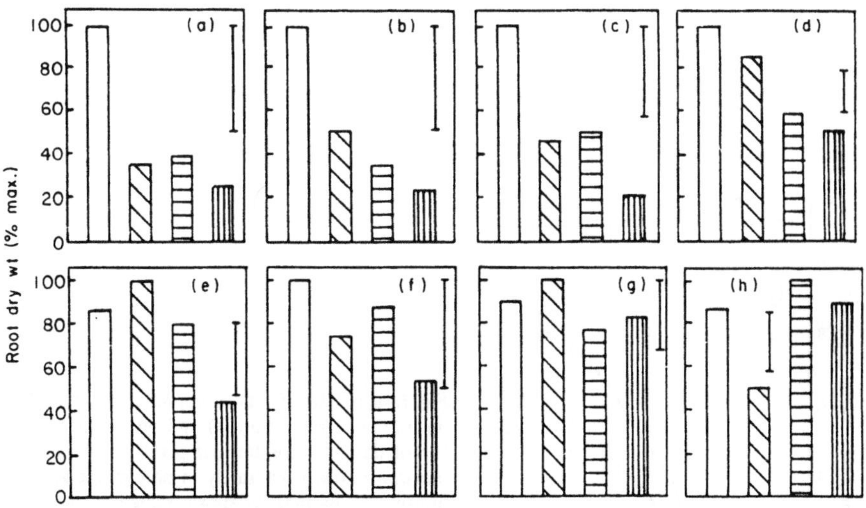

Fig. 7.2. The effects of waterlogging and salinity on the root dry weight yield of **a** *Festuca rubra*, **b** *Juncus gerardii*, **c** *Armeria maritima*, **d** *Plantago maritima*, **e** *Aster tripolium*, **f** *Triglochin maritima*, **g** *Puccinellia maritima* and **h** *Salicornia europaea:* ☐, drained non-saline; ▧, waterlogged non-saline; ▤, drained saline, ▥, waterlogged saline. *Vertical bars* indicate confidence intervals above which any two treatment means are significantly different. Results obtained after 2 months. (Cooper 1982)

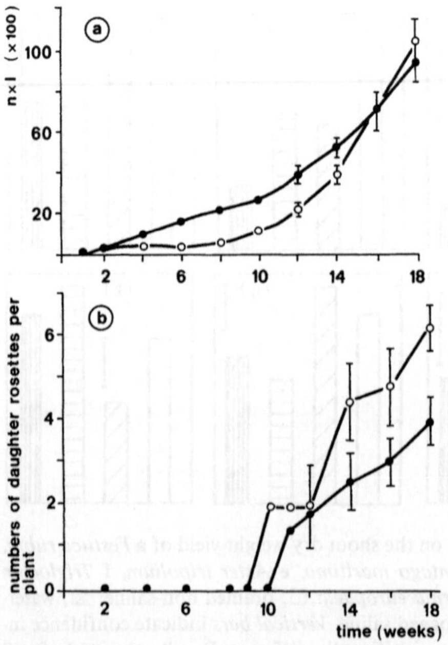

Fig. 7.3a, b. The effects of inundation with sea water as well as with fresh water on the biomass production (**a**) and the mean number of daughter rosettes per plant (**b**) of *Plantago maritima* grown separately in pots in the greenhouse. The biomass is approximated by the multiplication of the mean number of leaves per plant (n) and the mean length of the longest leaf (1). ○——○, inundated with sea water; ●——●, inundated with fresh water (controls). I 2S.E. n = 20. (Blom 1983)

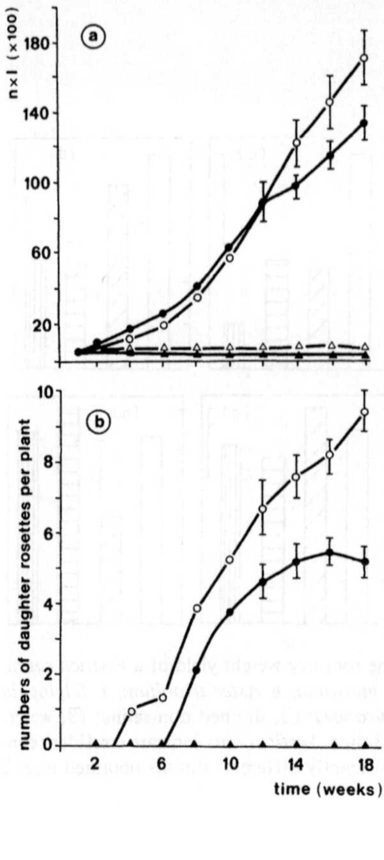

Fig. 7.4a, b. The effects of clipping and trampling on biomass production (**a**) and on the mean number of daughter rosettes per plant (**b**) of *Plantago maritima* grown separately in pots in the greenhouse. The biomass is approximated by the multiplication of the mean number of leaves per plant (*n*) and the length of the longest leaf per plant (*l*). △——△, clipping treatment; ○——○, trampling treatment; ▲——▲, combined clipping and trampling; ●——●, controls, I 2S.E. n = 20. (Blom 1983)

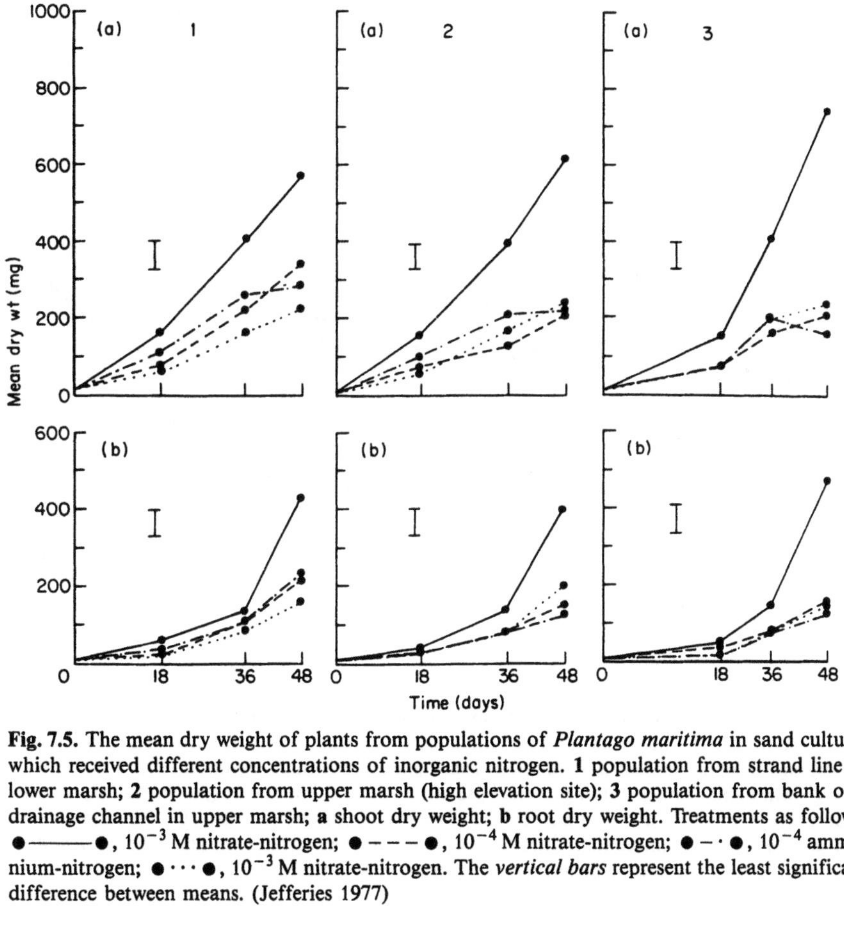

Fig. 7.5. The mean dry weight of plants from populations of *Plantago maritima* in sand cultures which received different concentrations of inorganic nitrogen. **1** population from strand line of lower marsh; **2** population from upper marsh (high elevation site); **3** population from bank of a drainage channel in upper marsh; **a** shoot dry weight; **b** root dry weight. Treatments as follows: ●———●, 10^{-3} M nitrate-nitrogen; ● – – – ●, 10^{-4} M nitrate-nitrogen; ● – · ●, 10^{-4} ammonium-nitrogen; ● · · · ●, 10^{-3} M nitrate-nitrogen. The *vertical bars* represent the least significant difference between means. (Jefferies 1977)

the upper levels of some coastal marshes the growth of certain species was limited by the availability of nitrogen. In a greenhouse experiment significant differences were found in growth response to nitrogen between populations of *P. maritima* (Jefferies 1977). Shoots of plants from the lower marsh populations showed significantly faster growth than those of plants from the upper marsh (Fig. 7.5). Significant differences in growth response were also found between populations within the upper marsh. These differences had developed between populations which were separated by a distance of less than 10 m (Fig. 7.6). Anylyses of variance showed significant population X treatment interactions, demonstrating genetic differentiation (Table 7.3). This phenomenon was also observed for other salt marsh species as *Aster tripolium*, *Salicornia europaea* and *Triglochin maritima*. Jerling (1985) reported on genetic differentiation in fitness-related characters in *Plantago maritima* along a gradient situated on a Baltic sea shore meadow. Grazing intensities and flooding frequencies varied along the gradient and differences in densities of *P. maritima* and in covering

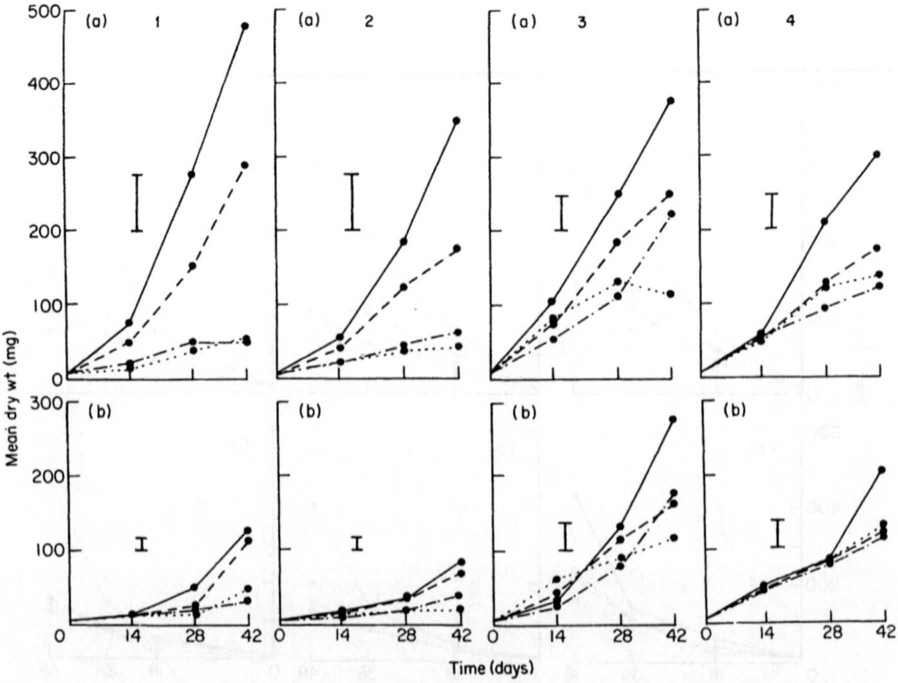

Fig. 7.6. The mean dry weight of plants from populations of *Plantago maritima* in sand cultures which received different concentrations of inorganic nitrogen. **1** population from bank of a drainage channel in upper marsh; **2** population from upper marsh (high elevation site); **3** plants with erect growth habit; **4** plants with prostrate growth habit; **a** shoot dry weight; **b** root dry weight. Treatments as follows: ●———●, 10^{-3} M nitrate-nitrogen; ●- - -●, 10^{-4} M nitrate-nitrogen; ●-·●, 10^{-4} M ammonium-nitrogen; ●···●, 10^{-5} M nitrate-nitrogen. The *vertical bars* represent the least significant difference between means. (Jefferies 1977)

of the adjacent vegetation were found. Results of demographic as well as experimental studies strongly suggest a genetically determined variation in seed size, germinability, growth patterns, and allocation of dry matter among plants of *P. maritima* occurring in the separated parts of the gradient. Selection due to grazing and competition resulted in genetic differentiation. Life-history traits of some genotypes appeared to be of adaptive value. Restricted gene flow and variation in male sterility also affected genetic differentiation in the gradient studied by Jerling (1988).

7.1.4 Conclusion

P. maritima is considered to be a highly variable species, especially in morphological characters as growth form (erect versus prostrate) and form of reproduction (generative versus vegetative). In spite of the evidence of a certain genetic differentiation in some physiological processes and the existence of

Table 7.3. Analysis of variance of the growth responses of *Plantago maritima* populations at Stiffkey Marsh (Britain) to different levels of nitrogen. For treatments see Figs. 7.5 and 7.6. (Results from Jefferies 1977)

Sources	*Plantago maritima*			*Plantago maritima* (upper marsh, high elevation site and drainage channel populations only)			*Plantago maritima* (erect and prostrate populations)		
	d.f.	Shoots significance level	Roots significance level	d.f.	Shoots significance level	Roots significance level	d.f.	Shoots significance level	Roots significance level
Treatments (T)	3	***	***	3	***	***	3	***	***
Harvests (H)	2	***	***	2	***	***	2	***	***
Population (P)	2	*	NS	1	*	***	1	***	***
TxH	6	***	***	6	***	***	6	NS	NS
PxT	6	**	NS	3	NS	*	3	NS	NS
PxH	4	NS	NS	2	NS	**	2	***	**
PxTxH	12	*	NS	6	NS	NS	6	NS	NS
Error	180			120			120		
Total	215			143			143		

*$P<0.05$; **$P<0.01$; ***$P<0.001$.
d.f. degrees of freedom.

ecotypes in separate areas, other results of experimental work on adjacent populations indicate that differences in morphological characters were due to phenotypic plasticity. Furthermore, the results described in the section showed that the reaction of *P. maritima* plants to flooding with sea water and trampling depended on the stage in the life cycle: relatively extreme variations in environmental factors may occur over distances which were too small for adaption by genetic differentiation. If selection acted on populations without reproductive isolation, selection favoured plasticity of *P. maritima*.

7.2 Ecotypic Differentiation in *Plantago major*

7.2.1 Introduction

A number of varieties of *Plantago major* have been described by taxonomists (e.g. Pilger 1937; see also Sagar and Harper 1964). Several papers deal with the heritable component of the characters in which these varieties differ (Groot and Boschhuizen 1970; Mølgaard 1976; Warwick and Briggs 1980; Van Dijk 1984). In order to evaluate these genetic differences between varieties as ecotypic differences, a relationship between characters and environment should be presented showing that these characters are adaptive for the genotypes under the reigning environmental conditions. Warwick and Briggs (1980) described differences in fitness of the lawn type and the roadside of type ssp. *major*, showing that each type is adapted only to its own specific environment.

A genetic analysis of some of the characters which are important for ecotypic differentiation has been made by Van Dijk (1984; Sect. 6.1). Genes, contributing to several of these characters, were found in the vicinity of the *Pgm*-1 locus, suggesting a gene complex around this locus.

Knowledge about the level of gene flow has been presented elsewhere (Sect. 6.1 and 6.5): in addition to short-distance gene flow, distinct long-distance seed transport occurs due to sticky seeds.

P. major is a predominantly selfing species and occasionally hybrids between different ecotypes will be formed. Because such hybrids may be important in the formation of genotypes adapted to habitats different from both parental habitats, the role of hybrids may be important in an evolutionary sense.

In a series of experiments the detailed adaptation of several ecotypes to their respective habitats was analyzed.

7.2.2 Origin and Characteristics of the Parental Plants

The 12 original parental plants were collected from various localities in The Netherlands (Van Dijk and Van Delden 1981). Six of the 12 plants: G_1, H_{19}, S_8, A_2, H_{44} and Z_2, were genetically analyzed (Van Dijk 1984) as far as leaf and inflorescence morphology and developmental characters were concerned. According to morphology and habitat the 12 plants were classified into five

Table 7.4. The parental types

Type	Mean number of seeds per capsule	Origin
ssp. *major*, lawn type		
G_1	9	Lawn, frequently mown
ssp. *major*, erect type		
G_7	9	Roadside, mown once a year, not trampled
H_{19}	11	Dike around sedimentary fields of a sugar factory, grazed by sheep
S_8	8	Sandy bank of a new-made lake
W_9	12	Dune valley near lake, poor in nutrients
ssp. *pleiosperma*, coastal type		
K_1	16	Coastal meadow, occasionally flooded by the sea
L_6	20	Sea dike, between basalt blocks
V_{13}	14	Muddy sea coast, frequently flooded
ssp. *pleiosperma*, inland type; small-seeded		
A_2	19	At the side of a newly made ditch, near the river Rhine
H_{44}	21	See H_{19}
Z_2	27	Riverbank, at the edge of a drinking pool for cattle
ssp. *pleiosperma*, inland type; large-seeded		
A_4	15	See A_2

groups (Table 7.4). The mean number of seeds per capsule (the main distinctive trait for the subdivision into subspecies) is also given in Table 7.4, together with a short description of the original habitat.

The plants were representative for the populations from which they were sampled. Only in H (a mixed ssp. *major*/ssp. *pleiosperma* population) and in A (containing two types, resembling A_2 or A_4) did a distinct intrapopulational variability for relevant morphological characters exist. In the Z population, seed number per capsule varied between 20 and 30. G_1 and G_7 were collected from different populations in the same area.

7.2.3 Results

Experiments were performed to analyze the genetically determined differences in strategy between the parental types. A greenhouse experiment showed differences in vegetative growth and reproductive output of the various types under optimal conditions. These conditions were more or less comparable with one of the field situations, a river bank, though the latter had a strongly variable growth season with a limited length.

In a transplantation experiment survival and reproduction were measured in the field situations by implanting seedlings. Additionally, a single experiment was carried out, starting with seeds: the garden trampling experiment. Finally, intra-specific competition under various conditions was studied in the greenhouse. In all cases the progenies of plants from different origin and of distinctly different morphology were used; in some experiments F_1's and F_2's of crosses between different types were included.

7.2.3.1 The Greenhouse Experiment

Pots with turf, each containing nine seeds of a parental type, F_1 or reciprocal F_2, were set at 25 °C and 16 h light a day. Germination occurred in 4 or 5 days (large and small seeds, respectively) and the germination percentage was 100%. Four weeks after sowing, plants were measured and transplanted to separate pots. After another 3 weeks they were measured again. During the rest of the experimental period, until 20 weeks after sowing, inflorescences were harvested as soon as the capsules were ripe.

Four weeks after sowing all plants were still in the vegetative juvenile stage. A measurement of growth was performed by taking leaf length and number of leaves. It was not possible to measure biomass dircectly, but available data (Kuiper 1983a) indicated that ssp. *pleiosperma* had a higher relative growth rate than ssp. *major*. Leaf length was considered relevant for the ability of competition for light; the ratio between leaf length and seed weight was considered relevant for relative growth rate. This ratio (Table 7.5) was considerably higher for ssp. *pleiosperma* than for ssp. *major*; in spite of the low seed weight the first subspecies showed the longest leaves after 4 weeks. A_4 had a high seed weight and a high growth rate and presented the largest type of the whole set. After four weeks the third or fourth leaf was usually the longest one (Table 7.5); in the lawn type G_1, the fifth leaf was the longest one. Obviously G_1 produced more and shorter leaves than the other types. Because the petiole of G_1 was also relatively short, one may conclude that G_1 was building up a close rosette. F_1 plants usually were intermediate in juvenile growth characters; the maternal effect of seed weight was strong in cases where the parents had distinctly different seed weights. The F_2 data are partially presented because no parental seed weights were determined.

Seven weeks after sowing almost all types were flowering; vegetative development had been completed and all types had produced their longest leaf. A considerable part of the effort was invested in reproduction. Only G_1, and to a lesser extent V_{13}, continued with leaf production and maintained their rosettes. The final maximum leaf length of the various types was not directly related with the relative growth rate. The moment of slowing down vegetative growth in favour of reproductive investment also was important for the ultimate leaf size: within their groups W_9, V_{13} and Z_2 were smaller than the other representatives (Table 7.7). Ssp. *pleiosperma* tended to be larger than ssp. *major*, especially for lamina length. Petioles of G_1 were very short. In the F_1's and F_2's the leaf morphology and dimensions were sometimes intermediate between their parents, sometimes close to one of them.

The slowing down of growth of the central rosette was associated with the formation of inflorescences or occasionally with the formation of secondary

Table 7.5. The greenhouse experiment: juvenile growth characters measured 4 weeks after sowing

Type	n	No. of longest leaf[a] 3	4	5	Length of longest leaf (mm)	Seed weight[b]	Growth rate[c]
G_1	9	–	2	7	39	0.285	137
G_7	9	–	9	–	56	0.275	204
H_{19}	10	6	4	–	44	0.242	182
S_8	9	1	8	–	46	0.281	164
W_9	9	–	9	–	40	0.250	160
K_1	9	2	7	–	58	0.177	328
L_6	12	5	7	–	74	0.204	363
V_{13}	12	–	12	–	63	0.236	267
A_2	10	9	1	–	53	0.188	282
H_{44}	9	–	9	–	67	0.152	441
Z_2	10	$3\frac{1}{2}$	$6\frac{1}{2}$	–	59	0.146	404
A_4	9	–	9	–	102	0.305	334
F_1's[d]							
G_1Z_2	5	–	5	–	70	0.300	233
Z_2G_1	3	–	2	1	42	0.199	211
$H_{19}H_{44}$	4	–	4	–	93	0.278	335
$H_{44}H_{19}$	4	–	4	–	86	0.264	326
A_2S_8	8	3	5	–	58	0.241	241
$V_{13}G_1$	6	–	5	1	49	0.272	180
L_6V_{13}	6	–	6	–	75	0.110	682
Z_2W_9	8	3	5	–	36	0.160	225
S_8W_9	9	$3\frac{1}{2}$	$5\frac{1}{2}$	–	60	0.472	127
F_2's[d]							
G_1Z_2	9	2	$6\frac{1}{2}$	$\frac{1}{2}$	48		
Z_2G_1	9	1	7	1	39		
$H_{19}H_{44}$	9	–	9	–	88		
$H_{44}H_{19}$	9	–	8	1	91		
A_2S_8	9	$4\frac{1}{2}$	$4\frac{1}{2}$	–	85		
S_8A_2	9	4	5	–	87		
$V_{13}G_1$	9	–	$7\frac{1}{2}$	$1\frac{1}{2}$	49		

[a] The oldest leaf (neglecting the cotyledons is no. 1, etc. If two leaves were equally long, $\frac{1}{2}$ has been given to both leaf numbers.
[b] Mean weight (mg) of the seeds in the whole batch of the appropriate cross.
[c] The ratio of length of the longest leaf and seed weight (mm mg^{-1}).
[d] Crossing progenies are indicated by the subsequent names of female and male parent.
Comparison of all means:

No. of longest leaf (F = 13.8)

A_2 H_{19} L_6 Z_2 K_1 S_8 G_7 W_9 V_{13} H_{44} A_4 G_1

Length of longest leaf (F = 28.1)

G_1 W_9 H_{19} S_8 A_2 G_7 K_1 Z_2 V_{13} H_{44} L_6 A_4

Table 7.6. The greenhouse experiment: inflorescences and secondary rosettes produced 7 weeks after sowing

Type	n	No. of inflorescences																				No. of sec. rosettes					
		Flowering										Developing															
		0	1	2	3	4	5	6	7	8	9	0	1	2	3	4	5	6	7	8	9	0	1	2	3	4	5
G_1	9	–	6	3	–	–	–	–	–	–	–	–	–	3	6	–	–	–	–	–	–	5	2	2	–	–	–
G_7	9	4	5	–	–	–	–	–	–	–	–	–	–	8	1	–	–	–	–	–	–	4	4	1	–	–	–
H_{19}	10	9	1	–	–	–	–	–	–	–	–	9	1	–	–	–	–	–	–	–	–	8	2	–	–	–	–
S_8	9	8	1	–	–	–	–	–	–	–	–	–	4	5	–	–	–	–	–	–	–	9	–	–	–	–	–
W_9	9	1	3	5	–	–	–	–	–	–	–	–	–	7	2	–	–	–	–	–	–	5	3	1	–	–	–
K_1	9	5	3	1	–	–	–	–	–	–	–	–	–	–	3	5	1	–	–	–	–	–	1	2	5	1	–
L_6	12	2	–	5	4	1	–	–	–	–	–	–	–	1	4	2	4	–	1	–	–	–	4	1	5	2	–
V_{13}	12	–	9	3	–	–	–	–	–	–	–	–	–	–	1	1	5	1	2	2	–	–	–	6	4	2	–
A_2	10	–	–	3	4	2	1	–	–	–	–	–	–	8	2	–	–	–	–	–	–	7	1	2	–	–	–
H_{44}	9	–	–	2	1	6	–	–	–	–	–	–	–	1	7	1	–	–	–	–	–	–	4	3	2	–	–
Z_2	10	–	1	2	3	4	–	–	–	–	–	–	–	5	5	–	–	–	–	–	–	8	2	–	–	–	–
A_4	9	–	–	–	–	–	–	4	1	2	2	–	–	–	4	1	2	–	1	–	1	2	3	4	–	–	–
F_1's																											
G_1Z_2	8	–	–	–	4	2	2	–	–	–	–	–	–	3	5	–	–	–	–	–	–	6	2	–	–	–	–
$H_{19}H_{44}$	8	–	–	–	6	2	–	–	–	–	–	–	–	6	2	–	–	–	–	–	–	–	4	3	1	–	–
A_2S_8	8	1	–	2	4	1	–	–	–	–	–	–	–	8	–	–	–	–	–	–	–	3	4	1	–	–	–
$V_{13}G_1$	6	1	2	2	1	–	–	–	–	–	–	–	–	1	5	–	–	–	–	–	–	2	1	3	–	–	–
L_6V_{13}	6	–	–	6	–	–	–	–	–	–	–	–	–	–	2	2	–	1	–	1	–	–	–	2	4	–	–
Z_2W_9	8	1	3	2	1	1	–	–	–	–	–	–	1	6	1	–	–	–	–	–	–	5	3	–	–	–	–
S_8W_9	9	–	2	5	2	–	–	–	–	–	–	–	–	7	2	–	–	–	–	–	–	4	3	2	–	–	–
F_2's																											
G_1Z_2	18	4	1	2	9	1	1	–	–	–	–	–	4	9	5	–	–	–	–	–	–	16	2	–	–	–	–
$H_{19}H_{44}$	18	2	1	4	3	7	1	–	–	–	–	–	–	10	6	2	–	–	–	–	–	2	5	9	1	–	1
A_2S_8	18	–	1	3	10	4	–	–	–	–	–	–	1	10	6	1	–	–	–	–	–	7	8	3	–	–	–
$V_{13}G_1$	9	1	3	5	–	–	–	–	–	–	–	–	–	6	2	1	–	–	–	–	–	6	1	1	1	–	–

Table 7.6 (continued)

Comparisons of all means:
No. of flowering inflorescences ($F = 54.5$)
H_{19} S_8 G_7 K_1 V_{13} G_1 W_9 L_6 Z_2 A_2 H_{44} A_4

No. of developing inflorescences ($F = 24.1$)
H_{19} S_8 G_7 A_2 W_9 Z_2 G_1 H_{44} K_1 L_6 A_4 V_{13}

No. of secondary rosettes ($F = 17.2$)
S_8 H_{19} Z_2 A_2 W_9 G_1 G_7 A_4 H_{44} L_6 K_1 V_{13}

rosettes. In Table 7.6 the number of flowering spikes, developing spikes and secondary rosettes are presented. After 7 weeks the inland *pleiosperma* types, especially A_4, were already abundantly flowering. The coastal *pleiosperma* types showed a considerably larger number of secondary rosettes than the other *pleiosperma* types. The coastal *pleiospermas* surpassed the other groups because the secondary rosettes also produced inflorescences. The inland *pleiosperma* types slowed down production of inflorescences, while A_4 still had a high production level. In the F_1's the characters of the reproductively most active parent dominated in most cases. In the F_2's, especially in $G_1 \times Z_2$ and $H_{19} \times H_{44}$, a return towards a midparent value of reproductive investment was noticed, probably due to gene segregation.

As expected, seed production started in the early flowering inland *pleiosperma* types (Table 7.8). In the first week of seed ripening, 12 weeks after sowing, Z_2 produced the largest number of seeds and A_4 the greatest seed biomass. A few weeks later the ssp. *major* and coastal *pleiosperma* types also started to ripen seeds on a large scale. At the end of the growing period 2000 – 2500 mg of seed biomass had been produced by most types. The early starting Z_2, which was rather small because of its low seed weight, could produce only 1000 mg of seed biomass per plant. G_1, which continued producing new leaves, also stopped at 1000 mg of seed biomass; to a lesser extent V_{13} also spent more energy in leaf production at the cost of seed formation. A delay in seed ripening of G_1 and V_{13} was noticed, due to shading of the prostrate spikes by new leaves.

The final number of seeds produced at the end of the growing period varied between 15000 per plant (ssp. *pleiosperma* types A_2, A_4, H_{44} and L_6) and less than 5000 per plant (G_1). The F_1's usually produced more seeds than the parents, the F_2's being again closer to the midparent value. Seed size differed up to a factor 3.6 between the largest (S_8) and the smallest (Z_2) type (Table 7.9). Ssp. *major* produced the largest seeds and maintained its seed weight reason-

Table 7.7. The greenhouse experiment: leaf dimensions (mm) measured seven weeks after sowing (mean values). Reciprocals of F_1 and F_2 have been combined

Type	n	Petiole length	Lamina length	Lamina width	Total length
G_1	9	28	73	52	101
G_7	9	62	88	64	150
H_{19}	10	55	88	62	142
S_8	9	52	83	61	135
W_9	9	46	74	55	120
K_1	9	80	106	68	186
L_6	12	83	126	76	209
V_{13}	12	55	95	64	149
A_2	10	74	118	65	192
H_{44}	9	70	108	76	178
Z_2	10	54	92	54	146
A_4	9	76	124	69	199
F_1's					
G_1Z_2	8	38	84	56	122
$H_{19}H_{44}$	8	86	118	75	204
A_2S_8	8	71	110	69	180
$V_{13}G_1$	6	37	72	52	109
L_6V_{13}	6	68	110	70	178
Z_2W_9	8	45	75	49	119
S_8W_9	9	49	90	67	139
F_2's					
G_1Z_2	18	35	75	50	109
$H_{19}H_{44}$	17	75	110	71	185
A_2S_8	18	74	108	69	182
$V_{13}G_1$	9	40	79	56	118

Comparisons of all means:

Petiole length (F = 56.0)

G_1 W_9 S_8 Z_2 H_{19} V_{13} G_7 H_{44} A_2 A_4 K_1 L_6

Lamina length (F = 44.0)

G_1 W_9 S_8 H_{19} G_7 Z_2 V_{13} K_1 H_{44} A_2 A_4 L_6

Lamina width (F = 18.2)

G_1 Z_2 W_9 S_8 H_{19} V_{13} G_7 A_2 K_1 A_4 H_{44} L_6

Total length (F = 54.8)

G_1 W_9 S_8 H_{19} Z_2 V_{13} G_7 H_{44} K_1 A_2 A_4 L_6

Table 7.8. The greenhouse experiment: cumulative amounts of seed production after 12,13, 14, 16 and 20 weeks [numbers (n) and biomass (mg); mean values per plant]

Type	Week 12		Week 13		Week 14		Week 16		Week 20	
	n	mg	n	mg	n	mg	n	mg	n	mg
G_1	–	–	–	–	41	11	866	205	4551	1012
G_7	–	–	219	60	1009	269	4495	1180	8463	1987
H_{19}	–	–	–	–	127	33	2362	606	12811	2635
S_8	–	–	45	16	408	146	3195	1114	6372	2123
W_9	–	–	364	92	1360	323	5764	1265	8644	1796
K_1	–	–	77	16	911	152	8106	1280	13211	1957
L_6	–	–	810	136	2361	397	9146	1441	16518	2510
V_{13}	–	–	–	–	168	32	3676	663	10621	1675
A_2	118	24	1232	235	3617	633	10167	1657	16702	2493
H_{44}	323	54	2220	390	4836	769	12152	1763	14687	2083
Z_2	1118	115	3109	312	4876	496	7926	814	12093	1188
A_4	716	206	2602	706	4726	1167	12805	2357	15108	2604
F_1's										
G_1Z_2	567	96	1779	298	2310	383	4959	781	10012	1533
$H_{19}H_{44}$	376	111	1292	500	3693	1156	8067	2179	11946	2935
A_2S_8	155	51	1000	355	2012	661	5703	1621	9830	2534
$V_{13}G_1$	–	–	111	37	111	37	1995	581	6073	1487
L_6V_{13}	–	–	–	–	98	29	2313	566	9972	2226
Z_2W_9	65	11	655	104	1266	202	4123	659	7167	1077
S_8W_9	105	44	398	142	1591	586	5358	1756	8203	2631
F_2's										
G_1Z_2	86	14	511	89	936	168	3611	635	6055	1021
$H_{19}H_{44}$	151	39	1166	322	2585	699	7462	1728	10208	2273
A_2S_8	143	45	1067	319	2917	791	7846	1965	11590	2631
$V_{13}G_1$	–	–	–	–	–	–	1786	429	5968	1358

Comparisons of all means:

n after 12 weeks ($F = 8.5$)

G_1 G_7 H_{19} S_8 W_9 K_1 L_6 V_{13} A_2 H_{44} A_4 Z_2

n after 13 weeks ($F = 18.3$)

G_1 H_{19} V_{13} S_8 K_1 G_7 W_9 L_6 A_2 H_{44} A_4 Z_2

n after 14 weeks ($F = 19.7$)

G_1 H_{19} V_{13} S_8 K_1 G_7 W_9 L_6 A_2 A_4 H_{44} Z_2

Table 7.9. The greenhouse experiment: mean seed weights (μg) over the whole production period, over the first 2 weeks of production and the last 2 weeks of production

Type	Whole period	First 2 weeks	Last 2 weeks
G_1	222	238	218
G_7	235	267	185
H_{19}	206	253	185
S_8	333	358	320
W_9	208	238	174
K_1	148	167	148
L_6	152	168	132
V_{13}	158	189	133
A_2	149	191	113
H_{44}	142	176	125
Z_2	98	100	74
A_4	172	271	107
F_1's			
G_1Z_2	153	168	136
$H_{19}H_{44}$	246	387	176
A_2S_8	258	355	201
$V_{13}G_1$	245	318	233
L_6V_{13}	223	268	210
Z_2W_9	150	159	143
S_8W_9	321	357	323
F_2's			
G_1Z_2	169	174	121
$H_{19}H_{44}$	227	276	188
A_2S_8	227	299	227
$V_{13}G_1$	228	240	234

ably well during the ripening period. In ssp. *pleiosperma* the weight was lower and it decreased in some cases to a considerable extent, especially in A_4 (Table 7.9; the first seeds were over 250 μg and the last seeds at the end of the period were less than 100 μg). In the F_1's and F_2's seed weights were usually higher than the midparent values and sometimes higher than the parents.

After finishing seed production, most plants exhibited a "winter rosette" appearance (a circle of small leaves at the basis of each rosette, while the large leaves died). Only G_1 and V_{13} still had fresh green leaves. Most A_4 plants died, possibly due to a lack of investment in winter survival. All plants released seeds easily after touching the ripened spikes except for W_9, whose capsules had to be cracked like nuts to release seeds.

7.2.3.2 The Transplantation Experiment

Three-week-old juveniles were implanted in May into four different plots in which *P. major* occurred naturally: a path, a lawn, a riverbank and a dune val-

Table 7.10. Survival in the transplantation experiment. Numbers implanted in May and numbers of survivors in the end of August; mean survival percentages for the two subspecies and F_1's and F_2's. In the lawn all plants survived

Type	Path		Riverbank		Dune valley	
	n	Surv.	n	Surv.	n	Surv.
G_1	24	12	24	14	32	27
G_7	24	4	24	18	32	12
H_{19}	24	9	24	14	32	10
S_8	24	8	24	16	32	17
W_9	24	11	24	13	32	16
K_1	16	0	24	13	32	11
L_6	12	0	24	9	32	16
V_{13}	24	1	24	6	32	22
A_2	24	2	24	14	32	7
H_{44}	24	2	24	10	32	20
Z_2	16	0	24	14	32	11
A_4	18	1	24	14	32	17
F_1's						
G_1Z_2	24	11	24	16	32	25
$H_{19}H_{44}$	24	10	24	15	32	21
A_2Z_8	16	1	24	13	32	17
F_2's						
G_1S_2	96	17	96	55	128	91
$H_{19}H_{44}$	96	18	96	58	128	90
A_2S_8	96	17	96	62	128	79
ssp. *major*		37%		62.5%		51%
ssp. *pleiosperma*		4%		48%		46%
F_1's		31%		61%		66%
F_2's		18%		61%		68%

Comparison of all group means:
Survival at the path ($F = 9.9$, $P < 0.001$)
ssp. *pleio.* F_2's F'_1s ssp. *major*

Survival at the riverbank ($F = 2.8$) and dune valley ($F = 1.7$): no significant differences between groups.

ley. The position of the plants (all types had a fixed position within each square meter) could be recovered by laying a plastic sheet with labelled holes (1 m^2) over the area. One month after implantation nearly all plants were found in good conditions. At the end of August, only a minor part of the plants on the path were alive; surviving plants were mainly ssp. *major* types and progenies of the crosses (Table 7.10). In the other three plots plants only died because of disturbance by cattle, rabbits, moles etc.; in the lawn all plants survived.

In the first year only a few plants produced a few, very small, spikes on the path and all plants died in late summer after long drought. In the garden trampling experiment an estimate for seed production in such a path situation is given.

Considerable survival was found in the lawn and the dune valley: 50 to 60%, no inflorescences were formed during the first season. The riverbank plants, on the contrary, flowered abundantly in the first summer. The production of spikes was more or less comparable with seed production in the greenhouse experiment. The relative production of inflorescences at the end of the season was about 1 : 2.3 : 3 : 2.1 for ssp. *major*, ssp. *pleiosperma*, F_1's and F_2's, respectively (Table 7.11). The similar ratios in seed weight (in the greenhouse), after

Table 7.11. Seed production in the transplantation experiment. Number of inflorescences at the end of the first season at the riverbank; number of surviving and flowering plants in the next year in the dune valley (quantitatively) and the lawn (qualitatively)

Type	Riverbank: mean no. of inflorescences per plant	Dune valley		Lawn[a]	
		Surv.	Flow.	Surv.	Flow.
G_1	0.6	25	3	+	+ +
G_7	1.2	9	–	–	
H_{19}	1.3	5	–	–	
S_8	2.0	10	–	–	
W_9	1.2	16	1	+	+
K_1	2.6	11	–	–	
L_6	2.0	6	–	–	
V_{13}	3.5	14	4	–	
A_2	3.6	5	1	–	
H_{44}	3.7	15	4	–	
Z_2	2.2	11	–	–	
A_4	3.6	2	1	–	
F_1's					
G_1Z_2	4.1	21	6	+	+
$H_{19}H_{44}$	4.0	15	–	+	+
A_2S_8	3.6	13	–	+	+
F_2's					
G_1Z_2	1.9	77	24	+	+
$H_{19}H_{44}$	2.7	44	2	+	–
A_2S_8	3.6	66	6	+	–

Means per group:		Comparison of all group means:
		No. of inflorescences at the riverbank
ssp. *major*	1.3	($F = 12.3$, $P < 0.001$)
ssp. *pleiosperma*	3.0	ssp. *major* F_2's ssp. *pleio.* F_1's
F_1's	3.9	Flowering in the dune valley ($F = 0.9$): no significant differences between groups.
F_2's	2.7	

[a] +, Only a part of the plants; + +, all plants present.

the 15th week, were 1 : 2.1 : 2.5 : 2.3. Assuming the same weight of individual seeds for greenhouse and riverbank plants, the ratio of seed number produced in the last situation would be about 1 : 3.3 : 2.4 : 2.5.

In the next year all riverbank plants had disappeared. A fraction of the lawn and dune valley plants were still alive and in some cases produced seeds. The number of inflorescences produced in the dune valley, together with a qualitative survey of the seed production in the lawn situation, is presented in Table 7.11. In the dune valley no clear differences between the types could be established, while in the lawn, the lawn type G_1 was distinctly most successful. The tendency of high survival and seed production of the F_1's and F_2's could also be noticed.

7.2.3.3 The Garden Trampling Experiment

A number of 100 seeds of each of the 12 parental types were sown as a mixture in summer. The area (2 m^2) was thoroughly trampled about twice a week during summer, autumn and the following spring. The established plants were identified by electrophoresis.

Several dozens of *P. major* seedlings were noticed on the experimental plot, togehter with a number of grass species, mainly *Poa annua*, and some herbs. The *P. major* plants which reached sufficient biomass to be used for electrophoresis were identified for allozyme genotype; they mainly belonged to ssp. *major*. Members of all ssp. *major* types, except G_7, survived and also some V_{13} plants. In the next spring and summer the most vital plants were H_{19} and G_1. H_{19} was the only type which produced seeds (Table 7.12); its spikes remained erect without severe damage, also after repeated trampling. The leaves also were relatively resistant to trampling. The spikes of G_1 remained small and were damaged by trampling; no seed production occurred. Other types as S_8, W_9 and V_{13}, survived as small non-flowering plants.

7.2.3.4 The Competition Experiment

In competition experiments in the greenhouse seeds of G_1, H_{19}, K_1, Z_2 and A_4 were germinated in containers with turf at regular distances of 2 cm in a 12×12 grid, in such a way that each type was surrounded by the other four types. To avoid margin effects the inner square of 10×10 plants was used (20 plants per type per container). Germination was 100%.

As expected (greenhouse experiment) A_4 was dominating under unlimited growth conditions. H_{19} and K_1 could slightly compete with A_4; the other 2 types were severely shaded and did not flower except for a single Z_2 plant. Next to shading, root competition could be important. The produced seeds mainly originated from A_4: H_{19} and K_1 also produced a considerable number of seeds (Table 7.13).

Table 7.12. Genotypes identified in the garden trampling experiment. Original number of seeds: 100 per type

Type	Number	Seed-producing
G_1	4	–
G_7	–	–
H_{19}	6	4
S_8	4	–
W_9	5	–
K_1	–	–
L_6	–	–
V_{13}	9	–
A_2	–	–
H_{44}	–	–
Z_1	–	–
A_4	–	–

Table 7.13. The results of the competition experiment: survival, shoot dry weight per plant, seed production per plant and product of survival and seed production

Type	Survival (of 20)	Mean shoot dry weight (mg per plant)	Mean seed production (numbers per plant)	Product of survival and seed production
1. Unlimited growth				
G_1	13	101	–	–
H_{19}	11	304	546	300
K_1	13	369	448	291
Z_2	1	330	363	18
A_4	14	289	1457	1020
2. Cut at 2 cm height				
G_1	20	414	283	283
H_{19}	18	18	–	–
K_1	19	53	2	2
Z_2	10	16	–	–
A_4	20	27	1	1
3. Trampled				
G_1	9	100	–	–
H_{19}	14	106	581	407
K_1	1	139	–	–
Z_2	–	–	–	–
A_4	1	22	–	–
4. High salt concentration				
G_1	8	126	200	80
H_{19}	6	81	67	20
K_1	14	529	1029	720
Z_2	5	80	200	50
A_4	13	702	1981	1288

The frequently clipped plant cover was dominated by G_1; it could retain enough leaf biomass to realize a reasonable seed production. A fraction of the other types survived but only a few small spikes with few seeds were produced.

As expected, in the trampled container G_1 and H_{19} survived reasonably well. Almost all plants of the ssp. *pleiosperma* died early. Only H_{19} produced seeds, due to the elasticity and toughness of the inflorescences; G_1 spikes were damaged fatally. The addition of NaCl to the soil had no strong influence on the competition ratios of the five types. A_4 was again dominating. The coastal type, K_1, was more successful than in the non-saline container. All types showed succulence and survived. These findings were in agreement with observations on a set of types implanted in a daily flooded coastal meadow; the plants survived during summer and became very succulent. The better tolerance for salinity of K_1, compared with the other four types, is consistent with the findings of D. Kuiper (1984), who established a better salt tolerance in K_1 than in G_1 and Z_2.

7.2.3.5 Physiological Ecotype Differentiation

Mineral Nutrition as an Environmental Factor. Plants of four inbred lines, A_3, G_1, K_1 and Z_2, were grown on two nutrient solutions, 100% (1/4 strength Hoagland solution) and 2%; part of the plants were transferred from one to the other growth condition, halfway through the time of the experiment (Kuiper 1982). Several physiological characteristics were followed during the experiment. The results indicated that the involved genotypes probably typified a large scale of ecotype differentiation in *P. major.* The 100% plants of the G_1 line (ssp. *major*) had a lower relative growth rate (ca. 135 mg g^{-1} day^{-1}) than the other genotypes, belonging to ssp. *pleiosperma* (ca. 185 mg $g^{-1} day^{-1}$. Low relative growth rates were accompanied by low root respiration rates (Kuiper 1983a), low root ATPase activities (Kuiper 1982), and low nitrate reductase activities in shoot and root tissue (D. Kuiper 1984). Similar results were obtained in the 2% plants. In conclusion, the noticed genotype differentiation could be translated to the field situation in terms of ecotype differentiation. Besides this observation, genotypic differences in phenotypic plasticity, as a response to transfer to the other nutrient condition, enabled the plants to cope with fluctuations in mineral strength. Plants of the slow-growing G_1 line were hardly capable in performing plastic responses. A positive correlation between relative growth rate and the rapidity of the appearance of a plastic response (and completion of the plastic response) was demonstrated (Kuiper 1982, 1983a).

Nitrogen Source as an Environmental Factor. The above-mentioned genotypic differences in phenotypic plasticity were underlined by results of experiments, in which plants of the four lines were subjected to different nitrogen sources (Fig. 7.7A). The lack of plastic response of the shoot to root ratio to different N sources was obvious in plants of the G_1 line. Plants of the other three lines

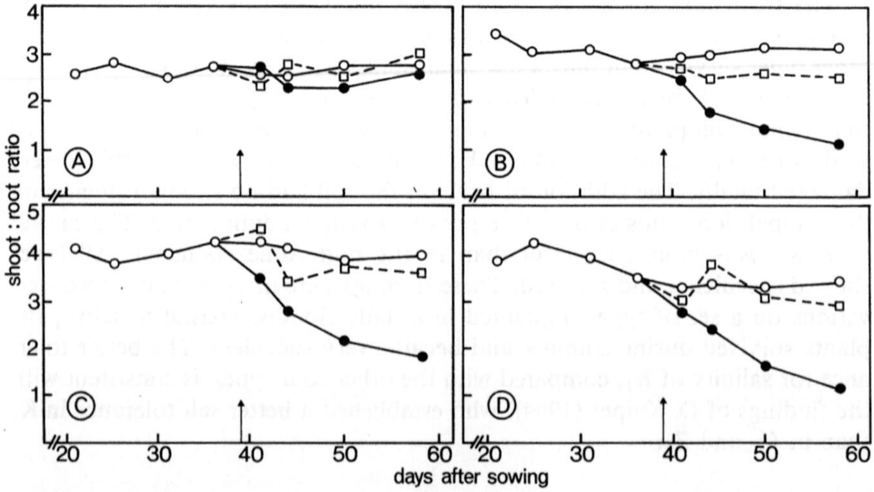

Fig. 7.7. Response of the shoot to root ratio of four lines of *Plantago major* (**A** ssp. *major,* line G_1: **B, C** and **D**; ssp. *pleiosperma,* **B** line K_1 **C** line A_3; **D** line Z_2) to a change in N nutrition, 39 days after sowing. ○, continued treatment with 2 mM NO_3^-; ●, change to no nitrogen; □, change to 2 mM NH_4^+

(Fig. 7.7 B, C, D) showed quick responses of the shoot to root ratio after transfer to a −N solution.

The responses of the nitrate reductase activity in the shoot (Fig. 7.8) were quite similar to the response of the shoot to root ratio, but the differences between the lines presented a more gradual response; the slow growing line G_1 (Fig. 7.8) had the slowest response and the quick-growing Z_2 (Fig. 7.8) showed the most rapid response.

Salt as an Environmental Factor. The responses of the Mg^{2+} (Fig. 7.9) and the Ca^{2+}-dependent root ATPase activities to 50 mM NaCl were compared with the previously described responses of physiological traits to a change in an environmental factor. In Fig. 7.9 A plants of the G_1 line again showed a lack of response. Plants of the other two tested lines (K_1, Fig. 7.9 B, and Z_2, Fig. 7.9 C) showed quick responses of ATPases to salinity.

From the results on the measurement of photosynthesis and chlorophyll (Kuiper 1984) and results on growth (D. Kuiper and B. Stuiver, unpubl. data) plants of line K_1 were more salt-resistant than individuals of the other lines.

Light as an Environmental Factor. Two genotypes belonging to ssp. *major* were grown from seeds, collected in an open and exposed habitat, and from seeds in a shaded habitat. Seeds were germinated and grown at two light intensities. Some of the plants were subjected to a transfer from one to the other light condition. Measurements on growth, respiration etc. (Kuiper and Smid 1985; Kuiper and Staal 1986) evidently showed ecotypic differentiation to light. The

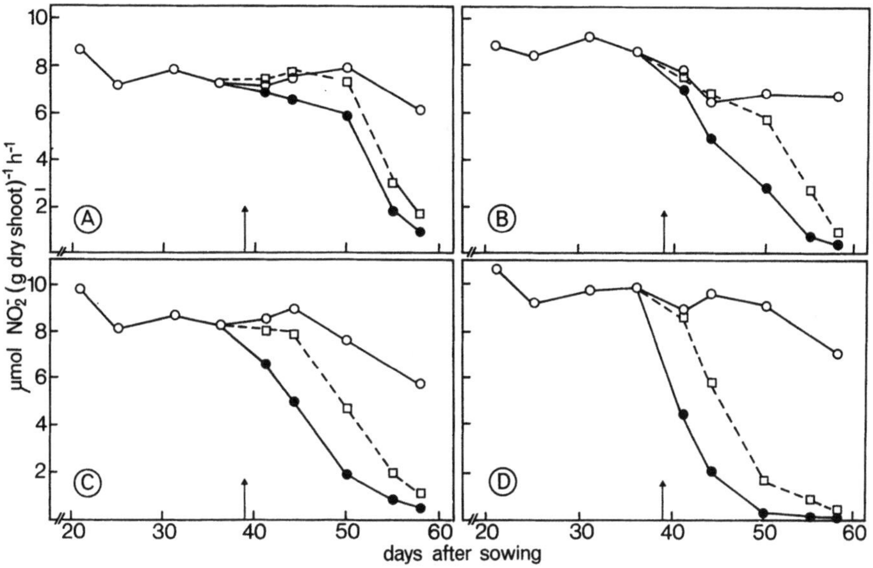

Fig. 7.8. Response of the nitrate-reductase activity of the shoots of four lines of *Plantago major* (see Fig. 7.7) to a change in N nutrition. Symbols as in Fig. 7.7

possibility to react plastically was constricted to the very young stage in these plants, confirming the results observed in line G_1, which also belonged to ssp. *major.*

7.2.4 Discussion

7.2.4.1 Alternatives in Life Cycle Strategy

A wide array of ecotypes occurred within the studied types of *P. major*; the two subspecies were readily recognizable and many other types were also distinctly adapted genotypes. A descriptive subdivision into numerous varieties makes no sense because each type was slightly different; an apparently continuous multivaried system was evident. A better approach is the analysis of alternatives which can be realized by selection pressure; alternatives to utilize limited resources in order to maximize fitness under the specific environmental conditions. Alternatives could be realized at successive developmental stages; sometimes they were qualitative, sometimes quantitative in nature. A list of the alternatives is shown in Table 7.14; the seven alternatives will be discussed in detail below.

1. Germination vs. Staying Dormant. *P. major* germinates at relatively high temperatures (Blom 1978); germination will occur only at open sites where

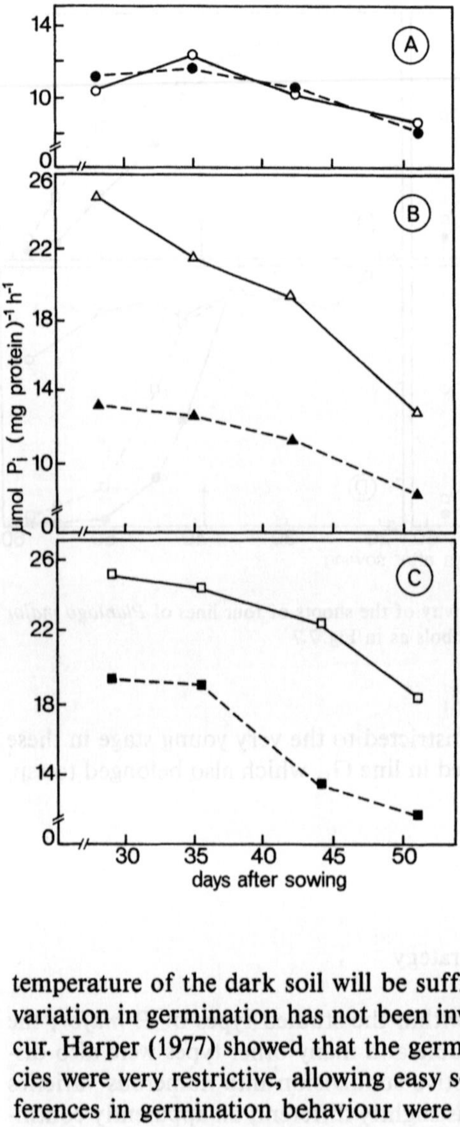

Fig. 7.9. Response of the Mg^{2+}-dependent root ATPase activity of three lines of *Plantago major* (**A** ssp. *major*, line G_1; **B** and **C** ssp. *pleiosperma*; **B** line K_1; **C** line Z_2) to 50 mM NaCl, added to the nutrient solution 39 days after sowing. *Open symbols* no NaCl added; *closed symbols* 50 mM NaCl added

temperature of the dark soil will be sufficiently high in early spring. Genetic variation in germination has not been investigated, but it is quite likely to occur. Harper (1977) showed that the germination conditions for *Plantago* species were very restrictive, allowing easy selection. In *P. lanceolata* genetic differences in germination behaviour were described earlier (Van Dijk 1985).

2. Low vs. High Juvenile Growth Rate. A bimodal distribution of growth rates was observed, coinciding with the two ssp. *major* and *pleiosperma*. Growth rates based upon leaf length over seed weight could be directly related to the growth rate of the maximum leaf length. Growth rate could be associated with adaptive properties. In ruderal conditions species with a high growth rate are observed, in a competitive situation a low growth rate is more frequent (Grime and Hunt 1975). Burden and Harper (1980) found a high level of variation in relative growth rate within populations of *Trifolium repens*; they concluded that differences in growth rate could be ascribed to genetic differences within

Table 7.14. Classification of the 12 parental types according to their strategy components

Alternatives	Condition	G_1	G_7	H_{19}	S_8	W_9	L_6	K_1	V_{13}	A_2	H_{44}	Z_2	A_4
1. Germination behaviour[a]		–	–	–	–	–	–	–	–	–	–	–	–
2. Juvenile growth	Slow,	+	+	+	+	+							
	quick						+	+	+	+	+	+	+
3. Leaves	Tall,		+	+	+	+	+	+		+	+	+	+
	closed rosette	+							+				
4. Switch to generative growth	Early,									+	+	+	+
	medium,		+		+	+	+	+					
	late	+		+					+				
5. Secondary rosettes	Few,	+	+	+	+	+				+		+	+
	many						+	+	+		+		
6. Leaf growth during flow. period	Much,	+							+				
	little		+	+	+	+	+	+		+	+	+	+
7. Survival at the end of the flow. period	High	+	+	+	+	+	+	+	+	+	+	+	
	low												+
8. Seed size	Very small,											+	
	small,						+	+	+	+	+		
	large,	+	+	+		+							+
	very large				+								

[a] No observations.

a species. In *P. major* genetic variation was mainly present between populations and between ecotypes and also for growth rate the ecotypes differed considerably.

Looking at the problem from a physiological viewpoint, one may ask whether a low growth rate is the consequence of a more adapted shoot (a cost-benefit situation) or to an inability to grow fast since metabolism was retarded (a lack of plasticity). Both explanations can be ecologically important: in the first case leaves could, for instance, become more trampling-resistant, the latter case could be important when a prostrate growth form was adaptive. A high growth rate combined with an early switch to flowering could present an alternative for the situation where growth was slow. In the present set of ecotypes H_{19} probably invested in trampling resistance at the cost of growth. G_1 had to be prostrate to maintain itself and to make more leaves. Two other ssp. *major* types, S_8 and W_9, were also better adapted to physically more extreme situations. G_7 and W_9 were types of this group which flowered relatively early and formed secondary rosettes (Table 7.6).

3. Forming Tall Leaves vs. a Closed Rosette. The preference for making more leaves with short petioles insteed of a few, but longer, erect leaves is easily understood from the occurrence in nature. The difference has also been described by Warwick and Briggs (1980); erect types in a lawn regime were severely defoliated by mowing or grazing. On the other hand, prostrate types produced less seeds in a high vegetation. It is remarkable that trampling resistance needs an erect type; trampling resistance is a matter of being elastic and tough at the same time. A better protection against loss of water after damaging the leaves may play a role in trampling resistance.

4. The Time of Switch to the Generative Phase. A counterbalance of advantages can be noticed: an early switch to seed production will ensure progeny when the season is short and the probability for surviving in winter is low. On the other hand, a long period of vegetative growth will lead to large quantities of seeds (ssp. *pleiosperma*), or will be of importance for a better winter survival and subsequent production of seeds in the following year (Hawthorn and Cavers 1978). The riverbank situation with a variable season length, and a variable position of the plant close to or more remote from the river, may favour different strategies and could even lead to a more or less stable polymorphism. In the inland *pleiosperma* group (A_2, H_{44} and Z_2) Z_2 seemed to be adapted to a very short season; its number of seeds produced during the first weeks of the greenhouse experiment was the greatest. In a longer season, the two other types produced more and larger seeds, which indicates a high fitness in long summer seasons.

5. The Initiation of Secondary Rosettes. A large investment in generative growth, as discussed in (2) and (4), can be realized by growth of the main rosette and by forming additional rosettes; in the last case seed production later in the season, or in later years, can be much greater because only one spike may arise from each node. The coastal types of ssp. *pleiosperma* clearly invests in secondary rosettes, which may be due to a long season, and a reasonable chance for survival in winter. Compared with the inland types of ssp. *pleiosperma,* seed production is delayed and compensated by additional inflorescences from the secondary rosettes for later in a sufficiently long season.

6. The Production of New Leaves During Flowering. In a situation without competition, the leaves at the beginning of flowering will be sufficient to complete the season as far as provision of energy and nutrients is concerned, and storage materials for a next season is saved; further investment in leaves would happen at the cost of seed production. However, in a competitive situation like a lawn in which maintenance of a closed rosette is decisive for the ensurance of a place in the vegetation on the long term, continuity in production of new leaves could be important. The lawn type G_1 and also V_{13} continued producing new leaves during flowering and their seed production was indeed consistently lower.

7. Winter Survival. *P. major* is perennial but there is a tendency towards annual behaviour in the inland *pleiosperma* types. The extreme type, A_4, often died after abundant flowering. It is possible to keep such plants alive at the end of the season, when kept under less optimal conditions, but investment in survival under optimal development is limited. The other inland *pleiosperma* types also looked less prosperous at the end of the season, which undoubtedly may be related to the habitat: survival in winter will be low at a riverbank situation, due to disturbance by the river water. Occasionally, survival is observed; such plants will have a relatively large seed production in the next year.

8. Seed Size. The alternatives here are producing many small seeds or a few large seeds; the highest fitness will be reached at the point where a further increase in seed number is compensated by a loss in seedling success. The compensation point is related to the juvenile growth conditions in the habitat: the large seeds in the ssp. *major* compared with ssp. *pleiosperma* are undoubtedly related to severe competition and physical stress, while at a riverbank or an open coastal vegetation these forces are weaker; equilibrium between number and size of seeds is observed at a lower seed weight level. Within both subspecies variation in seed weight exists. Although seed weight in ssp. *major* always exceeded 200 µg, S_8 had a very high seed weight: more than 300 µg. In ssp. *pleiosperma*, Z_2 produced small seeds of about 100 µg, while the other types ranged between 100 and 200 µg. A special position was occupied by A_4, which in the early seed production made seeds of 250 µg, like ssp. *major*. In view of the rapid growth in early generative development and the low ability to survive after flowering, A_4 was probably adapted to arable land conditions. Indeed the habitat of *P. major* ssp. *pleiosperma* has often been described as wet arable land (Mølgaard 1976).

7.3 Ecotypic Differentiation in *Plantago lanceolata*

7.3.1 Introduction

Plantago lanceolata is a highly variable species found in various grassland habitats (Pilger 1937; Böcher 1943; Haeck et al. 1985; Sect. 3.1). Genetic differentiation at the population level has been reported for morphological and life history traits (Böcher 1943; Cavers et al. 1980; Primack and Antonovics 1981; Wolff 1987; Wolff and Van Delden 1987), as well as physiological characters (Wu and Antonovics 1976; Teramura and Strain 1979; Teramura 1983; Smit, see Sect. 4.3). The species is also characterized by a large phenotypic plasticity (D. Kuiper 1984; Van Dijk et al. 1988). Warwick and Briggs (1979) and D. Kuiper (1984) emphasize that phenotypic plasticity is more prominent than genetic differentiation. This conclusion is supported, especially for life history traits, by data of Antonovics and Primack (1982). Experiments were performed to investigate the degree of genetic differentiation of *P. lanceolata* and answer the following questions:

- Does local adaptation occur and if so, on what scale?
- In which stages of the life cycle does adaptation occur?
- Which environmental factors are selective?
- Which plant traits and developmental mechanisms are involved in selection processes?

Differentiation was studied in relation to the type of grassland management which has an important effect on the structure of a sward. Indications for its importance as a selective factor were found for grassland species such as *Agrostis tenuis* Sibth. (Bradshaw 1959, 1960), *Potentilla erecta* (L.) Rausch. (Watson 1969), *Poa annua* L., and *Plantago major* L. (Warwick and Briggs 1978, 1980), but only weakly so for *Plantago lanceolata* (Warwick and Briggs 1979). In The Netherlands, large differences in the demographic behaviour of *P. lanceolata* were found in grasslands under management as hayfields (i.e. mown once or twice a year) and as pastures (i.e. grazed by cattle) (Mook et al. 1989; Sect. 4.1). In this section, differences between plants from such habitats will be analyzed further.

Firstly, a garden experiment will be discussed in which indirect evidence was found for ecotypic differentiation, together with indications for selective factors involved. Secondly, reciprocal transplant experiments will be reported, giving direct evidence for both adaptation and selection in the field. Thirdly, the effects of separate environmental factors on different populations will be demonstrated, using greenhouse experiments.

7.3.2 Garden Experiment

In this experiment the genetic variability between pasture and hayfield populations was assessed. The sites were geographically isolated from each other, in an area of $100 \times 160\ km^2$ in the central part of The Netherlands, except for two adjacent sites, that were separated by 30 m (populations 6 and 7). Management of these two sites by grazing or mowing was the only habitat difference for the last 22 years (Table 7.15). Measurements on morphology and seed production were made in the original habitat. Thereafter, plants were propagated by means of root cuttings, cultivated in a garden during the next year, and measured in a similar way as in the field.

In the garden, significant differences between hayfield and pasture populations were present in a large number of characters (Table 7.16). Hayfield plants, compared to pasture plants, had more erect, longer leaves and fewer side rosettes. They flowered earlier and had a small number of long inflorescences, heavy seeds, and a low seed production. This variability appears to be mainly due to genetic differences, although the root cuttings did not completely exclude pot effects and the condition of the mother plant. However, the genetic basis of a large number of morphological characters was confirmed by genetical analyses (Wolff and Van Delden 1987, Sect. 8.1). Differences in general morphology (leaf length, flowering date, total number of inflorescences) between hayfield and pasture types coincided well with field measurements (Table 7.16). As the expression of genes depended on the environment during plant development, and conditions in the garden were more or less optimal, some of the characters measured only represent optimal values. In traits related to seed production, a clear contrast between field and garden was present: in the garden a much higher number of mature inflorescences was found than in the pasture

Table 7.15. Management, soil type and vegetation height for *P. lanceolata* populations used in garden experiment (Nos. 1 – 10), transplant experiments 1 – 3 (Nos. 5, 9), transplant experiment 4 (Nos. 2, 4, 10), and transplant experiment 5 (No. 10). Duration of present management is indicated in years. Vegetation height was measured at the end of May

Population	Management		Soil type	Vegetation height (cm)
	Present	Previous		
1. Hasselt	H (15)	HPA	Clay	43
2. Heteren	H (10)	HPA	Clay	65
3. Veerslootlanden	H (>50)	H	Peat	19
4. Bruuk	H (30)	H	Peaty loam	35
5. Merrevliet	H (>50)	H	Peat	35
6. Ossenkampen hayfield	H (22)	PAH	Clay	21
7. Ossenkampen pasture	PA (22)	PAH	Clay	13
8. Loenermark	PA (10)	A	Sand	4
9. Westduinen	PA (>50)	PA	Sand	7
10. Junne	PA (>50)	PA	Sand	6

Type of management indicated as: H = hayfield, mown once a year (site 1 twice); HPA = alternate hayfield, grazed by cattle after yearly hay cut; PA = pasture, grazed by cattle during the growing season; PAH = alternate pasture, grazed by cattle, but every other year a hay cut; A = agricultural field.

sites, where many inflorescences were grazed. Moreover, these inflorescences had smaller spikes, and a lower seedset, possibly caused by a shortage of energy or nutrients as a result of the repeated removal of leaves and spikes (Van der Toorn, unpubl. data).

Most traits were correlated and data were further analyzed with multivariate statistics, using principle component analysis (PCA) for the field data and multiple discriminant analysis (MDA) for the garden data. By these analyses the set of observed characters was transformed to a number of variates, composite characters which were unrelated to each other and which were arranged in order of decreasing importance for discrimination of the populations. In both analyses a significant part of the total variance was explained by the first variate (PCA: 73%; MDA: 57%). This variate coincided with the contrast between hayfield and pasture populations: important morphological traits such as leaf length and total inflorescence number had high loadings (Fig. 7.10A, A'); hayfield and pasture populations had different positions, indicating their different morphology and growth habit (Fig. 7.10B, B'). This variate was related with the vegetation structure, since it was significantly correlated with vegetation height ($P<0.01$) and light transmission ($P<0.05$; Fig. 7.10C, C'). The second variate was correlated with the number of rosettes in the PCA, but its meaning in the MDA was less clear. In both analyses the second variate (explaining 12% of the total variance in the PCA and 19% in the MDA, respectively) was weakly correlated with soil factors. Evidently, the structure of the vegetation played an important role in the genetic differentia-

Table 7.16. Quantitative traits for hayfield (H) and pasture (PA) populations of *P. lanceolata* (mean values per group of populations) measured at the original site and in a garden. For the garden the significance (*P*) of the F-ratio for the difference between hayfield and pasture populations is indicated. (Van der Toorn, unpublished data)

Character	Field[a]		Garden[b]		
	H	PA	H	PA	*P*
Number of populations	5	3	6	4	
Leaf angle (degrees)	–	–	136	158	***
Leaf length (mm)	108	50	111	78	***
Leaf width (mm)	20	16	21	17	***
Rosette number	1.8	1.5	5.2	7.3	***
Start of flowering (day after 1 March)	88	101	108	114	***
Spike length (mm)	22	14	23	25	*
Inflorescence length (cm)	35	17	30	27	***
Total inflorescence number	8.1	11.2	11	17	***
Mature inflorescence number	3.3	0.7	4.3	8.4	***
Seeds/spike	68	20	64	68	
Seeds/mm spike	3.1	1.4	2.9	2.8	
Seed biomass/spike (mg)	73	13	96	86	*
Weight/seed (mg)	1.1	0.8	1.6	1.3	***
Seeds/plant	236	14	302	584	***
Seed biomass/plant (mg)	247	10	440	714	***
Aboveground biomass (g)	–	–	8.6	8.7	
Reproductive effort	–	–	0.53	0.58	***

Significance indicated as: *, **, *** = $P<0.05$, <0.01, <0.001.
– = Not measured.
[a] Populations 1–4 and 6–8.
[b] Populations 1–10.

tion of populations and chemical and physical soil factors were of minor importance.

As mentioned above, traits related to seed production showed an opposite behaviour in the field when compared to the garden. This also was apparent from the multivariate analyses, since these traits showed opposite positions in the PCA and MDA, in contrast to the other traits (Fig. 7.10A, A′). In pastures seed production was lower as a consequence of grazing, whereas under uniform conditions the potential seed output for pasture plants was higher than for hayfield plants. Therefore, grazing pressure in relation to vegetation structure, seems to be one of the mechanisms involved in the selection processes.

It is likely that in hayfields the time of mowing is a selection mechanism. A nearly significant ($P = 0.09$) correlation between flowering and mowing time was found for the populations 1 to 6. In all populations the majority of the inflorescences (>66%) did mature before the relevant mowing date (Van der Toorn, unpubl. data), a strong suggestion for a causal relationship between flowering and mowing time.

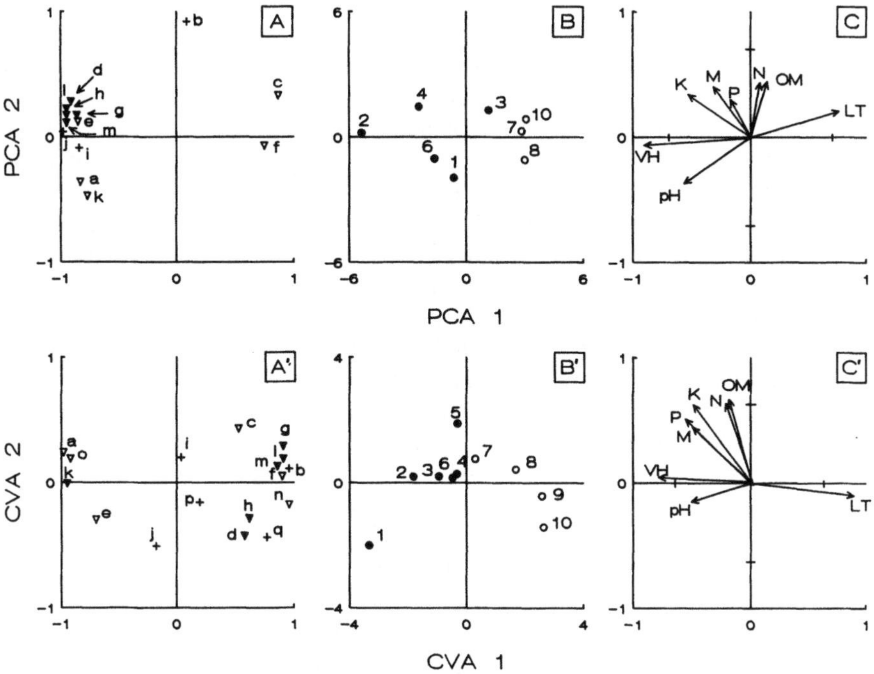

Fig. 7.10. Results of multivariate analyses for quantitative traits of *P. lanceolata* populations (garden experiment). **A, B, C** Principal component analysis based on population mean values measured in the field: **A′, B′, C′** Multiple discriminant analysis based on mean values per genotype (root cuttings) measured in a garden. Results are given as plot of plant characteristics (**A**), populations (**B**) and correlation coefficients of habitat factors (**C**), relative to the first two principal component (**A**–**C**) or canonical (**A′**–**C′**) axes. Plant characteristics indicated as: *a* leaf length; **b** rosette number; *c* start of flowering; *d* spike length; *e* inflorescence length; *f* total inflorescence number; *g* mature inflorescence number; *h* seeds/spike; *i* seeds/mm spike; *j* seed biomass/spike; *k* weight per seed; *l* seeds/plant; *m* seed biomass/plant; *n* leaf angle; *o* leaf width; *p* above-ground biomass; *q* reproductive effort. Plant characteristics are indicated according to their position in A and A′: ▽ = similar position; ▼ = opposite position; + = position somewhat different (but not opposite) or plant characteristic only presented in one of the analyses. Populations indicated as: ● = hayfield; ○ = pasture. Habitat factors indicated as: *LT* light transmission to soil level; *VH* vegetation height; *OM* organic matter content; *N* total nitrogen content; *K* potassium content; *P* total phosphorus content; *M* = moisture content; *pH* pH-KCl. *P* = 0.05 level of significance given as a transverse line on the axes (Van der Toorn, unpubl. data)

In the garden experiment similar differences as observed for hayfield and pasture populations (Table 7.16), were present between the adjacent Ossenkampen populations (Nos. 6 and 7 in Fig. 7.10 B′); significant differences were found in leaf angle, leaf length and inflorescence number (Van der Toorn, unpubl. data). These differences strongly indicate selection for erect, long-leaved plants in the hayfield and for prostrate, short-leaved plants in the pasture as these populations differed in management only during the last 22 years. Genetic differences seem to have arisen within about ten generations (Sect. 4.1), which indicates that selection pressure must have been strong.

7.3.3 Transplant Experiments

Between-Population Differences. Adaptive significance of genetic differences can best be proved by reciprocal transplants (Antonovics 1976; Antonovics and Primack 1982; Van Dijk 1989). Such experiments should involve plants in various stages of the life cycle, since selection pressures may be present at any stage. Transplantation was done in three life stages for two of the populations, 5 and 9, representing a wet hayfield and a dry pasture ecotype (Van Groenendael 1985 a, b). Firstly, survival of seeds in the soil was followed in time (Experiment 1). Seeds of five size classes were put in nylon mesh tubes and buried at three depths, reciprocally at each site; 1200 seeds for each ecotype and each site. Every 3 months seed viability was tested. Survival of the two ecotypes did not significantly differ, there was no origin by site interaction and no direct indication for adaptation to the local habitat was found (Fig. 7.11). Seeds of the pasture habitat were much smaller, with the biggest seeds of approximately the same size as the smallest seeds of the hayfield ecotype; the small seeds of the pasture had a higher mortality in the hayfield (site × origin × size interaction: $P<0.05$), giving some evidence for local differentiation. Indications for the latter were also found after correction for depth of burial: seeds from the pasture buried near the surface had a relatively higher survival in their own habitat (site × origin × depth interaction: $P<0.025$; Van Groenendael 1985 a). Germination and establishment of seedlings was investigated in another experiment (Experiment 2), using marked seeds which were sown in a 1-cm grid in spring; 400 seeds for each ecotype and each site. In the hayfield all seeds germinated in the first season: germination was delayed in the pasture, showing strong environmental effects on the time of germination. In both habitats the hayfield ecotype showed the highest germination and no significant interactions between site and origin could be demonstrated (Van Groenendael 1985 a). After germination, in both habitats the seedlings of the hayfield ecotype performed better than those of the pasture ecotype (Fig. 7.11) but interactions again were absent. After 2 years the pattern changed: in the pasture nearly all plants of the hayfield ecotype died after a severe drought; in the long run each ecotype survived best in its own habitat (site × origin interaction: $P<0.05$; Van Groenendael 1985 a).

Juvenile and adult survival was also investigated using shoots, propagated vegetatively from leaves collected from flowering plants in the field; ten genotypes were represented, each with ten individuals (Experiment 3). Plants from both ecotypes showed the best survival in their own environment (origin × site interaction present after 1 year: $P<0.01$). The high mortality for both ecotypes in the pasture was mainly due to drought (Fig. 7.11). These experiments indicate that selection was hardly, if not at all, present in the seed and seedling phases, but was confined to the later growing stage.

Reciprocal transplants in the juvenile stage were also made with a number of other populations (2, 4 and 10, Table 7.15; Experiment 4). Root cuttings from adult plants collected in the field were used: for each type and site 20 genotypes were represented by ten (sites Heteren and Junne) and five individuals

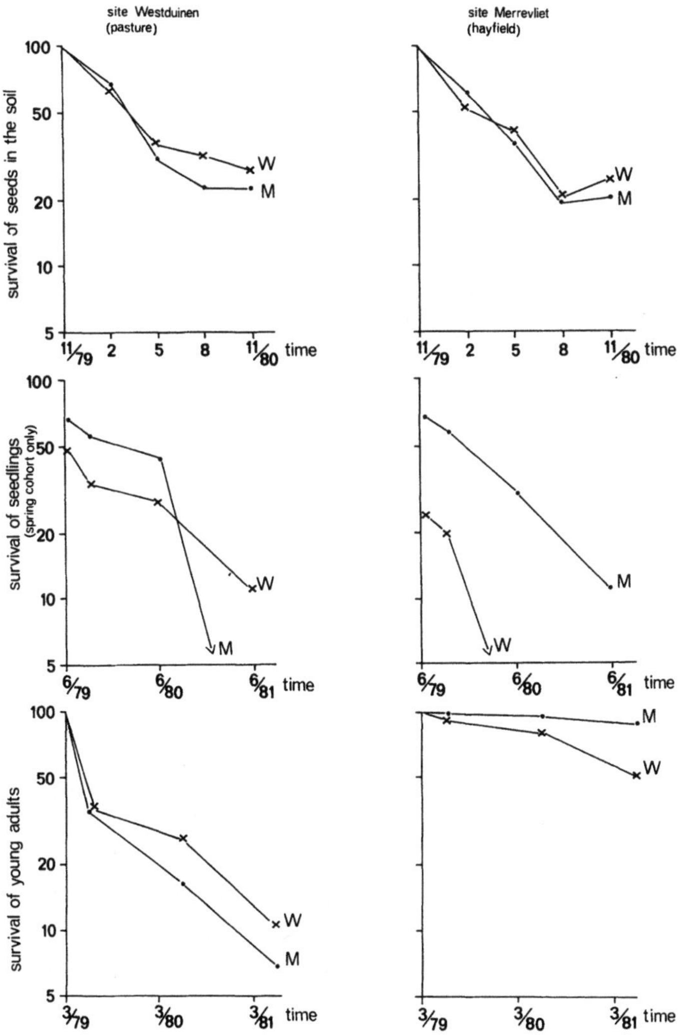

Fig. 7.11. Survivorship curves of various experimental cohorts of two *P. lanceolata* populations originating from Westduinen, a dry pasture (*W*), and Merrevliet, a wet hayfield (*M*), transplanted reciprocally into the relevant sites (transplant experiment 1, 2 and 3; Van Groenendael 1985 a)

(site Bruuk). The experiment was followed for 3 years and – in contrast to the foregoing experiment – more data on reproduction was obtained (Van Tienderen 1989, Van Tienderen and Van der Toorn 1991 a, b).

Between sites, distinct differences in survival were present (Fig. 7.12). In the hayfield Heteren a lower mortality occurred than in the other sites. Here plants from the pasture Junne showed a higher mortality than the native type (Table 7.17), especially in the third summer, which presumably was a consequence of a better growth of the surrounding grasses that season. The high summer

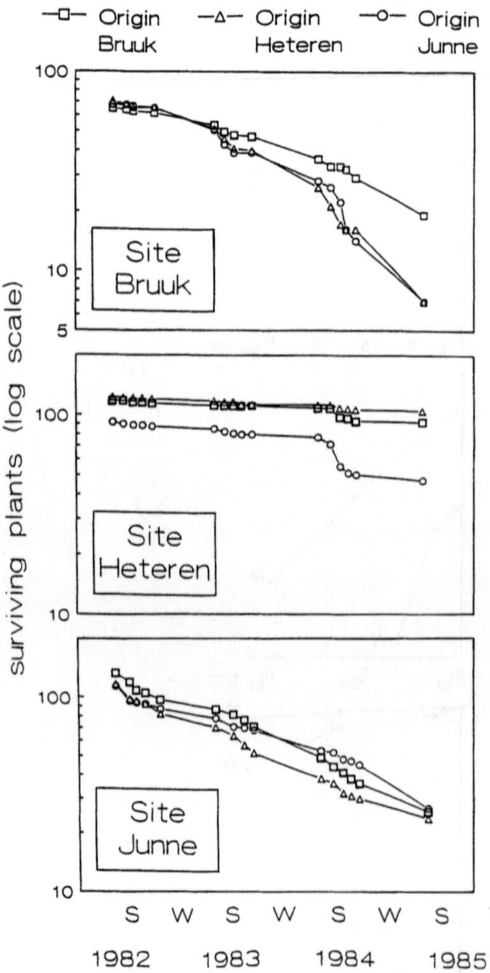

Fig. 7.12. Survivorship curves of transplants of three *P. lanceolata* populations originating from two hayfields (Bruuk, Heteren) and a pasture (Junne) transplanted reciprocally into the relevant sites. *S* summer; *W* winter (transplant experiment 4; After Van Tienderen and Van der Toorn 1991a)

mortality was in correspondence with the demographic behaviour of *P. lanceolata* in hayfields, where mortality was shown to be correlated with competition for light (Van der Toorn and Pons 1988; Mook et al. 1989). In the hayfield Bruuk, alien plants had a higher mortality than native plants; but only the difference between Bruuk and Heteren plants was significant (Table 7.17). The lower mortality of Junne plants, when compared with Bruuk plants, was not significant when initial fresh weight (lower for Junne plants) was taken into account. In the pasture Junne no significant differences in survival between origins were found. This was the case in both hayfields; when differences occurred, the native type showed the highest survival.

The success of native and alien plants was expressed as the total number of seeds produced by all plants of a certain origin in due course of time. The total seed yield is determined by a number of processes during the develop-

Table 7.17. Survival of transplants in three habitats (transplantation experiment 4). Results from log-linear analysis of survival data. Chi-squares for each effect estimated by maximum likelihood analysis (SAS procedure CATMOD). Initial fresh weight used as a co-variate. (After Van Tienderen and Van der Toorn 1991 a)

Site Source	df	Bruuk X^2	Heteren X^2	Junne X^2
Origin	2	7.18*	12.47**	1.17
Bruuk vs. Heteren	1	7.18**	0.24	0.14
Bruuk vs. Junne	1	1.65	7.38**	0.54
Heteren vs. Junne	1	1.34	9.83**	1.13
Time	6	9.73	52.37***	16.95*
Initial	1	3.52	47.09***	13.77***
Summer vs. winter	1	0.14	5.59*	0.00
Fresh weight	1	6.49*	0.05	0.04
Time × fresh weight	6	6.89	6.91	3.37
Origin × time	12	11.49	15.54	14.25

Significance levels indicated as: *, **, *** = $P<0.05$, <0.01, <0.001.

ment of a plant. To evaluate the cumulative contribution of these processes to the total seed yield, four components of seed yield were distinguished: the 1st component the number of growing seasons that a plant is present; the 2nd component the number of seasons that a plant is flowering; the 3rd component the number of successful seasons (i.e. when at least one ripe spike was formed); the 4th component the total number of seeds produced by all plants of a certain origin. In all seed yield components, the native type showed a better performance than the alien types, except in the Junne site for component 1, the number of seasons present (Fig. 7.13). The relative advantage of the native type was increased with component as each component included the effects of previous ones, resulting in large differences in total seed yield. In the pasture Junne a somewhat different pattern in the seed yield was found, as relative differences were smaller than in the number of successful seasons. Here, hayfield plants scored better than the native ones in the seed production of plants that formed a ripe spike (30, 26, and 22 seeds for plants originating from Bruuk, Heteren and Junne, respectively). Yet, their high seed output could not compensate for the low incidence of flowering and maturation of spikes, resulting in a low total seed yield compared to the native plants.

Hayfield plants formed few and long inflorescences that contained more seeds than pasture plants in all transplant sites. They also differed in growth habit: hayfield plants were characterized by few side rosettes and few and long leaves (Van Tienderen 1989).

Selection differentials were calculated for the total experimental population (i.e. all plants from the three origins put together) in each site (Table 7.18) in order to analyze in how far selection was acting on specific plant characteristics. For all traits the sign of the selection differential for the total experimental population (S_{tot}) appeared to be the same as for all traits in which it was cal-

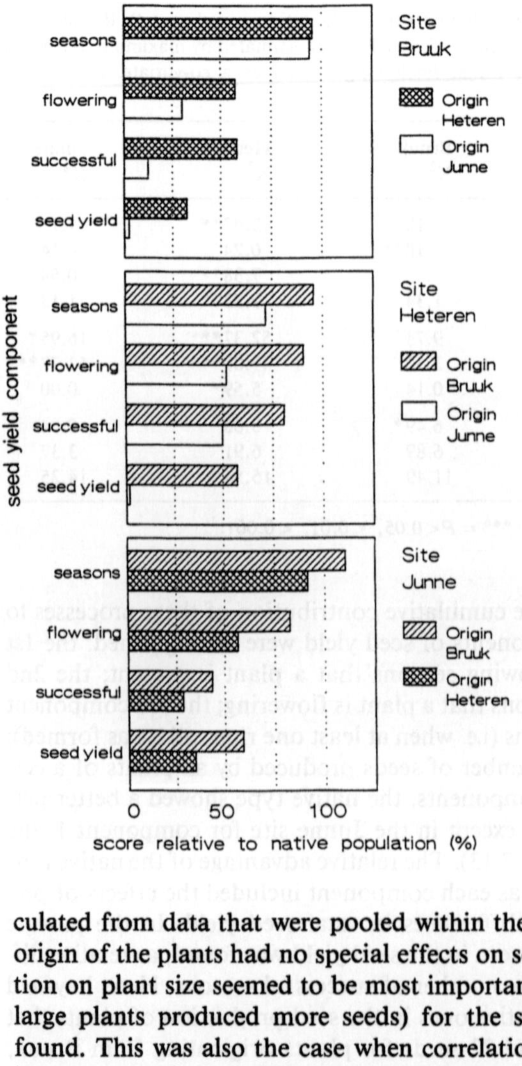

Fig. 7.13. Relative scores for the alien populations in four components of seed yield (each component includes the effects of the previous components), at the three sites of transplant experiment 4. The performance of the native population set at 100% for all components. (After Van Tienderen and Van der Toorn 1991 a)

culated from data that were pooled within the three populations (S_{within}): the origin of the plants had no special effects on selection. In all three sites, selection on plant size seemed to be most important; large and positive values (i.e. large plants produced more seeds) for the selection differential (S_{tot}) were found. This was also the case when correlations with other traits were taken into account ($S_{partial}$). Selection on plant size was not as trivial as it may seem because in each site the native plants showed the best performance. Variation in plant size was determined both by conditions for growth, and by genetic differences in the ability to acquire resources under the particular conditions the individual plants encountered. Unfortunately, a further analysis of these factors was impossible as growing conditions of each individual plant were not measured.

Selection on early flowering (expressed by a negative sign of the selection differential) occurred in the Heteren and Junne sites. For Heteren, selection for early flowering could be expected as this hayfield was mown relatively early. Here, the contribution of genetic factors to the variation in flowering was relatively high (35, 83 and 1% for Heteren, Bruuk and Junne, respectively). The

Table 7.18. Standardized selection differentials for traits and final seed yield for the sites in transplant experiment 4, for the total experimental population (S_{total}), pooled effects within the three origins (S_{within}, after removal of effects of population of origin) and the partial effect after correction for correlations with the other traits ($S_{partial}$). Plant size and growth habit were measured in the first year of the experiment. (After Van Tienderen and Van der Toorn 1991b)

	n	S_{total}	S_{within}	$S_{partial}$
Site Bruuk				
Plant size	189	1.25***	1.04	1.20***
Growth habit	189	−0.14	−0.04	−0.00
Onset flowering	32	0.07	0.02	0.14
Initial fresh weight	189	0.44	0.29	−0.30
Site Heteren				
Plant size	312	0.46***	0.38	0.41***
Growth habit	312	−0.28***	−0.20	−0.21***
Onset flowering	283	−0.36***	−0.22	−0.01
Initial fresh weight	312	0.19***	0.14	−0.09
Site June				
Plant size	247	0.75***	0.88	0.77***
Growth habit	247	−0.06	−0.29	−0.26
Onset flowering	140	−0.39*	−0.50	−0.28
Initial fresh weight	247	0.16	0.33	0.13

Significance levels indicated as: *, **, *** = $P<0.05$, <0.01, <0.001. Significance tests based on t-values for jack-knifed z-transformed correlations (S_{total}) and jack-knifed partial standardized regression coefficient ($S_{partial}$). No test available for S_{within}.

selection for early flowering in the pasture Junne was unexpected, since this population was characterized by late flowering. However, the value of the selection differential was doubtful because it was strongly influenced by a small number of late flowering plants that did not produce any seeds. In both sites the time of flowering depended primarily on plant size as larger plants flowered earlier. As a consequence, partial selection differentials were small.

Only at the Heteren site was a selection on growth habit found. Plants that scored high, i.e. with a pasture-like growth habit, apparently had a lower seed yield than expected in a dense and high vegetation. The amount of variation in growth habit that could be attributed to genetic factors also was higher at the Heteren site (29%), when compared to the other sites (Bruuk: 3%; Junne 11%).

In conclusion, results indicate that selection on timing of flowering as well as on growth habit may have caused differentiation between populations. Yet, the quantitative measurement of selective forces was hampered by the fact that populations differed in plant size at an early stage, and that variation in plant size dominated variation in fitness.

The performance of *P. lanceolata* plants in the experiment was strongly affected by the height of the sward. Vegetation height was measured around each plant in the first year of the experiment and its effects on the components of seed yield were evaluated. In each of the three sites a negative relationship be-

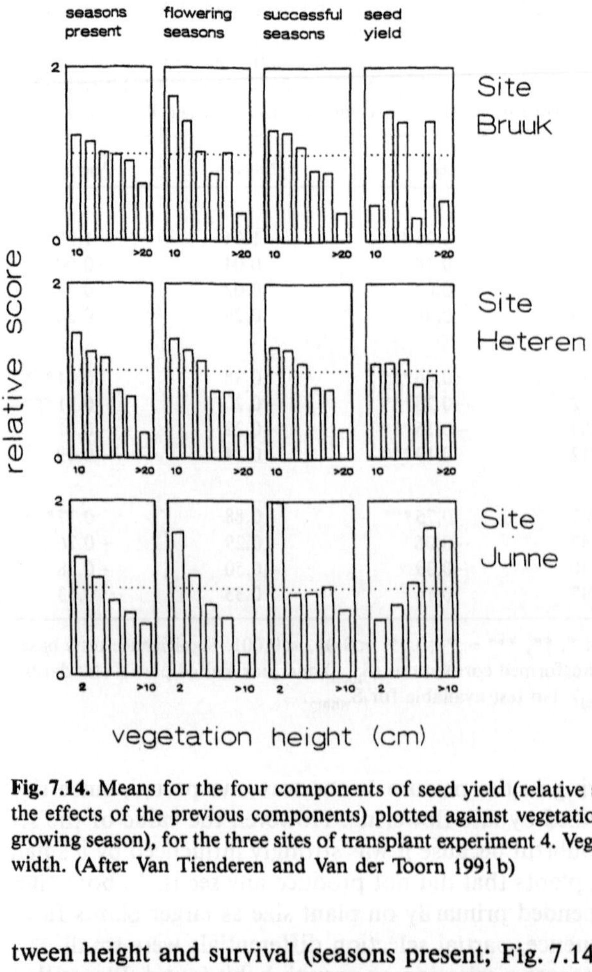

Fig. 7.14. Means for the four components of seed yield (relative scale, each component includes the effects of the previous components) plotted against vegetation height (measured in the first growing season), for the three sites of transplant experiment 4. Vegetation height in classes of 2 cm width. (After Van Tienderen and Van der Toorn 1991 b)

tween height and survival (seasons present; Fig. 7.14) was found, presumably as a consequence of competition increasing with vegetation height. In the Heteren site a negative relationship was also found in other seed yield components, as was the case in the Bruuk site (except for the total seed yield). Here, a change in water level may have been more important during the stage of seed setting. However, a different pattern was found for the pasture Junne: vegetation height was negatively related to survival and positively with total seed yield. Places with a high sward probably had a low grazing pressure and, consequently, plants had more spikes which compensated the low survival.

Within-Population Differences. Within populations of *P. lanceolata* a considerable amount of additive genetic variation was present in morphological traits (Sect. 8.1). To see whether this variability could be related to local environmental conditions, a transplant experiment (Experiment 5) was set up in a site with a gradient in environmental conditions. Within an extensively grazed field near a former river branch, a topographical gradient was chosen, ranging from a

Table 7.19. Description of the four sites of transplant experiment 5

	Dry top	River bank	Wet low	Wet rough
Water table in summer (cm)	−170	−64	−20	−15
Number of higher plant species	25	24	26	28
Dominant plant species	*Festuca ovina* *Agrostis tenuis* *Festuca rubra* *Cerastium arvense*	*Festuca rubra* *Holcus lanatus*	*Cynosurus cristatus* *Carex panicea* *Carex ovalis* *Trifolium repens*	*Glyceria fluitans* *Agrostis stolonifera* *Juncus effusus* *Carex panicea* *Carex nigra*
Density *P. lanceolata* (m^{-2})	0.2	0.4	0.6	0.4
Percentage of total area with similar vegetation	29	30	10	10
Approx. grazing pressure	Low	High	High	Low
Soil upper 5 cm				
pH (H_2O)	4.9	4.9	5.5	5.2
Total N (mg/100 g)	325.5	220.0	496.5	663.5
Olsen-P (mg/100 g)	1.1	1.7	1.2	1.1
Organic matter (%)	8.5	6.3	13.6	18.4

dry sand dune to a low and wet site. Within this gradient of about 100 m length four sites were selected (Dry Top, River Bank, Wet Low, Wet Rough) that showed clear differences in ground water level, plant species composition, grazing pressure, and soil characteristics (Table 7.19). Adult plants were collected in the field and propagated by means of root cuttings. In each site 16 mother plants were selected, that were replicated and transplanted back so that each mother plant was represented by eight transplants per site (32 transplants in total). Survival, growth and seed production was followed during a 2-year period.

It appeared that plants showed a higher survival in summer than in winter, except for the dry top site. This seems to be a usual trend for open and low grass vegetation (Mook et al. 1989, Sect. 4.1). In the dry top site many plants died during a drought in the summer of the second year. Here, native plants had a significantly lower mortality score than alien plants (Table 7.20). In both wet sites, mortality in winter was much higher than in summer, presumably as a result of a long-lasting inundation at the end of the winter period. At the wet low site, the native plants showed a lower mortality than plants from the dry top and the river bank site and in the following summer also from the wet rough site. In the latter period, the native type also showed a better growth (expressed as a higher total leaf area). Furthermore, at the rough site, plants of the dry top had a lower growth yield after the first winter (Van Tienderen 1989). No evidence was found for a better performance of the native type in terms of seed production. The large variability shown by this trait as a result of grazing may have obscured genetic differences. Therefore, few indications

Table 7.20. Percentage of plants that survived during four periods of a reciprocal transplant experiment (No. 5) with four sites in a single population, and significance of origin effects (between plants from different sites, df = 3). Significance inferred from the difference in deviance between log-linear models with and without the effect included, compared with the critical X^2. The initial number of plants was 128 per origin per site. (After Van Tienderen 1989)

Site	Period	Percent survival per origin per period				Origin effect
		DT	RB	WL	WR	
Dry top	1st summer	91	91	88	89	0.9
	1st winter	56	53	55	58	0.5
	2nd summer	17	8	0	8	15.8**
	2nd winter[a]	27	0	–	40	
River bank	1st summer	95	95	95	91	1.9
	1st winter	80	78	82	88	5.1
	2nd summer	91	87	97	89	7.5
	2nd winter	74	61	68	66	3.3
Wet low	1st summer	99	98	100	98	4.5
	1st winter	72	76	84	85	8.1*
	2nd summer	90	84	98	94	15.0**
	2nd winter	28	23	41	28	7.7
Wet rough	1st summer	87	87	92	90	2.8
	1st winter	33	35	42	46	4.9
	2nd summer	89	97	92	81	7.2
	2nd winter	18	26	29	23	1.3

[a]These percentages are based on 11, 5, 0 and 5 plants for the four origins, respectively, not tested.
Significance levels indicated as: *, ** = $P<0.05$, <0.01.

for a local adaptation to extreme environmental conditions in data on survival and growth were found; such indications were absent at the river bank site.

In the study area gene flow was not limited by barriers; reproductive isolation by differences in flowering time did not occur. Pollen and seed dispersal was supposed to be limited in *P. lanceolata* (Tonsor 1985b; Bos et al. 1986). However, in- and outstreaming flood water and grazing by cattle may have imposed large effects on seed dispersal. *P. lanceolata* seeds were covered by a mucilaginous layer and could adhere to cattle (Van der Pijl 1972). Seeds that were eaten by cattle may have also been transported (50% of ripe seeds appeared to germinate after passage through the digestive system, for references see Sagar and Harper 1964). Local adaptation was possible only if seed dispersal was limited and selection pressures were sufficiently strong. The results of the experiment indicate that at some sites local adaptation may have occurred. On the other hand, local adaptation will be hindered if all seeds were dispersed over the whole area, notwithstanding the selective pressures within the different sites. The river bank site, where no evidence for local adaptation was found, occupied a large part of the total area (Table 7.19) and plants at this site had the highest seed yield and survival rate of all four sites. Consequently, a good performance at this site contributed greatly to the success of a genotype in the total population provided that its seeds were dispersed over the entire population. The lack of evidence for local adaptation may thus be the result of the unequal contribution of the four sites to the seed pool, and adaptation to the predominant site, the river bank, may have played a more important role than local adaptation to the other, relatively marginal sites.

7.3.4 Greenhouse Experiments

Transplant experiments can indicate in how far plants are adapted to their own environment and they may give information on selection processes that act on separate plant traits. However, the identity of the factors which cause such a selection remains unclear. Therefore, experiments were set up to analyze the effects of various environmental factors on the performance of contrasting *P. lanceolata* populations.

In the experiments seedlings were used, originating from the pasture at Junne and the early-mown hayfield at Heteren (Table 7.15). In a first experiment ("Seedling experiment") the survival of *P. lanceolata* seedlings was followed in a vegetation of *Festuca rubra*. Two treatments were used: a "low" treatment in which the grass was repeatedly clipped down to 4 cm (a simulated pasture sward) and a "high" treatment in which the grass was clipped down to 10 cm (a simulated hayfield sward). Three-day-old seedlings (48 seedlings per origin per treatment) were planted into the grass and followed during a 70-day period. In the "high" treatment a much higher mortality (84% after 70 days) occurred than in the "low" treatment (6%). However, there were no differences in survival between both populations. It was concluded that competition for light resulted in high mortality of *P. lanceolata* seedlings and that this mortality was non-selective (Van Tienderen 1989).

In a second experiment ("Manipulation experiment"), *P. lanceolata* plants were subjected to various treatments that simulated aspects of their respective habitats. Forty plants per treatment per population were grown separately in pots and followed up to the fertile growing stage. After 3 weeks, three treatments were applied, labelled: Normal, Trampled and Clipped. The normal treatment was a simulation of the conditions in an early mown hayfield. Plants were left undisturbed and harvested at an early stage of seed ripening. There was competition for light as the pots were tightly packed. Trampling by cattle may damage plants by mechanical injury and soil compaction (Blom 1979). It was simulated by placing a cylindrical weight (diameter 5 cm, weight 2 kg) on rosettes, twice for 5s with an interval of 5s; this procedure was repeated three times a week. Clipping was used as a simulation for grazing by cattle; all leaves from a rosette were cut back repeatedly to 4 cm, and the spikes were not clipped.

Plants of both populations showed clear morphological differences before the various treatments were applied. These differences were similar to those in the garden experiment: hayfield plants from Heteren had long leaves and when

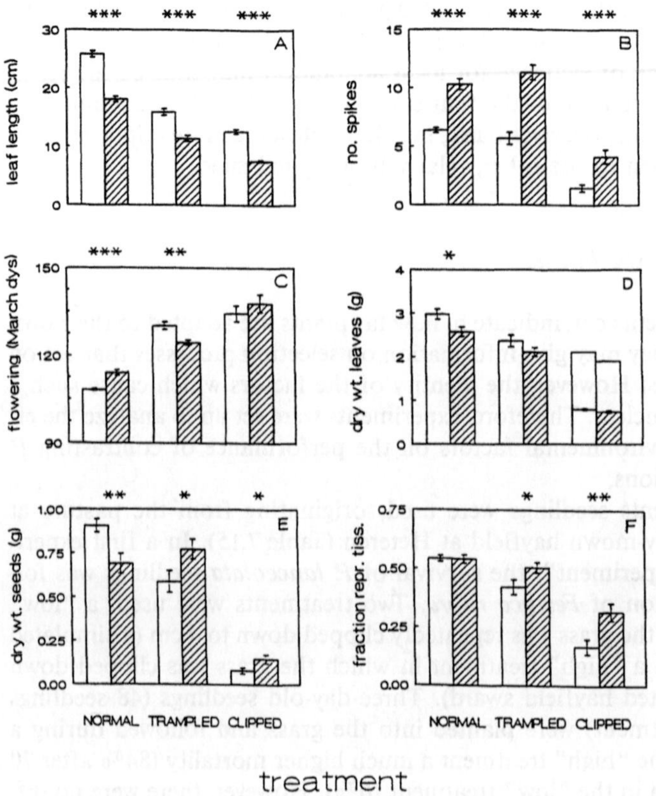

Fig. 7.15. Measurements of *P. lanceolata* plants from hayfield Heteren (*blank*) and pasture Junne (*hatched*) in three treatments of a manipulation experiment at the final harvest. **A** length of the longest leaf; **B** number of spikes; **C** start of flowering in days after March 1; **D** dry weight of leaves (the weight including clipped leaves is also indicated for the clipped treatment); **E** dry weight of seeds; **F** the ratio total reproductive tissue to total above-ground biomass. Standard errors indicated for all points. For each treatment significance levels from t-tests of population effects are indicated as: *, **, *** = $P<0.05$, <0.01, <0.001. (Van Tienderen 1989)

left undisturbed produced a low number of spikes that flowered early, contrary to pasture plants from Junne (Fig. 7.15). In the normal treatment, hayfield plants produced more leaf biomass (Fig. 7.15 D) and more total aboveground biomass than the pasture plants. Hayfield plants also had a higher seed production, but there was no difference in the fraction of reproductive tissue (Fig. 7.15 F). Pasture plants had a higher potential seed production, mainly by the higher number of spikes (Fig. 7.15 B). Yet, their later flowering date (Fig. 7.15 C) resulted in a later maturation of seeds; because of the early harvest, their seed output remained lower than that of the hayfield plants.

Plants from the hayfield Heteren that were trampled flowered later than the pasture plants from Junne, in contrast to the situation in the other treatments (Fig. 7.15 C). In the beginning, Heteren plants were damaged due to mechanical injury to the main rosette, and erect leaves developed. The pasture plants from Junne produced more, but slightly smaller, inflorescences than Heteren plants, resulting in a higher fraction of reproductive tissue and a higher seed biomass in contrast to the normal treatment (Fig. 7.15 F, E).

Clipping resulted in a delay of flowering, compared to the other treatments. No difference in the onset of flowering occurred between both populations (Fig. 7.15 C), but more Heteren plants remained in the vegetative growth stage: only 21 out of 40 plants from Heteren flowered, compared to 31 for Junne (X^2-test for difference; $X^2 = 5.5$, $P < 0.05$). Clipping also reduced the number of spikes, compared to the other treatments, but Heteren plants still produced relatively fewer spikes than Junne plants (Fig. 7.15 B). This resulted in a lower fraction of reproductive tissue and a lower seed production for the Heteren plants (Fig. 7.15 F, E).

It appears that consistent differences between both populations were maintained under different environmental conditions: hayfield plants had longer leaves, a higher shoot weight, and a lower number and taller spikes. A reversal in the ranking of the two populations occurred only for traits associated with reproductive output. Hayfield plants dropped flowering or reduced spike production when shoot growth was limited by clipping or trampling. Under these conditions, more energy was allocated to shoot growth, compared to pasture plants. Large differences between treatments occurred in the plastic response of various traits.

7.3.5 Discussion

Adaptation. The results of the transplant experiments give strong evidence that populations of *P. lanceolata* are uniquely adapted to their native environment, resulting in ecotypic differentiation. Adaptation to local heterogeneity in environmental conditions may also occur, but adaptation for subsites in the pasture Junne was less pronounced than that for geographically isolated populations. It is likely that local adaptation was prevented in the Junne site, because seeds could be transported over fairly large distances as a result of flood water and grazing. In the case of the Ossenkampen populations (garden experiment), seed exchange was limited between both sites, which may have enhanced the differentiation of both populations.

The transplant experiments show that adaptation was manifest in the later growing stages, especially the reproductive phase. The reduction in life time seed yield of aliens could be up to 95%. These high selection coefficients approach the levels observed in extreme environments, e.g. in the case of metal tolerance (McNeilly and Bradshaw 1968; Hickey and McNeilly 1975; McNeilly 1979).

In an experiment with F_2 plants, originating from a hayfield and a pasture site, Wolff (1988) found that in the hayfield habitat both survival and plant size was related to genotype; the genotypes originating from the hayfield almost always performed better. In the pasture habitat, survival was nearly genotype-independent and environmentally determined; plant size was genotype-dependent. A similar difference between a hayfield and a pasture habitat existed in transplant experiment 4 (but not in transplant experiment 2 and 3): in two hayfield sites adaptation was found both in survival and fecundity, whereas in the pasture site adaptation was only expressed in fecundity. It is likely that the difference between both habitats is related to the degree of environmental stability: the hayfield habitat is a relatively stable and homogeneous environment, whereas the pasture habitat is much less uniform in time and space, and characterized by unpredictable disturbances, caused by grazing and/or drought (Van Groenendael 1986; Van Tienderen 1989).

Selective Factors and Life History Strategy. The structure of the vegetation seems to have been of prime importance for the differentiation of *P. lanceolata* in the grasslands studied. One of the factors involved is competition with surrounding grasses (Fig. 7.14). Competition for light seems to play a critical role, as selection for growth habit occurred in the hayfield Heteren. Plants with long, erect leaves were favoured, compared to plants with short and prostrate leaves (Table 7.18, Fig. 7.13). Also, hayfield plants had a higher fitness under simulated hayfield conditions than plants from a pasture (Fig. 7.15). Probably differences in photosynthetic properties were also involved: hayfield plants had a lower light compensation point than pasture plants (Smit, Sect. 4.3). Also, plants from a shaded site had a lower rate of light-saturated photosynthesis than plants from an exposed site (Teramura and Strain 1979).

The time of mowing seems another selective force in hayfields, as it is related to the timing of flowering (garden experiment). This is confirmed by the finding that selection on early flowering occurred in an early-mown hayfield (Table 7.18). In agreement, a similar selective pressure was found under simulated conditions (Fig. 7.15).

Grazing pressure also is of selective importance, as indicated by the results of the manipulation experiment: the pasture type showed a higher resistance to trampling and produced more seeds when trampled or clipped than the hayfield type. The pasture type was characterized by a higher fraction of reproductive tissue, resulting in an excessive number of spikes. This strategy probably originated in order to avoid the risk that all spikes were removed by grazing. In contrast, the hayfield type allocated more biomass into shoot growth when trampled or clipped. This strategy, combined with the formation of long and erect leaves, is favourable under conditions of a high vegetation with intensive competition by grasses. In the biomass allocated into reproductive tissue, the difference between the hayfield and the pasture ecotype fits reasonably well in the r-K hypothesis; a selection for K-strategists with a lower reproductive effort may be expected under stable environmental conditions (Harper and Ogden 1970; Gadgil and Solbrig 1972).

The strategy of pasture and hayfield type corresponds well with the importance of life history traits, as derived from matrix models describing the dynamics of populations in the field (Van Groenendael and Slim 1988): for the pasture type especially seed production, and for the hayfield type adult survival was important. The timing of germination in spring (pasture type) or in autumn (hayfield type) also appeared to be significant for population growth. Further investigations are needed to assess the selective importance of seed dormancy and timing of seedling emergence and their relation with other characteristics (Van der Toorn and Pons 1988).

The smaller seed size of pasture plants (Table 7.16) may be a consequence of a trade-off between seed size and seed production. A small seed size probably is not disadvantageous in a pasture, because competition for light is low and seedling survival was shown to be high (Van Groenendael 1985 a, b; Mook et al. 1989). Silvertown (1989) has questioned the importance of selection as an explanation for variation in seed size. Arguments, supporting his view for *P. lanceolata*, are the large phenotypic plasticity (cf. Table 7.16). On the other hand, at least some genetic variation in seed size is present in most populations, although heritabilities are low (Wolff and Van Delden 1987). A further argument against a direct selection on seed size is the absence of a consistent relationship between seed size and fitness. Van der Toorn and Pons (1988) reported that *P. lanceolata* seedlings in a dense grass vegetation were selected for time of emergence. Selection on the length of cotyledons was unimportant, indicating that seed size, which is closely related to cotyledon length (Van Groenendael 1985 a), did not play a significant role in determining seedling success. Furthermore, transplant experiments showed ambiguous results on a possible selection in the seed and seedling stages: in experiment 1 and 2 little or no evidence for selection in these stages was found (Van Groenendael 1985 a, b). This conclusion was supported by the results of the seedling experiment and data of Antonovics and Primack (1982) and Van Damme (1983 a). Yet, in a reciprocal transplant experiment between three sites Van Tienderen (1989) found local adaptation of seedlings to their own environment in a pasture site and not in two hayfield sites. Therefore, it is concluded that for *P. lanceolata* seed size may be adaptive, but selective pressures under natural conditions probably are weak, and responses to selection slow, given the low heritabilities for seed size.

Development. Plants of hayfield and pasture ecotype differ in the number of daughter rosettes that develop from axillary meristems (Table 7.16). In *P. lanceolata*, axillary meristems may remain dormant or develop into a daughter rosette or an inflorescence. Each new leaf that is formed by a daughter rosette also has an axillary meristem and may develop into an additional potential inflorescence. Teratological malformations indicate that the control of the shoot apex on development of meristems is weaker in secondary rosettes. Also, when many but relatively small leaves are formed, more axillary meristems are available which in turn may develop into inflorescences (Van Groenendael 1985 c). Van Tienderen (1989) hypothesized that the hayfield and pasture ecotype differ in the degree to which plant development is controlled by the main rosette via apical dominance. This hypothesis is based on the similarity in the pattern of genetic differences and plastic responses, such as found in the transplant and manipulation experiments: individual genotypes adopted either a "hayfield" or a "pasture" type growth habit when grown in either a dense or an open vegetation. Furthermore, in the garden experiment a close similarity between genetic and environmental correlations between traits was also found (Van Tienderen 1989). There is evidence that hormones that affect apical dominance play a role in the plastic reactions that determine "pasture" and "hayfield" type growth (Soekarjo 1980; Van Tienderen 1989). The correspondence between plastic and genetic responses therefore suggests that selection has acted on genes that regulate the control of the apical meristem over plant development. Such a mechanism facilitates a rapid change in growth habit, because many traits are affected when gene frequencies change. Wolff and Van Delden (1989) indeed demonstrated that the morphology of *P. lanceolata* can be changed by an artificial selection on a separate trait: selection on leaf angle resulted in correlated responses in leaf length, spike length, and scape length. This finding seems to support the above-mentioned hypothesis on an internal mechanism causing variation in growth habit. Such a mechanism facilitates quick adaptations to environmental changes by selecting new genotypes, given a large within-population variability in traits. It also means that adaptations are possible to changing conditions during the lifetime of the individual, due to plastic reactions in the regulation of rosette development.

7.4 The Seed Bank in Relation to Distribution, Abundance and Life History of *Plantago* Species

7.4.1 Introduction

Although the major part of the mortality in plants occurs in the seed stage (Hickman 1979), the emphasis of the study on demography of plant populations has been on the dynamics of seedlings and adult plants (Cavers 1983). Besides, most papers concerning the fate of seeds in soils are dealing with weeds in arable systems. Studies on seed banks of grassland species are rare.

A seed bank has been defined as a reserve of viable seeds present in the soil and on its surface (Williams 1983). Numbers of viable seeds in or on the soil are strongly dependent on environmental factors such as disturbance and soil-chemical and physical conditions (Harper 1977). The dynamics of the total seed population may be described in terms of the fraction of seeds in dormancy, the loss due to germination and the loss of seeds that have become inviable (Roberts 1972). Seed dormancy has often been expressed as the degree of germination response to dormancy-modifying conditions (Richter and Switzer 1982). Following this procedure the effect of temperature on germination of seeds of *Plantago* species has been determined in laboratory experiments using moist filter paper. Germination of seeds of *P. major* and *P. media* is stimulated, for example, by increasing temperatures in the range of 15 to 25 °C (Sagar and Harper 1960; Blom 1978; Soekarjo 1982b), while this was not the case for seeds of *P. lanceolata.* Schat (1981) reported a pronounced polymorphism of seeds of *P. coronopus* with regard to the temperature requirements for rapid germination. A distinct temperature requirement of individual seeds produced by a single plant may result in a spreading of the risk of seedling mortality (Schat 1981; Soekarjo 1982b).

Vegetation cover may also be important in determining whether a seed germinates or remains dormant in the seed bank (Harper 1977; Pons and Van der Toorn 1988). It has been suggested (Cohen 1966; Harper 1977) that seed dormancy may have adaptive significance, since it generally helps to maintain a reserve of viable seeds in the event of catastrophes.

In this chapter experiments on seed banks of *Plantago* species in relation to surface vegetation and life history will be reviewed.

7.4.2 The Seed Bank in Relation to Surface Vegetation

Harper (1977) pointed out that a close correlation between the surface vegetation and the seed flora of the soil beneath in general is rare: species abundant in a certain vegetation type may be very poorly represented in the seed bank. The distribution of adult plants of *P. maritima* and the numbers of seeds present in the soil was the subject of several studies. Seed banks of *P. maritima* in a salt marsh in which the distribution of *P. maritima* was restricted to two sites (Blom 1983) were investigated. Soil samples (layer: 0–6 cm, volume: 29 cm^3, n = 184) taken along a gradient over the whole area contained only seeds of *P. maritima* if 20 or more adult plants per m^2 were present in the surface vegetation. Jerling (1983, 1985a) reported that along a successional gradient of a salt marsh very few seeds of *P. maritima* were present in the seed bank, whereas seeds of other species in this area were well represented in the seed bank. Jerling (1985a) also demonstrated that dispersal of seeds of *P. maritima* was very low. Bakker et al. (1985) also studied the dispersal, germination, and establishment of halophytes and glycophytes along a salt marsh gradient. In the latter study, seeds of *P. maritima* also occurred in the seed bank at sites higher on the gradient than where they were found as adults in the vegetation. Bakker and his colleagues showed that in this last salt marsh, seeds of *P. maritima* could be dispersed by flooding along the whole gradient: driftline material contained viable seeds. Seedling establishment and not dispersal and germination appeared to limit the occurrence of *P. maritima* and other halophytes.

As part of a study in which differences in life-history traits between (Lotz and Blom 1986, Lotz 1990) and within populations of *P. major* (Lotz et al. 1990) were investigated in relation to environmental factors, the correlation between numbers of seeds in the soil and the abundance of the species at the surface area was studied. At five sites where *P. major* plants occurred, soil samples (layer: 0–2.5 cm and 2.5–5 cm, volume: each 125 cm^3, n = 10 per site) were taken in early spring when seedlings were not yet present. The selected sites were more than 5 km apart. In August, numbers of adult plants (at least 1 year old) and seedlings were counted at the same sites. It was previously observed that in these five populations germination of seeds was very rare in autumn. Considering the

Table 7.21. The relationship between the abundance of *Plantago major* and the seed bank. Pearson correlation coefficients and levels of significance. Numbers of viable seeds were determined after 6 months' incubation (12 h light/12 h dark, 22 °C). See text for further information

	No. of viable seeds	
	Per soil sample (0–2.5 cm)	Per soil sample (2.5–5 cm)
A) Data over five populations, n = 5		
Density of adult plants in August	0.531	0.961**
Density of seedlings in August	0.216	0.947**
B) Data within one population, n = 9		
Density of adult plants in August	−0.132	−0.144
Density of seedlings in August	0.298	0.247
Density of seedlings emerged in spring-August	0.568	0.621*

*$P<0.05$; **$P<0.01$. No indication means not significant.

data of five populations, the density (means over three sample plots of 0.125 m^2 per site) of both adult plants as well as seedlings was linearly correlated with the mean number of seeds in the soil layers of 2.5 to 5 cm depth (Table 7.21); such correlations were absent at the surface layer (0–2.5 cm depth). The lack of correlation between the abundance of *P. major* plants and the density of seeds in the top layer of the soil may possibly be explained by the fact that small seeds may move down the soil profile with percolating rain water or by animal activity (Harper 1977). The ability of seedlings of *P. major* to emerge from depths down to 5 cm below the surface (Froud-Williams et al. 1984b) corresponds with the correlation between seed density in the layer of 2.5 to 5 cm and the number of *P. major* plants per unit area at the surface.

The relationship between the abundance of *P. major* and the size of the seed bank was further investigated within a single population. On a former beach plain *P. major* occurs in three microhabitats which differ strongly in nutrient supply, water saturation of the soil and vegetation cover (Lotz and Spoormakers 1988). In early spring soil samples (layer: 0–2.5 cm and 2.5–5 cm, volume: 125 cm^3, n = 3 per site) were taken next to three plots in which the demography of adult plants and seedlings was followed. Numbers of seeds in the soil (both layers) did not correlate with the mean densities of adult plants and seedlings as determined in August (Table 7.21). Nevertheless, within this beach plain the total number of seedlings emerged from spring to August correlated significantly with the number of seeds in the soil layer 2.5 to 5 cm. Similar to the results over five populations of *P. major*, this correlation was not found with numbers of seeds in the surface layer of the soil. The fact that the contents of the seed bank correlated with the total recruitment over a whole season and not with the number of seedlings present at the end of the growing season (August) may be explained by the dependence of survival of seedlings on the characteristics of the microsite in question (Lotz et al. 1990).

It can be concluded that in many stages of the life cycle, besides reproduction and seed dispersal, other factors may influence the relation between the size of the seed bank and the abundance of the species in the surface vegetation. These factors probably are soil characteristics and mortality during seedling establishment.

7.4.3 The Seed-Bank Dynamics in Relation to Life History of Plantago

Thompson and Grime (1979) distinguish between transient and persistent seed banks. In the transient seed bank, all seeds produced in the previous year either germinate or die. In the persistent seed bank, seeds with enforced dormany are present throughout the year. The latter seeds will germinate after transfer to favourable conditions. Long-lived perennials often have transient seed banks, whereas short-lived plants may have more persistent seed banks (Weaver and Cavers 1979). Seed-bank dynamics may be closely related to the life history of *Plantago*, and seed dormancy may be of adaptive significance (Cohen 1966; Harper 1977; Cook 1980).

The relationship between seed-bank dynamics and habitat characteristics was studied by Van Groenendael (1985a). He compared life-history characteristics of two populations of *P. lanceolata.* The population dynamics of the population in a dry dune grassland (Westduinen) with an open vegetation was under extreme abiotic control: the soil was poor in nutrients and water supply also was frequently limited. Plants were short-lived. Seeds were already produced by 2-year-old plants and had a short period of innate dormancy. Throughout the year a substantial amount of viable seeds was present in the soil (Fig. 7.16), indicating a persistent seed bank. The second population was situated in a wet meadow (Merrevliet); the soil was completely organic with high levels of nutrients. As a result of interaction of *P. lanceolata* with neighbouring plants, the population dynamics probably were under strong biotic control. A relatively small percentage of adults flowered, producing a large amount of seeds. The seeds only had a short period of innate dormancy. The seed bank at this wet meadow was of a more transient type (Fig. 7.16). According to Bakker (1983), *P. lanceolata* and other species of hayfields without fertilization have transient seed banks; species of trampling sites and sites with periodically dry and wet soils have persistent banks. Clearly, populations of *P. lanceolata* demonstrate a similar variation in seed-bank dynamics related to environment. It may be questioned whether the difference between both populations was induced by environmental factors or whether it was genetically determined. To answer this question, seeds of both populations were stored in nylon bags in the soil of both study sites and viability of the buried seeds was tested during the following year (Van Groenendael 1985a). In the soil of the wet meadow seeds of both populations had about the same half-lives, whereas in the dry dune grassland seeds originating from the dry site had a higher half-life than seeds of the wet meadow population (Table 7.22). The higher longevi-

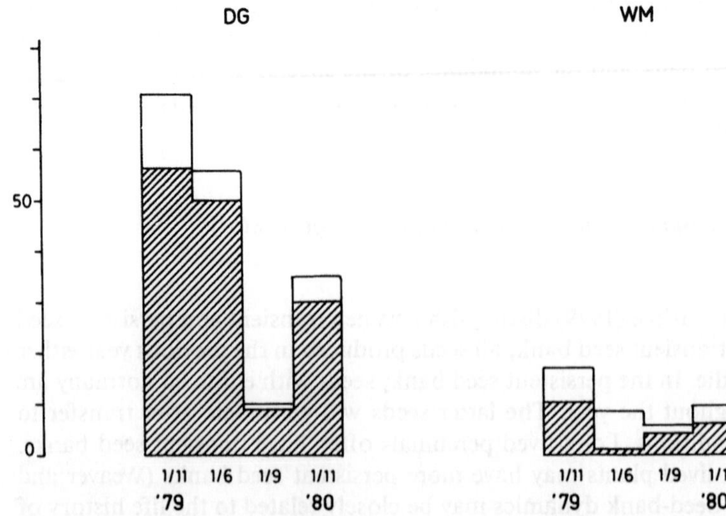

Fig. 7.16. Number of seeds in the soil for *Plantago lanceolata* in a dry dune grassland (DG) and a wet meadow (WM) at four moments in the year, based on 50 soil samples each. The clustering of seeds in the samples is indicated by the *variance* over mean ratio. The number of seeds present in the top layer (0–1 cm) is indicated by *shading*. The remaining seeds were found between 1 and 10 cm. (After Van Groenendael 1985a)

Table 7.22. Half-lives (days) of seeds of two populations of *Plantago lanceolata* stored in nylon bags in the soil. (Data derived from Van Groenendael 1985a)

	Study sites	
	Wet meadow	Dry dune grassland
Population		
Wet meadow	108	111
Dry dune grassland	107	133

ty of seeds is probably genetically determined, and it yields a greater contribution to the persistent seed bank.

Seeds of the dry dune grassland were smaller than those of the wet meadow population, even when plants were grown in the greenhouse (Van Groenendael 1985a). Large seeds probably have other dormancy-breaking requirements than smaller seeds (Harper 1977; Cook 1980; Silvertown 1981). Seed size is said to be negatively correlated with germination date. The delay of the germination of small seeds is caused by the fact that smaller seeds have relatively thicker seed coats than large seeds (Karssen 1970c). Thick seed coats may form a mechanical barrier which should be overcome by the seed embryo.

Table 7.23. Percentage of dormancy of seeds from two populations of *Plantago major* ssp. *pleiosperma* after incubation on moist filter paper (12 h light/12 h dark, temperature 22 °C) during 28 days. Means are given over ten plants with each ca. 40 seeds. (Data from Lotz 1989)

	Size class (mg)	River-bank population	Shore-meadow population
A	<0.2	79	–
	0.2–0.3	42	2
	>0.3	–	1
B	<0.2	54	–
	0.2–0.3	43	11
	>0.3	–	6

A, Seeds were harvested 1 day before the start of the experiment; B, Seeds were stored for 7 months in the dark at 4 °C.

The relationship between seed size and dormancy was also studied in two populations of *P. major* ssp. *pleiosperma* (Lotz 1989). The habitat of the first population, a sandy river-bank, was characterized by a high temporal variability due to inundation. Plants of this population had an annual life cycle, a short vegetative period and produced a high number of small seeds, all characteristics of an r-strategist. The second population was located on a shore meadow. The vegetation was dense and highly productive. The coastal plants may be characterized as K-strategists, having a more perennial life cycle, with a long vegetative period and a high single-seed weight (Van Dijk 1985; Lotz 1989). Seeds were harvested at the natural habitats one day before the start of the germination experiment. They were sorted into three size classes, <0.2 mg (riverside population), 0.2–0.3 mg (both populations), and >0.3 mg (shore-meadow population). A low percentage of seeds from the shore-meadow population was still dormant after 28 days of incubation, whereas this percentage was considerably higher for seeds from the river bank (Table 7.23, $P<0.001$). In addition, the percentage of dormancy decreased with increasing size, irrespective of a stratification treatment during 7 months ($P<0.05$). The mean germination time of seeds from the river-bank population was 1.7 and from the shore-meadow population 1.3 weeks (difference between populations significant at $P<0.001$). It is concluded that seeds from both populations differ in germination pattern in time, and that this difference is, at least partially, associated with differences in seed size. On the shore meadow, mortality of *P. major* ssp. *pleiosperma* until first flowering was extremely high in more or less intact vegetation, compared to that of plants that had emerged in gaps in the vegetation (Lotz 1990). Early germination may increase fitness of seedlings in these gaps that are also colonized by grasses (Naylor 1980; Venable 1984). The ability, associated with a relatively large seed size, to germinate as soon as a gap has been created by e.g. cattle, will probably stay under strong selective pressure in environments such as a dry dune grassland (Van Groenendael 1985) and the river bank, where dormancy of seeds may be of adaptive significance by spreading the risks in time (Cohen 1966).

7.4.4 Discussion

In general, most seeds enter the soil in the close proximity of the parent plant (Werner 1975; Bos, Section 6.4). A relation between the size of the seed bank and the surface vegetation may be influenced by at least three factors. Firstly, the environmental conditions, and indirectly the species composition of the surface vegetation might have been changed. Seeds produced by a parent plant of the early vegetation stage may still be present dormant in the soil (Harper 1977). Secondly, because of spatial variation in biotic and abiotic factors as seed predation and permeability of the soil, losses of the seed bank may vary. The result may be that only if many plants produce seeds and the seeds are present in certain soil layers, then the size of the seed bank may reflect the composition of the surface vegetation. Long distance seed dispersal is the third factor that may influence the correlation between seed bank and surface vegetation. Seeds may enter new areas by dispersal. Successful establishment in new habitats will depend on seed dispersal and germination and seedling survival.

An important demographic property of seeds is the capacity to remain dormant in the soil (Cook 1980). Seeds of *Plantago* populations from highly variable environments such as dry dune grasslands or river sides, had higher degrees of dormancy than seeds of meadow populations of the same species. Differences in seed-bank dynamics between populations have been rarely reported on before: Cavers (1983) mentioned that an annual species, *Chaenorhinum minus*, was represented by a transient seed bank on a river side and by a more persistent seed bank in a railway track.

There is evidence that differences in the degree of dormancy may depend on seed size, as in *P. lanceolata* and *P. major*. In his review, Cook (1980) concluded that selection for persistence in the soil may favour a decrease in seed size and a relative increase in seed-coat thickness. Besides, plants producing many small seeds as *P. major* plants from the river-side population, have higher dispersal probabilities, but are probably limited to growth conditions of low interspecific interference (Harper 1977; Silvertown 1981). The reverse holds true for plants producing few seeds such as *P. major* plants from the shore meadow. Thus, seed size presents constraints on spreading risk in a variable environment, on a successful colonization, and on establishment (Silvertown 1981). In studies on life history of plant populations, analyses of seed-bank dynamics are essential.

8 Genetic Strategies in *Plantago*

8.1 The Genetics of *Plantago* Species

8.1.1 Introduction

The presence of genetic variation is a prerequisite for evolution. Particular genes or sets of genes are involved in adaptation to specific environmental stresses. Phenotypic variation, partly based on genetic variation, can be found on the molecular, biochemical, physiological and morphological levels as well as for fitness and life history traits. Contrary to the situation in most animal species, plant phenotypes are often not as rigidly canalized; in plants it is often hard to analyze the genotypic basis (if any) for the phenotypic variation observed in natural populations. This implies the separation of phenotypic and genotypic effects and the arduous task of quantification of the genetic contribution. The ability of a particular genotype to be plastic, that is to express different phenotypic states, can be of adaptive value itself; and it may be under genetic control. For studies on populations in an evolutionary context, it is of paramount importance to elucidate the contributions of both genetic and non-genetic variation to adaptation. It should be clear, however, that the ability of a species to colonize a particular habitat and to maintain itself in that habitat (in comparison with other species which are unable to do so) must be entirely ascribed to genetic reasons (Bradshaw 1984). It is the purpose of this chapter to discuss the data on genetic variation in the *Plantago* species which occur in the Netherlands against this background.

8.1.2 Breeding System and Amount of Genetic Variation

Knowledge of the breeding system is essential for the interpretation of data on genetic variation, present in populations and species, and more specifically for the understanding of their genetic structure. In comparing the five plantain species in The Netherlands, it appears that considerable differences occur (Table 8.1): *P. major* and *P. coronopus* are self-compatible and as a result self-fertilization is possible; *P. lanceolata, P. media* and *P. maritima* cannot self-fertilize due to a self-incompatibility system of the gametophytic type (see also Sect. 2.6). All species are protogynous (see Sect. 2.5) and predominantly wind-pollinated. *P. lanceolata, P. coronopus* and *P. maritima* differ from the other, completely hermaphrodite species, in being gynodioecious.

Table 8.1. Breeding systems in five *Plantago* species

	P. major	*P. lanceolata*	*P. coronopus*	*P. media*	*P. maritima*
Self-compatibility	+	−	+	−	−
Gynodioecy	−	+	+	−	+
Protogyny	+	+	+	+	+
Diploidy	+	+	+	− [a]	+

[a] Tetraploid in The Netherlands.

It is obvious that the breeding system has a profound impact on the amount of genetic variation present in populations, on interpopulation differences, and on the level of individual heterozygosity. In a predominantly self-fertilizing species, such as *P. major*, it is expected that the level of intra-population polymorphism will be low while the degree of homozygosity will be high. On the other hand, in species with self-incompatibility and gynodioecy such tendencies will be counteracted.

The determination of allozyme polymorphism by screening a number of enzyme loci provides an estimate of the amount of genetic variation (Sect. 6.1). Three *Plantago* species, *P. major*, *P. lanceolata* and *P. coronopus* were extensively surveyed for allozyme variation (Van Dijk and Van Delden 1981; Van Delden 1985; Van Dijk 1985a, b; Van Dijk et al. 1988). The results for the remaining two species are based on less loci and less samples and should be considered as preliminary. A summary of the results is given in Table 8.2, where P is the mean fraction of polymorphic loci over populations, H_e the mean gene diversity per locus ($H_e = 1 - \Sigma p_i^2$, where $p_1, p_2 \cdots p_i$ are the allele frequencies). Also presented are the F-statistics, F_{IS} and F_{ST}. F_{IS} describes the relation between the observed fraction of heterozygotes and the fraction expected from Hardy-Weinberg ratios; F_{ST} describes the degree of population differentiation. *P. major* was characterized by a low fraction of polymorphic loci; the P values presented are averaged over populations. The fraction of variable loci in *P. major* on the species level was much higher. This difference is to be attributed to the high level of homozygosity, found in *P. major* populations. In *P. lanceolata* and *P. coronopus*, the population means ($\bar{P}_p$) were nearly as

Table 8.2. Levels of allozyme variation and F statistics of five *Plantago* species (see text)

	$\bar{P}$	$\bar{H}e$	F_{IS}	F_{ST}
P. major	0.15	0.047	0.79	0.14
P. lanceolata	0.33	0.127	0.12	0.04
P. coronopus	0.31	0.088	0.07	0.11
P. media	0.42/0.50[a]	0.16/0.32[a]	−	−
P. maritima	0.30[b]	0.10[b]	−	−

[a] Diploids: first figures; Tetraploids: second figures (P.J. Van Dijk, pers. commun.).
[b] Preliminary data.

high as the species means ($\bar{P}_S$) and two to three times higher than in *P. major*. Although both *P. major* and *P. coronopus* are self-compatible, it was found that the fraction of self-fertilization differed considerably: Van Dijk (1985 a, b) calculated that the outcrossing rate in a representative population of *P. major* amounted to 0.165, while – surprisingly – in *P. coronopus* the outcrossing rate was about 1. The low outcrossing rate in *P. major* may partly explain the low values of both $\bar{P}_p$ and $\bar{H}$. More direct estimates of outcrossing rates (t_m) were obtained by Wolff et al. (1988) and Wolff (1991 a, b) by means of a multilocus procedure (see Sect. 6.1.4). *P. major* had t_m values of 0.02 to 0.08, while for *P. coronopus*, t_m values ranged from 0.34 to 0.93. The association between mating system and amount of genetic variation agrees well with the general picture emerging from the literature: a positive correlation between outcrossing rate and amount of polymorphism (Hamrick et al. 1979; Brown 1979; Nevo et al. 1984). Nevo et al. (1984), for example, give the following P values, averaged over species, for selfers, mixed species and outcrossers: 0.16, 0.32 and 0.55, respectively. The corresponding H values are: 0.037, 0.056 and 0.076, respectively.

For morphological variability the amount and distribution of genetic variation in our populations was calculated for *P. major*, *P. coronopus* and *P. lanceolata* (Wolff 1990b). The effects of the mating system on morphological variability are approximately identical to those on electrophoretic variation. The outcrossing species *P. lanceolata* possesses the highest total variation and the inbreeding species *P. major* has the strongest differentiation between populations. In *P. coronopus*, however, morphological differentiation among populations is low in comparison with electrophoretic variation. The reason for this will be discussed in Section 8.1.4.3.

8.1.3 Heterozygosity and Environment

A plant species can cope with the challenge of a variable environment in two ways: by phenotypic plasticity or by genetic variation. In the latter case several genotypes are present which possess different maximum fitnesses in different parts of the habitat (as with spatial heterogeneity) or in different seasons (as with temporal heterogeneity). Such an array of genotypes may enable a species or a population to live in a greater variety of environmental conditions. This relation between genetic variation and ecological amplitude has been referred to as the "niche width variation" hypothesis (Van Valen 1965; Soulé 1971; Babbel and Selander 1974). Babbel and Selander (1974) found for two *Lupinus* species an inverse relationship between the level of genetic variation and the degree of edaphic restriction. On the other hand, two *Hymenopappus* species, which occurred on similar soil types, also showed similar levels of genetic variation. A positive relationship between heterozygosity and niche breadth has also been claimed for animal species. In 18 Hawaiian *Drosophila* species, for example, genetic variation was directly related with the number of plant species used as breeding substrates (Steiner 1977). Comparison of *Plantago* species for the relationship between genetic variation and niche breadth seems at first sight not adequate in view of the profound differences in their breeding systems and the corresponding differences in the amount and distribution of genetic variation. At the intra-species level, however, such comparisons appear relevant.

It is often assumed that a relationship exists between fitness and heterozygosity: an increase in heterozygosity should be accompanied by an increase in fitness. In F_1's from crossings between different strains, increased values for various fitness components are often associated with a rise in heterozygosity, a phenomenon known as heterosis. It has also been repeatedly found that a decrease in heterozygosity, as found with inbreeding, is accompanied by loss of fitness (Charlesworth and Charlesworth 1987). The generality of this relationship in natural populations is unknown at the moment. Furthermore, it is not clear whether fitness is a linear function of heterozygosity or more complex, e.g. following an asymptotic curve (Frankel and Soulé 1981).

Intra-species comparisons for heterozygosity can be done at the population level and at the individual level. Several reports on animal and plant species, at the individual level, describe the relationship between heterozygosity and survival, sexual maturity, and growth rate. Besides the described relation between fitness components and heterozygosity, a negative correlation between morphological variability and heterozygosity is often reported. This has been interpreted as a relationship between developmental stability and heterozygosity (Lerner 1954). For summaries see Frankel and Soulé 1981; Mitton and Grant 1984; Livshits and Kobyliansky 1985. A convincing case for a plant species, *Lyatris cylindracea*, has been provided by Schaal and Levin (1976): the average heterozygosity for allozyme loci increased with increasing age class, suggesting that survival was positively correlated with heterozygosity. Positive correlations were also found between individual heterozygosity and fecundity, longevity and growth rate. Ledig et al. (1983) found a positive relationship between allozyme heterozygosity and growth rate in populations of *Pinus rigida*. A positive relationship between heterozygosity and fitness at the inter-population level has been reported for a number of *Drosophila* species (Powell 1971; McDonald and Ayala 1974; Steiner 1977; Powell and Wistrand 1978). However, the generality of the relationship between genetic variation and fitness has been questioned: negative results have been also reported (see above reviews). Although many of the characters studied in association with heterozygosity can be considered as fitness components, the relationship does not seem to be restricted to these traits; often several kinds of quantitative traits seem to be involved, and their relation with fitness cannot easily be demonstrated. Most estimates of heterozygosity are based on a randomly chosen number of allozyme loci; it does not seem likely that these loci themselves realize the effects. The allozyme loci more likely reflect general heterozygosity at the individual or population level.

To study whether a positive relationship between heterozygosity and fitness occurred in *P. lanceolata*, Wolff and Haeck (1990) undertook an allozyme analysis of 181 plants from a natural population. Demographic data of these

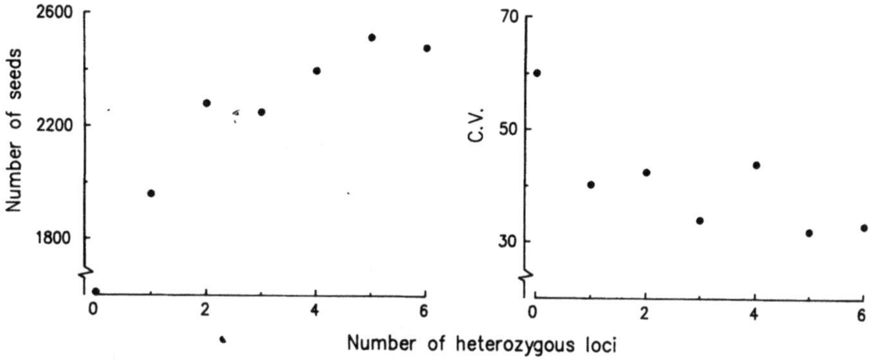

Fig. 8.1. The relationship between heterozygosity and both the number of seeds and the coefficient of variation for seed number

plants had been collected over a period of 10 years. In this outbreeding species a positive correlation was found between heterozygosity and some characters associated with generative output, while no relationship existed with age of plants. Further evidence was found by Wolff (1987), who screened individual *P. lanceolata* plants from an F_2 cross for allozyme variation at six loci and for 17 characters, related with morphology and reproduction. The plants were classified into seven groups, from 0 to 6 heterozygous loci; means and coefficients of variation (CV) were calculated for each group and for each quantitative character. A significant positive correlation existed between the number of heterozygous loci and a number of character means. This relationship existed only for generative but not for vegetative characters. The degree of phenotypic variation of the characters, expressed as C. V., was determined for each of the seven groups, differing in the number of heterozygous loci. A negative correlation between CV and heterozygosity was found (Fig. 8.1). It thus appears that the level of heterozygosity in *P. lanceolata* is related with generative fitness characters and variability, a finding which fits well with comparable results in other animal and plant species.

With some prudence it can be concluded that a decrease of genetic variability in *P. lanceolata* will lead to loss in fitness. In the predominantly selfing species *P. major*, probably quite different mechanisms must operate because heterozygosity is extremely low in this species. *P. coronopus* is an interesting case in this respect: outcrossing occurs frequently, although selfing may also occur, especially in isolated populations. Two studies have been done to answer the question whether inbreeding effects are present in this species. Wolff (unpubl.) did a garden experiment in which plants from a first and a second selfed generation (S1 and S2) were compared to plants from biparental origin (F_1) which were then selfed (F_2). Significant inbreeding effects occurred for the amount of generative output (weight of scapes, spikes and seeds together) but not for vegetative growth (weight of leaves). A second study on inbreeding effects in *P. coronopus* was done by Van Damme and Van der Reest (unpubl.).

They showed a stronger decrease in fitness in crosses with plants originating from a closer distance, compared to plants from a wider distance. Generative characters were positively correlated with heterozygosity in the two *P. lanceolata* cases mentioned above, involving a natural population and an artificial cross, and in the *P. coronopus* crosses. The reason for this phenomenon is not clear.

8.1.4 Genetic Variation for Quantitative Characters

Quantitative traits in plants, like plant height, leaf length, flowering date, number of seeds, seed weight and many others are closely connected with survival and reproduction and may consequently contribute significantly to individual fitness. The phenotype is the outcome of both genetic and environmental influences. In the following paragraphs some genetic aspects of quantitative variation in three *Plantago* species, *P. major*, *P. lanceolata* and *P. coronopus*, will be discussed; especially those aspects associated with ecotypic differentiation will be considered.

8.1.4.1 Ecotypic Differentiation in *P. major*

The two subspecies of *P. major*, ssp. *major* and ssp. *pleiosperma* (Pilger 1937; Mölgaard 1976), differ in a number of characters, number of seeds per capsule being the most distinctive. The subspecies also differ in the type of habitat in which they occur: ssp. *major* is predominantly found on roadsides and lawns, while ssp. *pleiosperma* is found in unstable sites like riversides and arable land (see Sect. 7.2).

Within both subspecies different ecotypes can be distinguished: in ssp. *major* lawn and roadside ecotypes are among the most prominent types (Warwick and Briggs 1979, 1980; Warwick 1980); in ssp. *pleiosperma* riverside and ruderal land types differ from each other. Van Dijk and Van Delden (1981) found in a study of allozyme variation in *P. major* that two allozyme loci, *Pgm-1* and *Got-1*, possessed subspecies-specific alleles; the $Pgm\text{-}1^{S_2}$ and $Got\text{-}1^{F}$ alleles were specific for ssp. *major* and the $Got\text{-}1^{I}$ allele for ssp. *pleiosperma*. At least for particular loci, the two subspecies were genetically differentiated at the allozyme level.

Allozyme variants are highly useful as markers for the study of the localization of genes controlling quantitative traits. For this purpose Van Dijk (1984) made crosses between plants which differed both for allozyme alleles at several loci and for ecotypic traits. The latter group comprised many morphological traits like leaf length, leaf blade length, leaf blade width, petiole length, inflorescence length and position, spike length, scape length and ratios of these traits and further quantitative traits like number of seeds per capsule, number of inflorescences and flowering time. It turned out that in the F_2's correlations between allozyme variants and quantitative traits existed. This is shown

Table 8.3. The F_2's $G_1 \times Z_2$: mean and standard deviation of nine characters for each genotype. Eventual significant differences between homozygotes are indicated per locus

Linkage group	Locus	Genotype	No. (max.)	Leaf length (mm)		Leaf blade length/width		Petiole length/blade length	
				Mean	SD	Mean	SD	Mean	SD
1	*Pgm*-1	S_2S_2	59	120.0	25.4	1.633	0.178	0.495	0.077
		S_2N	93	130.8	28.7	1.794	0.205	0.530	0.098
		NN	33	144.3***	37.3	1.875***	0.203	0.635***	0.087
2	*Got*-1	SS	50	128.7	29.6	1.597	0.135	0.531	0.093
		SI	99	130.1	32.3	1.758	0.171	0.539	0.109
		II	36	128.7	26.0	1.981***	0.218	0.543	0.093
3	*Me*-1	NN	54	126.4	30.7	1.753	0.219	0.545	0.107
		SN	93	131.3	30.0	1.745	0.211	0.539	0.100
		SS	38	129.4	30.7	1.796	0.222	0.523	0.098
3	*Shdh*	NN	48	122.2	28.8	1.755	0.228	0.535	0.100
		S_2N	106	132.5	30.1	1.755	0.215	0.542	0.102
		S_2S_2	31	130.1	31.9	1.769	0.202	0.525	0.105
4	*6Pgd*-2	OO	51	135.8	30.0	1.739	0.202	0.543	0.103
		NO	85	134.4	29.2	1.752	0.216	0.538	0.105
		NN	49	112.0***	26.5	1.787	0.228	0.531	0.100
4	*Got*-2	NN	48	136.0	30.2	1.752	0.217	0.551	0.105
		NF	77	133.3	29.2	1.746	0.223	0.534	0.101
		FF	60	117.6**	29.1	1.776	0.207	0.531	0.100
4	*Est*-4	NN	52	132.9	27.0	1.792	0.238	0.548	0.092
		NF	75	135.9	31.7	1.749	0.209	0.541	0.121
		FF	58	116.5**	27.8	1.738	0.202	0.524	0.081
Entire F_2			185	129.4	27.0	1.758	0.215	0.537	0.102
F_1's $G_1 \times Z_2$			9	149		1.78		0.55	
$G_1 \times G_1$			9	121		1.395		0.44	
$Z_2 \times Z_2$			10	166		2.41		0.71	

in Table 8.3 for a particular cross $G_1 \times Z_2$. G_1 is a ssp. *major* lawn ecotype with round leaves, short petioles, bent inflorescence position, high spike/scape ratio, low number of seeds per capsule and late flowering time. Z_2, on the other hand, is a typical ssp. *pleiosperma* ecotype from a river bank with oval leaves, long petioles, double bent inflorescence position, low spike/scape ratio, high number of seeds per capsule and early flowering time. For each particular trait the association, if any, with the various allozyme variants can be determined. The F_2 showed, e.g., a correlation of leaf length with *Pgm-1*, *6Pgd-2*, *Got-2* and *Est-4*, but not with *Got-1*, *Me-1* and *Shdh*. This fits well with the data on the linkage relationships between the allozyme loci (Sect. 6.1.2). In the same way, the correlations between allozyme markers and the other quantitative traits were determined. When this procedure was repeated for other crosses

Table 8.3 (continued)

Inflorescence length (mm)		Inflorescence position		Spike/ scape		Seed no. per capsule		Flowering time class		No. of inflorescences	
Mean	SD	Mean	SD	Mean	SD	Mean	SD	Mean	SD	Mean	SD
132.9	35.9	4.20	0.65	1.633	0.358	13.44	2.39	3.37	0.79	0.64	1.37
146.4	39.7	4.03	0.60	1.533	0.330	15.69	2.86	2.33	0.95	3.02	2.50
154.4*	36.2	3.90	0.70	1.292***	0.255	17.54***	3.09	1.22***	0.49	4.94***	2.00
151.3	42.8	4.22	0.64	1.569	0.331	15.42	2.92	2.28	1.09	2.66	2.70
142.5	35.8	4.01	0.65	1.506	0.364	15.52	3.25	2.44	1.11	2.70	2.61
134.0	37.9	3.92*	0.58	1.506	0.316	14.45	2.82	2.83*	1.08	2.19	2.40
141.9	42.0	4.07	0.63	1.504	0.346	15.49	3.19	2.45	1.17	2.23	2.40
145.4	38.3	4.07	0.67	1.563	0.368	15.25	3.16	2.55	10.4	2.73	2.70
140.1	34.0	3.97	0.56	1.452	0.279	15.07	2.87	2.32	1.21	2.76	2.57
137.6	40.7	4.08	0.62	1.492	0.333	15.06	3.18	2.45	1.19	2.21	2.37
146.5	38.3	4.06	0.67	1.553	0.365	15.47	3.09	2.54	1.03	2.73	2.71
140.9	35.7	3.96	0.56	1.469	0.295	14.99	3.05	2.29	1.27	2.71	2.52
155.8	38.4	3.89	0.64	1.632	0.341	14.65	3.06	2.59	1.08	2.37	2.32
147.0	37.2	4.04	0.62	1.551	0.344	15.50	3.25	2.48	1.16	2.52	2.59
119.9***	30.8	4.23*	0.64	1.354***	0.295	15.57	2.81	2.33	1.06	2.96	2.87
157.2	37.9	3.81	0.67	1.607	0.333	14.86	2.45	2.48	1.13	2.50	2.30
146.1	37.6	4.04	0.56	1.571	0.353	15.45	3.23	2.49	1.11	2.53	2.52
125.4***	33.9	4.25**	0.63	1.389***	0.311	15.41	3.41	2.45	1.11	2.75	2.92
148.4	34.6	3.90	0.65	1.588	0.309	15.24	2.83	2.42	1.11	2.73	2.48
148.3	39.2	4.03	0.62	1.542	0.389	15.55	3.40	2.46	1.10	2.57	2.67
130.1*	39.0	4.20*	0.62	1.437*	0.307	14.98	2.96	2.53	1.14	2.50	2.61
143.2	38.4	4.05	0.64	1.522	0.346	15.28	3.10	2.47	1.11	2.59	2.59
156		4.5		1.12		20.3				4.1	
122		3.5		2.45		8.4		4.0		1.6	
114		5.0		0.84		34.5		1.0		3.3	

*, **, *** Significant differences between homozygotes, $P<0.05$, 0.01 and 0.001, respectively.

involving parents with different ecotypic characteristics, a composite picture was obtained concerning the correlation of allozyme markers and relevant quantitative traits. Table 8.4 shows the final outcome; it appears that most of the quantitative characters involved in specific ecotypic traits in *P. major* were localized in linkage group 1; a few others are associated with linkage groups 2 and 4. The genes controlling these characters were apparently clustered and linked with particular allozyme loci, especially with *Pgm-1*; pleiotropic effects of the allozyme loci themselves were not probable (van Dijk 1984). In this way, different characters which contribute to the composition of a specific ecotype were combined into a gene complex. The lawn ecotype in spp. *major*, for example, had short blades giving rise to a round leaf shape, while petioles and scapes were short and the inflorescence positions was bent. The resulting phenotype was prostrate and highly resistant to mowing; leaves and inflorescences

Table 8.4. The association of quantitative traits with linkage groups in *P. major*

Character	Linkage group		
	Pgm-1	*Got*-1	*6Pgd-2/Got-2/EST*-4
Leaf length	+	−	−
Blade length/width ratio	+	+	−
Petiole length/blade length ratio	+	−	−
Inflorescence position	−	−	−
Spike/scape ratio	+	−	+
Seed number/capsule	+	−	−
Flowering time	+	−	−

+, Indicates a significant correlation; −, indicates no correlation.

were to a considerable degree protected from cutting. A ssp. *pleiosperma* river bank ecotype was erect, it had a high relative growth rate and an early flowering time and many (but small) seeds were produced. This ecotype is adapted to its highly unstable and unpredictable habitat; floodings may occur irregularly, destroying the plants. Rapid growth and early seed production will diminish the risk of not leaving offspring.

The gene complexes in the different ecotypes will contain particular sets of alleles for several loci, providing different phenotypes. Frequent recombination would disrupt such a gene complex, destroying the co-adapted group of genes. Tight linkage will decrease the probability of disruption of the complex and although no data are available concerning the number of loci determining the various quantitative traits, the results point to a rather close localization of a group of genes responsible for coherent traits around the *Pgm-1* locus. For *P. major*, as a predominantly self-fertilizing species (Van Dijk 1987), however, such complexes are well preserved because the effects of high selfing rate are identical to those to tight linkage (Hedrick 1980). Thus *P. major* is able to maintain highly genetically differentiated and specialized genotypes. The adaptive significance of these specialized genotypes has been confirmed by transplantation and other experiments (Van Dijk 1989; Chap. 7, Sect. 7.3).

8.1.4.2 Ecotypic Variation in *P. lanceolata*

In *P. lanceolata* the majority of important quantitative characters related with fitness appears to be polygenetically determined. Wolff (1987) performed an extensive quantitative genetic analysis of morphological variability in two hayfield (He and Me) and two pasture (Wd and Pa) populations. The populations studied differed considerably in quite a number of phenotypic traits when measurements under field conditions were performed. The hayfield populations were characterized by long leaves, cotyledons, spikes and scapes and further by small leaf angles and high seed weight. The pasture populations had short leaves, cotyledons, spikes and scapes, a wide leaf angle and low seed weight. The two hayfield populations also differed from each other for particular characters, as did the two pasture populations. When cultured under controlled experimental conditions, in a greenhouse or in an experimental garden, the characteristic differences proved to be conserved, pointing to underlying genetic differences. Wolff and Van Delden (1987) used full-sib families from the populations to estimate the between and the within population components of variability. Principal component analyses were performed on means of families and it was shown that three components explained 70% of the total variation. The first component was interpreted as a contrast between leaf angle and both leave and scape length, providing an index of plant height. The second component was indicated as an index for vegetative biomass and included number of leaves and of rosettes, and flowering date. The third component was interpreted as an index for generative biomass, especially total seed weight.

To estimate the genetic component in the total phenotypic variation, the variation was partitioned into variation between populations, between families and within families. It appeared that substantial variance was present both within and between populations. Table 8.5 gives (for the greenhouse data) between-population and between-family variance components, expressed as the percentage of total variance, derived from a netted analysis of variance. The characters measured show large differences in the distribution of the variance components. For some characters, like leaf angle, leaf and scape length and cotyledon length, most of the variation was found between populations; for other characters the variance was equally distributed over populations and

Table 8.5. Between population and between family variance components of quantitative characters in *P. lanceolata* presented as percentages of total variance (see text)

Character	Between-population	Between-family
Leaf angle	61***	21***
Number of rosettes	2	24***
Flowering date	6	51***
Number of leaves	18*	14*
Leaf length	55***	15***
Leaf width	18***	25***
Number of scapes	37***	12*
Scape length	48***	13**
Spike length	27**	16**
Leaf weight	30**	33***
Generative weight	10	15*
Seed weight	11	15*
Reproductive effort	11	26**
Leaf length/width ratio	18*	38***
Root weight	28	35***
Shoot/root ratio	13*	18*
Cotyledon length	55***	16***

$^{*}P<0.05$, $^{**}P<0.01$, $^{***}P<0.001$.

Table 8.6. Average heritabilities (h^2) for a number of characters in *P. lanceolata*. Heritability estimates were obtained from full-sib families from four populations

Character	h^2	Character	h^2
Leaf angle	0.99	Leaf weight	0.84
Number of rosettes	0.71	Generative weight	0.34
Flowering date	0.60	Seed weight	0.44
Number of leaves	0.54	Reproductive effort	0.61
Leaf length	0.56	Leaf length/width ratio	0.66
Leaf width	0.68	Scape/spike ratio	0.59
Number of scapes	0.35	Root weight	0.55
Scape length	0.56	Shoot/root ratio	0.43
Spike length	0.47	Cotyledon length	0.66

families. The last category of characters, including flowering date, number of rosettes and leaf length/width ratio, had a between-family component exceeding that of the between-population component.

The analyses of variance of full-sib families within each population provided the data for the estimation of the heritabilities (h^2) for each of the characters studied. Despite the large variances of the h^2 estimates, substantial genetic variation was indicated for most characters. Table 8.6 gives the h^2 values, averaged over populations for each of the test environments (greenhouse and garden).

Wolff (1988, 1990) also studied correlations between characters; such phenotypic correlations may have genetic and/or environmental causes. In a hayfield population, most characters had positively genetic correlations, except for flowering date; the latter was negatively correlated with leaf length, number of scapes, seed weight etc. This points to the existence of a hayfield ecotype, characterized by early flowering, combined with long leaves and scapes and heavy seeds. Pasture ecotypes had, among other characteristics, large leaf angles with short leaves. The characteristics are summarized in Table 8.7, which again shows that the populations within each category (either hayfield or pasture type) were still distinguishable; the differences could be related to differences in the respective habitats (Wolff and Van Delden 1987). Wolff and Van Delden (1989) showed in a bidirectional selection experiment for leaf angle that sufficient genetic variation is present in populations for a rapid response to selection.

Despite a great phenotypic plasticity, it follows from the genetic analyses (Wolff and van Delden 1987) that considerable ecotypic differentiation exists within this species. At least two prominent ecotypes could be distinguished, a hayfield and a pasture type, each with characteristics specific for the habitat (Table 8.7). In hayfields with high vegetations, light transmission is impaired (Slim and Van der Toorn 1983) and plants compete for light. A number of characteristic traits of hayfield populations can be associated with this habitat condition: long cotyledons will contribute to a high survival during the early life stage, while adults with long, erect, leaves will be better competitors for light.

Table 8.7. Genetic differentiation in *P. lanceolata*. Summary of population-specific traits for four populations: Merrevliet (Me), Heteren (He), Pannerden (Pa) and Westduinen (Wd)

Character	Hayfield populations		Pasture populations	
	Me	He	Pa	Wd
Leaf angle	Small	Small	Great	Great
Number of rosettes	Low	Low	Low	High
Number of leaves	Low	Low	Intermediate	High
Number of spikes	Low	Low	Low	High
Scape length	Long	Long	Intermediate	Short
Flowering date	Early	Early	Late	Late
Individual seed size	Great	Intermediate	Intermediate	Small
Total seed weight	About equal	About equal	About equal	About equal
Cotyledon length	Long	Long	Intermediate	Short

Long scapes will improve pollen capture and pollen dispersal, and thus improve reproductive success. For pasture plants light is not a limiting factor; the plants have to deal with the defoliating effect of grazing; pasture ecotypes have been selected for many side rosettes, large leaf angles and numerous and short leaves and scapes. The degree of ecotypic differentiation varied, depending on the features of the habitat and its temporal and spatial variability. For the pasture ecotypes, for example, the Wd population was more extreme in this respect than the Pa population. The apparent reason is that the latter had a more ruderal appearance with a higher vegetation. In view of the great amount of genetic variation shown to be present for these characteristics, it is not surprising that fine-tuning of the phenotype to the specific environmental demands of a particular habitat often occurs. The importance of the genetic differences between the various ecotypes is demonstrated by Van Groenendael (1985a), who did reciprocal transplantations between hayfield and pasture populations; it appeared that the genotypes showed the highest performance in their own environment.

For the analysis of the localization of some of the quantitative characters involved in the ecotypic differentiation in *P. lanceolata*, Wolff (1987) made crosses between a hayfield (He) and a pasture (Wd) population. The individuals used as parents differed for morphological and life-history trait characteristics, and for six allozyme markers. Analysis of the F_2's showed associations of allozyme markers and quantitative traits in a number of cases. However, unlike the situation in *P. major*, the traits were not clustered in a few major gene complexes marked by particular allozyme markers. Closely related characters were localized in different linkage groups. In *P. lanceolata*, outcrossing probably prevents the development of gene complexes as found in *P. major*. Recombination will enable the formation of a great array of genotypes in *P. lanceolata*.

8.1.4.3 Ecotypic Differentiation in *P. coronopus*

Populations of *P. coronopus* from different parts of the Netherlands were screened for quantitative variation (Wolff 1991 a, b). For this purpose plants were collected in the field and eight plants per population were selfed giving eight selfed families (F_1). From each F_1 eight plants were grown under greenhouse conditions and measurements were performed on a number of traits like leaf angle, number of leaves, leaf length, number of scapes, scape length and flowering date (Wolff 1991 a, b). Contrary to the findings in *P. lanceolata* and *P. major* populations, differentiation was nearly absent in *P. coronopus*. The between-family variation, however, was about equal in *P. coronopus* and *P. lanceolata*. The heritabilities found for the characters under study were significant in most cases. Although the *P. coronopus* population sampled occurred in quite different habitats, which, for example, varied considerably in vegetation height, no ecotypic morphological differentiation could be demonstrated. The reasons for the lack of ecotypic differentiation are unknown.

8.1.5 Strategies

The three *Plantago* species, which have been studied for genetic variation thus far show remarkable differences. Figure 8.2 gives an overview of the amount of genetic variation both for allozymes and for quantitative characters (Wolff 1990b). Total genetic variation is split up for the intra-population, the inter-population and the family level.

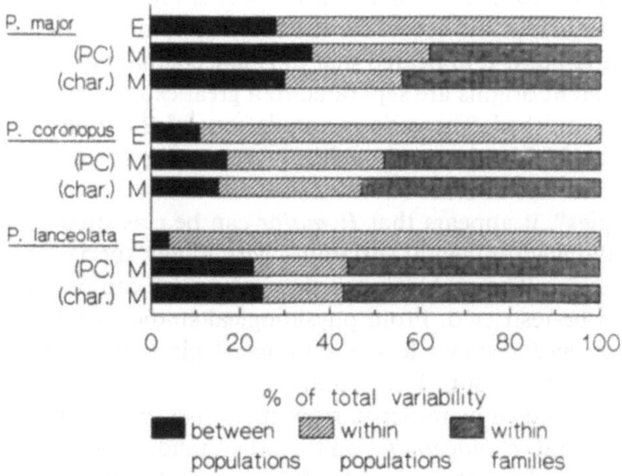

Fig. 8.2. Partitioning of electrophoretic (*E*) and morphological (*M*) variability into components of variability between and within population, expressed as percentages of total variability. Morphological variability is presented both as mean variability over a character set common to all three species (*char.*) and as principal component scores (*PC*) for the same character set. For morphological variability, the within-family component is also indicated

P. lanceolata shows considerable genetic variation, both for allozymes and for quantitative characters. Also, within a population, much variation is found. Populations are hardly differentiated for allozyme variation but they show considerable – ecotypic – variation for morphological and life-history characters. The allozyme data suggest, assuming neutrality of the variants, that gene flow (probably mainly through pollen dispersal) is effective in creating rather uniform allele frequency distributions in different populations. Selection pressures, on the other hand, are sufficiently strong to cause inter-population differentiation for quantitative characters, which are related with fitness. It has been shown (Van Tienderen 1989) that also within populations, on a microscale, genetic differentiation for such characters occurs associated with spatial habitat heterogeneity.

Compared to *P. lanceolata*, a quite different situation is found in *P. major*. Within-population variation is low for allozymes and for quantitative traits. Considerable inter-population differentiation occurs for allozyme polymorphisms, while differentiation for quantitative traits is strongly associated with specific habitat types. Genetic differentiation has substantiated two different subspecies in *P. major*, which are easily distinguishable by morphology and by allozyme markers. The chromosomal clustering of ecotype-specific traits in this species, associated with the high degree of selfing, leads to highly specialized phenotypes with specific combinations of morphological traits.

P. coronopus exhibits genetic variation at the intra- and inter-population and the species levels. Surprisingly, there is some interpopulation differentiation for allozymes, but more for quantitative traits. The differences between the three *Plantago* species in morphological differentiation are visualized in Fig. 8.3, which shows the results of principal component analyses for several quantitative characters (Wolff 1991a, b). Mean factor scores for each family in each population were calculated and scores for the first two components are plotted for each species separately. It appeared that in *P. lanceolata* and *P. major*, populations from different origins are separated to a great extent and that ecotypes can easily be distinguished. The various populations of *P. coronopus*, however, showed a considerable overlap and could not be distinguished from each other under greenhouse conditions.

In considering "strategies", it appears that *P. major* can be classified as a species encountering spatial environmental variability with highly specialized, nearly homozygous, genotypes. The range of environments which can be mastered in this way seems to be restricted. From physiological studies it follows (D. Kuiper 1983b) that *P. major* shows less physiological plasticity than *P. lanceolata*. The species *P. major* is split into a limited number of specialized genotypes which can be considered as rather narrow specialists, each capable of occupying a different habitat. *P. lanceolata*, on the other hand, shows much more physiological plasticity than *P. major* (D. Kuiper 1983b, 1984). The ecotypes form a more or less continuous range with pronounced hayfield and pasture ecotypes at the extremes. An average *P. lanceolata* individual seems to be able to cope with a greater variety of habitats than an average *P. major* individual. Thus, *P. lanceolata* is, comparatively speaking, more a generalist than

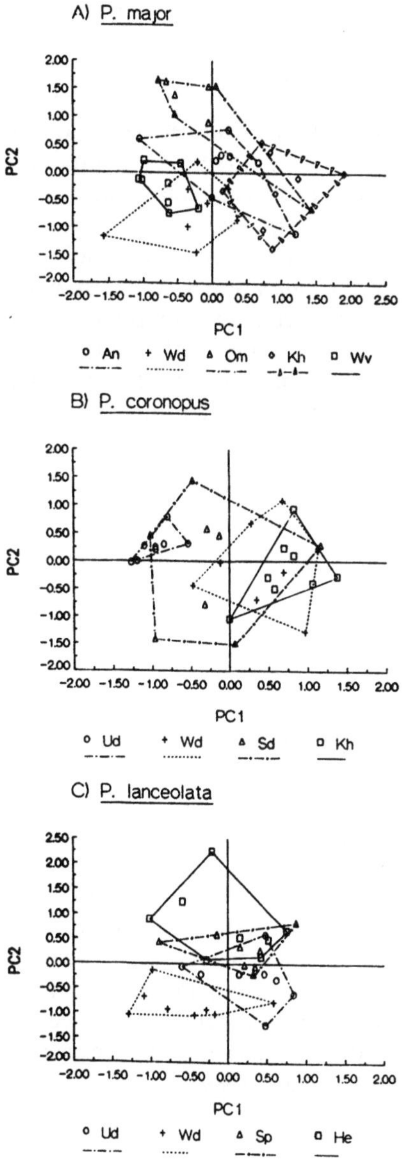

Fig. 8.3. Family means of individual component scores on the first and second axis derived from principal component analysis of populations of *P. major* (**A**), *P. coronopus* (**B**) and *P. lanceolata* (**C**). Family means of a particular population are indicated by *different symbols* and the most extreme family means of each population are connected by *lines*. The populations are indicated *below the PC 1 axis*

P. major. The situation for *P. coronopus* is not yet clear. In the populations studied so far, hardly any differentiation for quantitative characters was observed, though interpopulation differentiation for allozymes was found and genetic variation for quantitative characters was present. Since *P. coronopus* plants often show in situ morphological differences between habitats which disappear for the greater part under greenhouse conditions, it is possible that high plasticity is important for survival in this species.

It is evident that even within a *Plantago* species great diversification of life history strategies may occur. *P. major* forms an excellent illustration of this phenomenon (Van Delden 1988). The two subspecies differ considerably in whole sets of coherent important fitness components (Van Delden 1985). *P. major* ssp. *major* has a low relative growth rate, flowers relatively late and produces big seeds: ssp. *pleiosperma* is fast-growing, early-flowering and produces great amounts of small seeds. These life history properties can be associated with the diverse habitats the subspecies encounter. Ssp. *major* occurs in rather stable habitats, while ssp. *pleiosperma* has to meet highly unstable conditions. These differences lead to the classification of ssp. *major* as a K-strategist, while ssp. *pleiosperma* follows an r-strategy. Thus, the genetic variation present within the species has been sufficient for the evolution of highly different complexes of physiological and morphological traits. The emphasis in future research should be directed to the decisive role of the breeding system in the accomplishment of particular evolutionary solutions in the genus *Plantago*.

8.2 A Population Biological Model of *Plantago major*: an Attempt to Integrate Genetic, Demographic and Physiological Knowledge

8.2.1 Introduction

The integration of the disciplines genetics, ecology and physiology of plants was an important aim of the *Plantago* project. In the participating institutes research at the interfaces of two of the three disciplines had already been carried out. In this section an attempt is made to integrate knowledge from all three disciplines simultaneously, as far as this knowledge has been gathered within the *Plantago* project, supplemented, when necessary, with assumptions where data were not available.

This is done by means of a model designed for computer simulation. The model has been tested for being realistic: the outcome is that the obtained simulation results are very close to the reality in nature in many respects. The model was constructed by combining and integrating an existing model on population structure, dealing with neutral genetic variation and genetic drift (see Van Dijk 1987) with a model on the action of natural selection upon ecotypes of *P. major* (see for experimental data Van Dijk 1985a, 1989).

8.2.2 The Ingredients of the Model

The core of the model is a smaller submodel: that of the plant. This submodel describes the production and allocation of biomass. The result depends on the genotype of the plant, and on the prevalent environment. Finally, random processes play a part (the simulations work completely at a stochastic basis).

Let us start with the environment: four environmental variables are chosen, which are of importance for ecotypic differentiation within *P. major* (Van Dijk 1985a). These four variables are:

1. Trampling intensity, The more intense trampling is, the more severely plants are damaged, unless they are "resistant". Instead of trampling also mowing or grazing could have been chosen, with similar implications for the model.
2. (Effective) season length. This length varies between 90 and 160 days. Short effective growth season lengths are met at river banks (where also a considerable year-to-year variation will be present) and on arable land.
3. Winter severity. This factor causes the loss of vegetation and of a large part of the seeds at river banks.
4. Ploughing (whether or not).

   Finally, there is an additional fifth variable: the time, in years, during which a growth place already exists, starting with a completely open situation, such as can be brought about by human influence.

Many of environmental factors which are typical for the occurrence of *P. major* (like compact soil, rather high amounts of water and nutrients), but do not play a part in ecotypic differentiation, are not included in the model. An extended model could comprise various levels of light intensity, salinity and mowing or grazing.

The five variables determine together the fate of a *P. major* population in the model, but also that of the rest of the vegetation: the fraction of the soil surface available for establishment of *P. major* is small if the vegetation is not kept open by trampling, winter damage or ploughing. Along with the amount of disturbance of the area, the opportunity for establishment of *P. major* seedlings increases, until at a strong disturbance level also for this species the environment will become too bad: no vegetation is left beyond a certain level. Within a particular range in the five-dimensional environmental space, higher densities of *P. major* are possible. This range will be enlarged by the fact that different ecotypes are adapted to different subranges. In the model, loci are present for four quantitative traits which are important for these ecotypic adaptations. For each trait three loci are present, each having a "high" and a "low" allele. The alleles work additively: there is no dominance and no interaction. The phenotype is determined by these genes and by random environmental influences.

If a habitat exists for a long time, and the environment is constant, finally the best-adapted available genotype(s) will occur at that place. This requires two more characters: firstly, a certain ability for colonization to provide for the possibility of arrival of the species at that particular place and also for the arrival of the most favourable alleles, and in the second place the possibility of recombination of traits, which is the more important if the environment is deviating from the original environments of the arriving seeds. Colonizing ability in this species may be correlated with stickiness of the seeds; recombination ability depends on the level of outcrossing.

If habitats generally exist during relatively short periods, or if environmental conditions change in the course of time, the importance of long-distance transport of seeds (the ability to escape from geographical restrictions) and the importance of recombination (the ability to escape from ecological restrictions) will increase. Gene flow and breeding system characteristics are, therefore, of essential importance for the model. Genetic variation (again three loci per trait) has been introduced in the model for both long-distance seed transport and outcrossing rate. Genetic variation for these characters is, however, not based on research data.

8.2.3 How the Model Works

8.2.3.1 General Aspects

In Fig. 8.4 the whole model is shown, as it is constructed using the smaller model of production and allocation of biomass; the numbers of individuals in the various life stages; the environmental variables and the genetic variables. The computer programme based on this model works as follows.

Let us start in spring when the establishment of new plants takes place: in Fig. 8.4 at the point with N_5. For each of the (in this computer programme: 10000) sites where an adult plant fits in (areas of 1 dm^2), is checked whether or not:

- a *P. major* plant which survived the last winter occupies that site;
- a plant of any other species occupies that site;
- seeds of *P. major* are present at that site.

Winter survival will be discussed later on. The presence of plants of other species depends on the time of existing of the habitat in years (y), and on the level of disturbance given by trampling intensity (i) and winter severity (w). The probability of open places being occupied by such plants is:

$$1-(4y+1)^{-[(1-0.5w)(1-1.5i)]}, \quad (1)$$

with the restriction that the maximum values that can be substituted for y, w and i are 5, 2 and 2/3, respectively. The ranges of w and i will be indicated later on. Initially (y = 0) no plants are present; beyond w = 2 or i = 2/3, no plants other than *P. major* are ever present. In the case of ploughing, 80% of the sites are occupied by other plants (crop plants and other weeds) in every year.

Seeds of *P. major* have been shed in the previous year by plants growing in that area, with a probability decreasing with distance from the parent's site to the site of the seeds. That probability is also directly proportional to the number of seeds produced by the particular parent. Occasionally, seeds may immigrate from other areas by means of long-distance seed transport.

If no surviving *P. major* plant, or a plant of any other species is occupying the site, but seeds of *P. major* are expected to be present, than 0, 1, 2 or more seedlings of *P. major* will appear. In the case of more than one seedling, seedling competition takes place. A seedling originating from a larger seed has a better chance of winning the competition than a seedling from a smaller seed. Only one seedling will survive. Of this winning seedling the genotype is determined by drawing a set of genes from the mother (which has already been

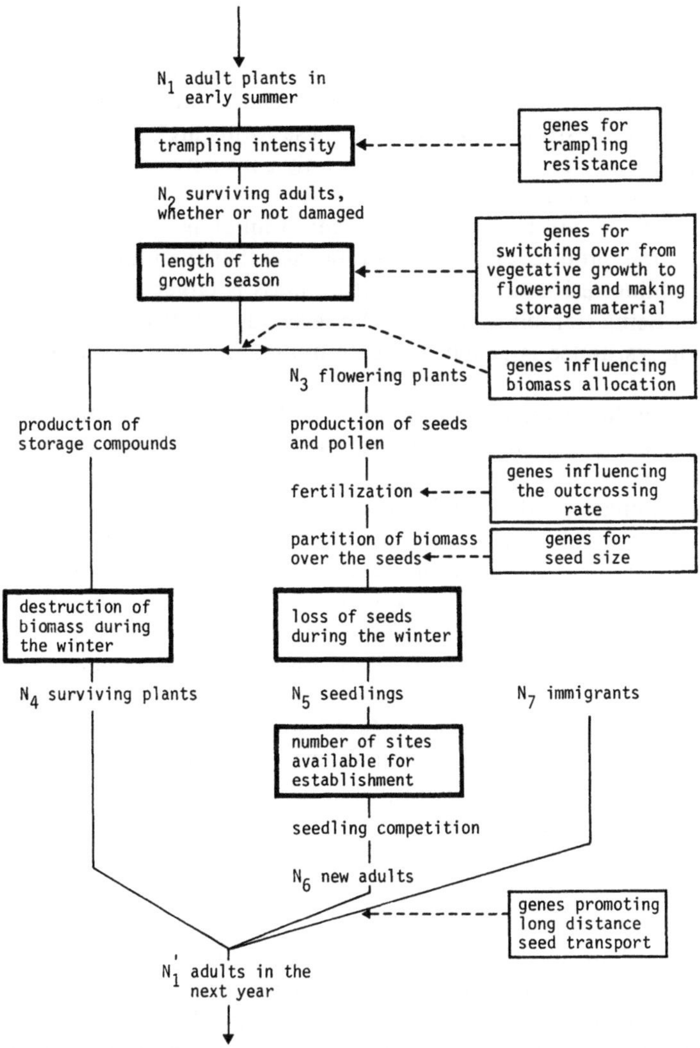

Fig. 8.4. The model. Environmental variables are in *bold rectangles*, genetic variables are in *light rectangles*

chosen, because it is the female parent plant which determines seed size) and from the father. This male parent is drawn from the pollen cloud, which is built up by the contributions of the plants in the area. The pollen cloud at the place of the female parent will usually contain a large fraction of own pollen of that female parent, responsible for selfing.

A newly established seedling starts growing, and at a certain moment it switches from vegetative growth (which is supposed to be proportional to the square of the time since germination) to investment in either seeds (flowering will start then) or in underground storage material. From the moment of the switch-over, the plant keeps the same size, while seed or storage production takes place linearly until the end of the growth season. Seeds need 20 days to ripen. The moment of the switch is genetically determined and will be discussed later on. At the end of the growth season, biomass will be present in the form of seeds and/or storage compounds: these forms are further called "biomass"; other tissues (leaves, roots, etc.) are only considered as vehicles in order to produce this "biomass".

During the growth season trampling can occur. Trampling causes damage at the cost of biomass production, at least in the case that the individual did not invest in trampling resistance. The loss of biomass is larger as trampling intensity is larger. Being trampling-resistant, however, also costs biomass. The level of trampling resistance is genetically determined and will be discussed later.

The ratio between the production of both forms of biomass: seeds and storage compounds, is also genetically determined. First-year plants have a threshold for flowering: below a certain investment in reproduction, flowering will not start and all biomass produced will be storage biomass.

The production of seed biomass is accompanied by a proportional production of pollen. The number of seeds produced depends, given a certain seed biomass, on individual seed weight. Seed weight is genetically determined. Another genetically determined trait is the ability of seeds to be (occasionally) tranported over long distances instead of falling down around the female parent plant.

Certain numbers of seeds have been produced now and are laying down at each of the 1 m^2 sites. This is the autumn situation. Winter severity determines the loss of seeds during the winter. Especially at river borders, a large proportion of seeds will disappear. With a winter severity parameter w (w varies between 0 and 3) a fraction of

$$10^{(-2-0.5w)} \tag{2}$$

of the seed remains. This means that normally only 1% of the seeds will survive (the others disappearing after being fallen down at unfavourable places, being eaten, getting affected by moulds, etc.). At a larger than zero w value, the remaining number of seeds may be reduced by a further factor up to about 30. Seeds being transported over long distances have an additional risk of arriving at unfavourable places: an additional proportion of 90% disappears.

A plant which has invested in underground storage biomass has a chance of winter survival, depending on the magnitude of the investment (r mg) and on winter severity (w). The survival probability is:

$$k = \frac{r}{400} \cdot 10^{-0.5w}, \tag{3}$$

with a maximum of k = 0.8 (there is always 20% chance of dying by aspecific causes like moles, rabbits, pathogens, etc.). In the case of ploughing, k = 0.

After having survived the winter, a plant has an advantage in the next year(s) over new established plants: in spring a large individual is already present which is able to flower relatively early. Therefore, an older plant will produce many more seeds than a seedling in the first year. Additionally, the perennial plant may become larger every year, if the circumstances and the genotype are favourable; seed production will increase proportionally.

In the model genetic variation is present for each of the six quantitative traits (see the light rectangles in Fig. 8.4) having three loci each. Also three neutral loci are included, together 21 loci, all segregating independently.

Passive gene flow is determined by the parameters σ_S, σ_P and t, the (axial) standard deviation of seed or pollen dispersal and the outcrossing rate, respectively. The outcrossing rate is variable, both through genetic variation for factors influencing this rate and through density (this will be discussed later on). In the model $\sigma_S = 10$ cm and $\sigma_P = 40$ cm, which implies that almost all seeds stay within a radius of 35 cm, and almost all pollen within a radius of 140 cm.

8.2.3.2 The Production of Biomass

As mentioned, not plant biomass itself is relevant for the model, but the "useful" biomass which is ultimately found in seeds and therefore determines the number of offspring. Storage compounds are an investment for future seed production and are useful in that respect.

In the model a seedling grows during a number of days (g) only vegetatively, and after the switch (at moment g) during the remaining of the season useful biomass is produced (with 20 days lag time for the ripening of the seeds and the deposition of storage materials). If (effective) season length = s, a period of $(s-g-20)$ days remains for the production of useful biomass. At the moment of the switch, the size of the plant is proportional to g^2 (because the growth rate is supposed to be proportional to the square of the time since germination). The amount of biomass produced is now equal to:

$$0.003\, g^2(s-g-20)\ \text{mg}. \qquad (4)$$

The maximum result is obtained if $g = 2/3\,(s-20)$, as can be calculated by differentiation of Eq. (4).

The value of g is determined by three loci, each possessing a "low" and a "high" allele, and by a certain environmental influence. Each high allele enhances the expectation of g with 8 days. If all six alleles are low, g = 46 days; if all alleles are high, g = 94 days. Environmental variation has a standard deviation of 5 days, and causes a random variation around the expectation set by the genotype. In the same way all further quantitative traits are evaluated.

To each season length belongs an optimal g value, and one or more optimal genotypes causing an expectation of g closest to the optimal value. In Fig. 8.5, for each even number of high alleles and for every possible season length it is shown how large the resulting biomass production is. Odd numbers of high alleles are intermediate (but are rare because of the high level of homozygosity

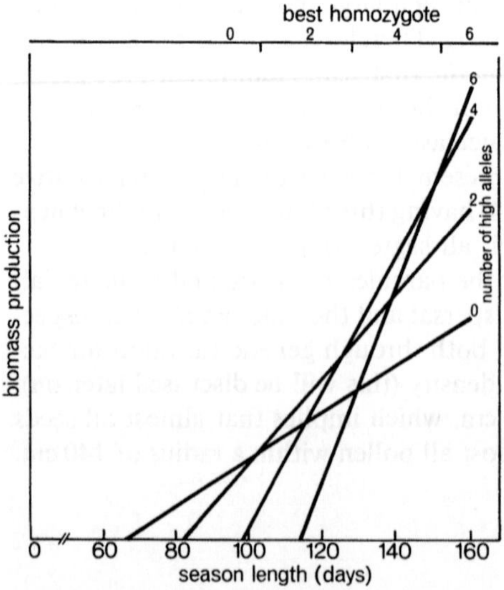

Fig. 8.5. Biomass production as a function of season length for several "switch-over" moments

due to selfing). Additionally is shown which (even) number of high alleles causes the highest biomass production for a particular season length.

Individuals having survived the winter, possessing r mg of storage material will start early with their biomass production. In the model their yield is equal to:

$$0.05\,r(s-60)\ \text{mg}. \tag{5}$$

Notice that the switch at moment g is a juvenile character only; in later years the switch is invariably early, independent of the genotype, and is 40 days.

Trampling resistance implies investment in stronger leaves and scapes at the cost of useful biomass. Trampling resistance (b) ranges from 0 to 0.6 at the cost

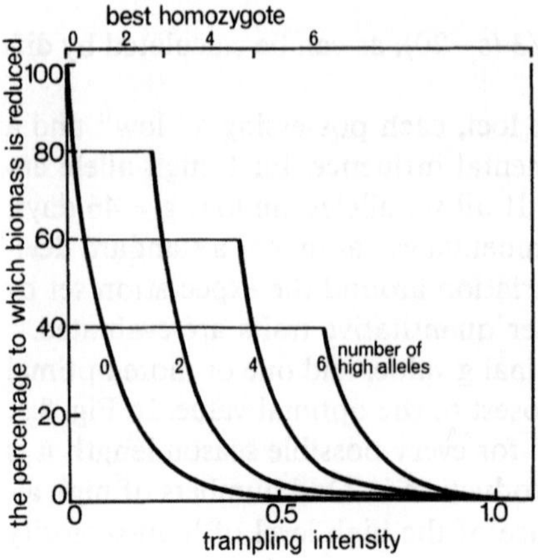

Fig. 8.6. Biomass reduction as a consequence of trampling resistance and trampling intensity at different levels of trampling intensity

of a fraction b of useful biomass. Trampling intensity i ranges from 0 to 1. The damage as a consequence of trampling is related to b and i; biomass is reduced to a fraction:

$$3^{10(b-i)}\,, \tag{6}$$

but if $b > i$ no damage will result.

The quantitative "trampling resistance" has an expectation of 0 with six low alleles, and an expectation of 0.6 with six high alleles. Standard deviation of environmental variation is 0.1. In Fig. 8.6 the relation is shown between the number of high alleles and the deleterious effect on biomass production in the range of i between 0 and 1. Also indicated is the most favourable (even) number of high alleles for each i value.

8.2.3.3 The Allocation of Biomass

The alternatives of investment of biomass in seeds or in storage materials deal with an annual, biennial or perennial strategy. Annuals produce nothing but seeds in their first (and only) year. Biennials produce storage materials but no seeds in their first year (or in later years) until, by whatever mechanism, flowering starts, and the effort is completely spent on seed production. Perennials often do not flower in the year of germination and invest both in seed and storage materials in their adult phase. Values for both the ratio of the two ways of investment and an threshold effort below which a plant will not flower, can describe all these possibilities.

In the present model there is a threshold value of 50% investment in reproduction below which a plant does not flower in its first year. In all other cases the ratio between investment in seeds and in storage compounds is determined by the familiar set of three loci with a low and a high allele each. An individual possessing six low alleles spends all on reproduction, while a plant possessing six high alleles spends 60% on storage materials and 40% on reproduction. The standard deviation due to the environmental variation is 5%.

What is the best strategy under particular circumstances: being an annual or a perennial? The answer depends on the ratio in mean yearly seed production by annuals and perennials under those circumstances, and of interest is that particular ratio at which both strategies can coexist. Perennials are thought to be represented by all age classes in a stable age distribution. The following parameters are defined:

k = the mean probability of winter survival (perennials);

v = the fraction of the annual number of empty sites that are available for establishment by *P. major* [a fraction of $1-v$ of the empty places will be occupied by other species, see Eq. (1)];

d_A = the density of annuals (fraction of the total number of sites occupied by annuals);

d_P = the density of perennials (the same);

f = the mean annually produced number of seeds by perennials divided by the mean number of seeds produced by annuals.

Each year a fraction of $v(1-kd_P)$ places is available for *P. major* (whether or not other species are perennials and occupy the same places as the year before is not considered). The probability that annuals occupy these sites is $d_A/(fd_P+d_A)$, and that perennials do the same is $fd_P/(fd_P+d_A)$. In this case of equilibrium (coexistence of both strategies):

$$d_A = \frac{d_A}{fd_P+d_A} \cdot v(1-kd_P) \qquad (7)$$

and

$$d_P = \frac{fd_P}{fd_P+d_A} \cdot v(1-kd_P)+kd_P. \qquad (8)$$

Equation (7) implies that $v(1-kd_p) = fd_P+d_A$, and substitution into Eq. (8) leads to $d_P = kd_P + fd_P$, or $f = 1-k$. (Each ratio between d_A and d_P is possible on condition that $d_A/d_{A,MAX}+d_P d_{P,MAX} = 1$ with $d_{A,MAX} = v$ is the density if only annuals were present and $d_{P,MAX} = v/(1-k + kv)$ is the density if only perennials were present).

Mean lifetime is directly connected with mean winter survival probability and is $L = 1/(1-k)$. This means that in the case of coexistence $fL = 1$, leading to the conclusion that in such a situation a perennial plant produces, during its whole lifetime, as many seeds as an annual plant does in 1 year. It does not matter when a plant produces seeds; the only important point is how many seeds per lifetime are produced. This result holds for a population which is nor growing (early reproduction would be favoured) nor diminishing in numbers (late reproduction would be favoured).

A second aspect of biomass allocation is seed size. Would it be better to produce (less) larger or (more) smaller seeds? Three loci again play a role in this character. Six low alleles lead to an expected weight of 100 µg per seed; six high alleles to an expected weight of 400 µg per seed. The standard deviation due to environmental variation is 15 µg. The optimal individual seed weight is related to the intensity of intra-specific competition between seedlings (inter-specific competition is not considered in relation to individual seed weight, but has an intensity related only to environmental parameters).

Thus far the fitness of a genotype could simply be defined as being proportional to the expectation of the seed production during a plants's lifetime (when population size is stable). The fitness of seed weight genotypes, on the contrary, appears to be very complex, and is both density-dependent and frequency-dependent. This will be elaborated only briefly. To start with, a few parameters will be defined:

z = the mean number of seeds per plant leading to seedlings, depending on the mean (annual) seed production per plant and the reduction due to Eq. (2);

d = the density of *P. major* (the fraction of the total number of sites occupied by this species);

k = the mean probability for winter survival;

v = the fraction of the annual number of empty sites that are available for establishment by *P. major* [a fraction of 1 − v of the empty places will be occupied by other species, see Eq (1)];

In the next year the mean number of seedlings at one site is zd. There is, however, only place for one adult. Assuming that seeds are randomly distributed over the area, the probability of finding at least one seed at a particular site is, using the formula for the Poisson distribution,

$$1-e^{-zd}\ . \tag{9}$$

The number of available sites for establishment of *P. major* seedlings is v(1 − kd). After seedling competition the total density of *P. major* plants is:

$$d' = kd + v(1-kd)\,(1-e^{-zd})\ . \tag{10}$$

In the case of equilibrium d′ equals d, and

$$-\ln\left(1-\frac{d(1-k)}{v(1-kd)}\right) = zd\ . \tag{11}$$

The straight line zd and the curved line of the left part of Eq. (11) can intersect in two points: in d = 0 and in a second point (see Fig. 8.7), representing the equilibrium density, which is situated closer to the maximum density $d_{MAX} = v/(1-k+kv)$ as z is larger. The tangent of the curved line in d = 0 is

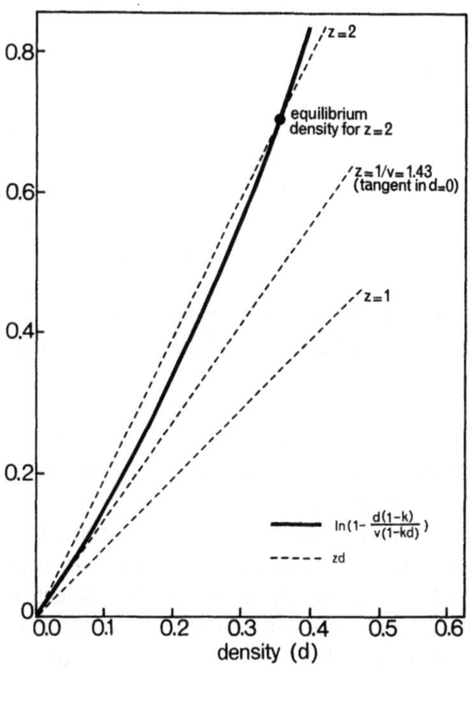

Fig. 8.7. Seed weight and plant densities (example with v = 0.7 and k = 0).

$(1-k)/v$ [because the left part of Eq. (11) approaches $d(l-k)/v$ when $d\to 0$], which means that the product of z and v should be greater than $1-k$ for a stable maintenance of a genotype leading to the production of the number of seeds associated with z.

Let us consider now the situation of a number of different seed weight classes (e.g. 400 µg, 300 µg, 200 µg and 100 µg, with an increasing value of z if seed biomass production is constant). The largest seed(ling) class (class 1) for which holds that $z_l v > 1-k$ will reach the equilibrium density d_l according to Eq. (11), not being influenced by smaller seed(ling) classes which are always outcompeted by the largest seed(ling)s. The remaining places, can be occupied by the next smaller seed(ling) class, etc. In this way it is possible that in an equilibrium situation several seed weight genotypes coexist. Relative fitness of each genotype is 1 in that case. A higher frequency will lead to a lower fitness, a lower frequency to a higher fitness (frequency-dependent selection).

So far a discontinuous seed weight distribution has been considered. In reality, a continuous distribution will occur. Given certain environmental conditions, a particular range of seed weights will coexist.

8.2.4 Simulation Results

Computer simulations using the model described can be carried out in different ways, depending on the geography and the stability of the terrain. Three possibilities will be discussed:

A) there is only one area, with constant environmental conditions both spatially and from year to year;

B) there is only one area, but environmental conditions may vary both spatially and between years;

C) a system of small areas exists, between which migration is possible. Each of these areas is spatially homogeneous, but areas differ from each other. The time of existence of an area is limited, after which it is replaced by a new, empty, area with new environmental parameters. A variant is that the areas persist, but that the environmental conditions will change gradually in a random direction.

When a simulation of type A is carried out, natural selection will favour an increase in frequency of the better adapted genotypes. This process can be split up into three partial processes:

1. the better-adapted genotypes which are present will increase in frequency;
2. as a result of recombination better-adapted genotypes are created;
3. favourable alleles and/or genotypes enter the population (by means of migration; mutation is not included in the model).

When process 1 is involved, population structure plays an important role. If genotypes are non-randomly distributed over the population area, and intermingling is slow because of restricted gene flow, there will be less selection against less fit genotypes, because these genotypes will often be surrounded by similar genotypes. An example will make this clear: let there be a population with 50% better-adapted genotypes, leaving a mean offspring number of 10, and 50% less fit genotypes, leaving a mean offspring number of 5. Total population size is constant, numbers are reduced independent of the genotype, and there is no cross-fertilization. In a random mixture of genotypes, the better-adapted genotype will be represented in the next generation by a frequency of 2/3. This is the maximum speed of selection possible under the given conditions. In the other extreme situation, each of both genotypes occupies a separate part of the area, constant in size, and there is no migration between the parts. In that case there is no selection at all. If a neighbourhood structure exists, the speed of selection is in between both extremes, dependent on the level of local differentiation for both genotypes (see Van Dijk 1987).

Essential for the partial processes 2 and 3 are recombination rate and the level of long-distance gene flow, respectively. These features lose their advantageous effects, however, if no new favourable alleles enter the population and no new genotypes arise. There will even be selection against characters promoting recombination and long-distance gene flow, if genetic variation for these characters is present. This will be directly clear considering long-distance seed transport: sending away seeds will occur at the cost of the local probability of getting offspring. The reason why also the ability of cross-fertilization is selected against will be discussed in the next section.

Simulations of type B are carried out using a terrain containing the environmental conditions for three well-known ecotypes of *P. major*: the "riverbank type" (small-seeded annuals), the "roadside type" (trampling-resistant perennials) and the "arable land type" (large-seeded annuals). A map of the terrain (10×10 m, containing 10000 sites of 1 dm^2 each) is shown partially in Fig. 8.8. At the riverbank the winter severity w decreases and season length s increases with the distance to the river. The values of these parameters also change every year. At the path, trampling intensity is the strongest in the middle, and also varies every year.

The simulation result is very realistic: at the riverbank small seeded annuals occur in strongly fluctuating densities. Higher at the river bank, the species *P. major* is hardly present: vegetation is too tall because the level of disturbance is too low. At both sides of the path, trampling-resistant perennials occur, and on the arable land a large-seeded annual dominates. In spite of a larger environmental variability and instability, selection against outbreeding and long-distance seed transport is constantly working, and these characters will finally disappear.

Simulations of type C are the most useful for an analytical approach. Problems which can be studied are, for instance, how genetic variation is adjusted to the environment and which role is played by cross-fertilization and long-distance gene flow in obtaining this adjustment. If the mean lifetime of a population is limited, strong selection appears in favour of long-distance gene flow if the system of populations (the metapopulation) is considered. Also cross-fertilization is maintained at a low level in the metapopulation in some of the simulations carried out. Genetic variation for these characters tends to disappear, facilitating the maintenance of a certain level of long-distance gene flow and outcrossing in the metapopulation.

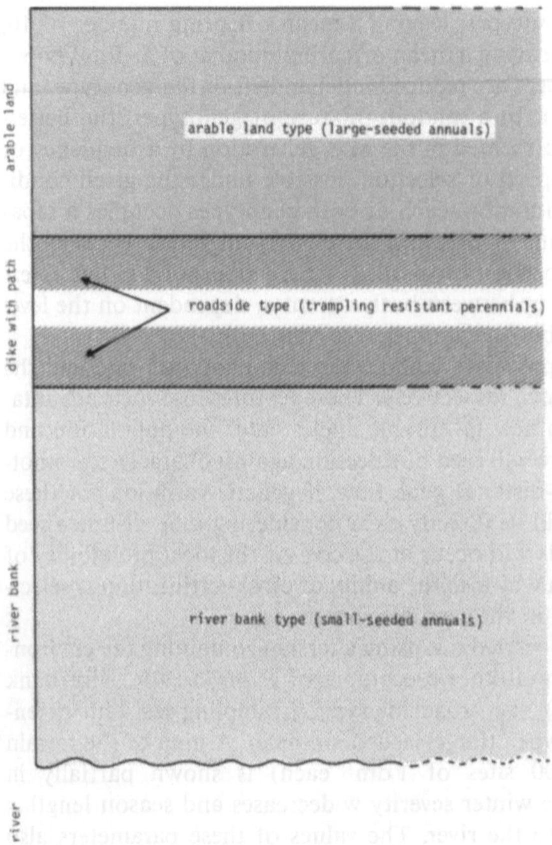

Fig. 8.8. The terrain in simulation B (partially)

8.2.5 Cross- and Self-Fertilization

An interesting topic in plant biology is the balance between selfing and outcrossing. When a species is self-incompatible (like *P. lanceolata*), a barrier has to be taken by a mutant gene which eliminates the self-incompatibility system. A strong inbreeding depression will then arise because deleterious recessive genes may become homozygous. In a predominantly inbreeding species like *P. major* such an inbreeding depression is practically absent (van Dijk 1989). If in a species like *P. major* a mutant gene appears, making cross-fertilization impossible but leaving the production and dispersal of pollen unaffected, a mechanism already described by Fisher (1944) will lead to the fixation of that mutant allele at the cost of the original allele, until the population is completely selfing. This process is visualized, in its most extreme form, in Fig. 8.9. The

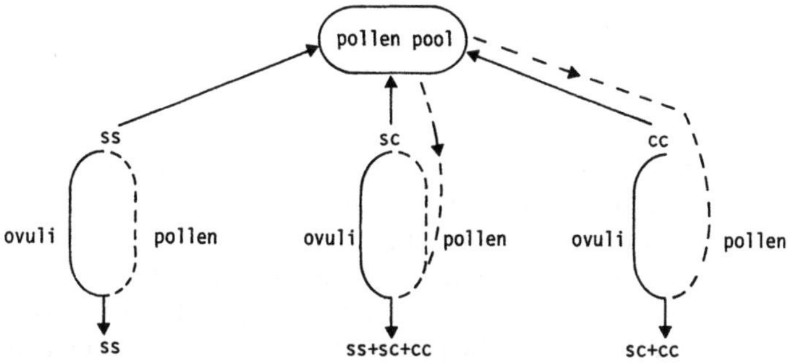

Fig. 8.9. The Fisher mechanism, causing a decrease in frequency of the c allele (promoting outcrossing)

mutant allele is indicated by *s*, the original one by *c*. The genotype *ss* is an obligatory self-fertilizer, *cc* an obligatory outbreeder, and the heterozygote can operate in both ways. An *ss* plant will produce only *ss* seeds, but will additionally spread *s* genes through its pollen. An increase in the *s* frequency will be the consequence. This mechanism will also promote mutant alleles which lead to a higher level of selfing than the original allele does. The mechanism of Fig. 8.9 does not work if the increase in selfing is accompanied by a loss of contribution to the pollen pool, which happens in the case of cleistogamy or when insect pollination is prevented.

In *P. major*, protogyny is the character which most strongly influences the outcrossing rate. Without protogyny, cross-fertilization would be nearly impossible, because new pistils would almost certainly be pollinated by native pollen being already present in large quantities. The genetic variation for outcrossing rate in the model may be considered as genetic variation for protogyny. The question why cross-fertilization is maintained in the species *P. major* in spite of Fisher's mechanism can also be formulated as: why is protogyny maintained?

In the model for self- and cross-fertilization, as elaborated by Van Dijk (1987), the rate of cross-fertilization t depends on the pollen dispersal parameter σ_P, density d and a factor Z which describes the probability of an own pollen grain to succeed in fertilization in proportion to that probability of an alien pollen grain:

$$t = \frac{2\pi\sigma_P{}^2 d}{Z + 2\pi\sigma_P{}^2 d} . \tag{12}$$

An equal pollen production for each plant is assumed. If pollen production varies, but remains proportional to the seed production, and has variance V, t will be smaller than in Eq. (12):

$$t = \frac{2\pi\sigma_P{}^2 d}{Z(1+V) + 2\pi\sigma_P{}^2 d} . \tag{13}$$

By way of protogyny, genetic variation is present for the Z factor. Z = 0 will lead to complete outcrossing (t = 1), whereas an infinitely high Z will lead to complete selfing (t = 0).

In the maintenance of outcrossing, the speed with which alleles for a lower Z tend to disappear, due to the Fisher mechanism, play an important role. If Z (or protogyny) is considered as a quantitative trait, genotypes for a low outcrossing ability will occur. In a low density situation complete selfing will be approximated, and the Fisher mechanism no longer works. But if (local) density increases, outcrossing will still be possible (because a larger value of d gives a higher t value in equation 13). Thus, the ability for outcrossing may remain "silent" during long periods without disappearing.

A second factor slowing down the loss of ability for outcrossing is population structure, as was mentioned earlier. The influence of both retarding effects implies that an equilibrium between breakdown and promotion of outbreeding ability is much less unlikely than in the Fisher model (but only at low outcrossing levels). In nature, outcrossing level is estimated to be about 10% under regular densities. In simulations of type C, the level decreases to about 5–10%. This could be either the equilibrium level in the natural situation of the species or the equilibrium level in the past, with a subsequent loss of genetic variation for outcrossing rate. The latter possibility raises the question whether genetic variation for protogyny is present in *P. major*, and, if not, how the species will react if alleles are introduced (by mutation) for weaker protogyny. Data concerning genetic variation for protogyny are not yet available for *P. major*.

8.2.6 What's the Use of the Model?

A model like the one presented here can be useful in three respects. In the first place, as a descriptive model, it generates a number of questions, stimulating experimental research in the fields where knowledge is insufficient. During the construction of the model, many assumptions have been made which evoke curiosity in order to establish how things really are. The introduction of more and more experimentally based knowledge into the model will lead to increasingly realistic simulation results. The present lack of knowledge largely holds for biomass production and allocation, and particularly the way in which genetic variation is involved. Summarizing, the model in its present state would be an excellent starting point for another period of multi-disciplinary research on *Plantago major*.

A second function of the model is the generation of theoretical questions which are of general interest, apart from the species *P. major*. A simplification of such a problem may enable a mathematical treatment, after which the solution can be tested on correctness, realism and robustness by replacing it into the large model which stands closer to reality. A few examples of such attempts to discover general rules are already presented: the advantage of an annual or a perennial life history; producing small or large seeds; being cross- or self-fertilizing. Particularly the latter topic is interesting in view of the colonizing ability of a species.

A third function of the model is its usefulness for educational purposes. Several aspects of population biology can be illustrated in such a way that also the interrelationships will become clear.

9 Conclusions

In the project the relationships between the demographic, physiological and genetic properties of the five *Plantago* species and the characteristics of their environment were studied at the species, population and individual plant levels. This allowed an analysis of phenotypic and genetic responses of populations and species to environmental conditions, the individual plant being the level at which selection takes place. Some conclusions of this analysis will be discussed below.

9.1 Comparison of Species

Each species occurred in at least several locations. Particularly, *Plantago lanceolata* was found over a wide range of different habitats whereas that for *Plantago media* was quite small. To establish common features in the locations of each species, phytosociological, soil chemical and microclimatological factors were considered. In general, the five species were found in different plant alliances but *P. lanceolata* sometimes occurred together with one of the other species. Soil and climatic values of the habitats of the species generally showed a considerable overlap, be it that the amplitudes in *P. lanceolata* were highest and that also those of *Plantago major* with its two subspecies and distinct ecotypes varied considerably. The salt content of the environment discriminated most between the species, *Plantago maritima* being on the halophytic side, and *P. media* on the glycophytic. The latter was restricted to locations rich in calcium carbonate in contrast to the other species.

By ordination of the plant communities according to their species composition the impression was obtained, that although they occurred over a wide range of soil conditions, *P. major* was found more frequently in nutrient-rich environments than *P. lanceolata*. However, no relationship could be established between the occurrence of the species and the separate chemical soil factors of their locations. Also, the chemical analysis of plant material and the C-A values did not suggest a relationship between chemical soil conditions and the occurrence of the species. Such a direct relationship cannot be expected from a comparison at this level, because nutrient availability is greatly influenced by soil physical factors, such as moisture content and compaction, and by the presence or absence of competition with other plant species. Therefore, with regard to nutrient availability, the fact that *P. major* often occurs in trampled, open sites (without competition), and *P. lanceolata* in hayfields (with competition), should be taken into account.

Field studies, common garden experiments and studies in the laboratory showed considerable differences between the species in their reactions to trampling, soil compaction, grazing and competition. The species reacted in a different way in their various life stages, e. g. radicle emergence, seedling establishment, growth and reproduction. *P. major* ssp. *major* can stand trampling and soil compaction much better than *P. lanceolata* and it even is stimulated by these factors, e. g. by a better moisture supply. On the other hand, germination and seedling establishment of *P. major* ssp. *major* in a dense vegetation was inferior to that of *P. lanceolata*, particularly in dry soils. This different reaction could be ascribed to a different response to the red/far-red ratio in the light conditions of the hayfields. *P. coronopus* was susceptible to trampling in the seedling stage, but not as an adult plant. This may explain why the species occurs in habitats that are only irregularly trodden by man and cattle. Experiments with *P. media* in Ca-rich dune soils show that its absence from the dunes is due to inferior germination and seedling establishment so that no plants reach the adult stage; however, planted adults survive for years.

From the above it emerges that separate factors can hardly be related to the occurrence of the species; due to the specific combinations (clusters) of these factors, the species in general are bound to different phytosociological plant alliances.

The demographic and life history properties of the five species had much in common but characteristic differences did exist. *P. coronopus* and *P. major* ssp. *pleiosperma* with their low seed weight, high seed production and low adult survival could be considered as r-strategists, whereas *P. lanceolata, P. maritima* and *P. media* were more K-strategists, i. e. with a comparatively high seed weight, low seed production and high adult survival; *P. major* ssp. *major* was intermediate. *P. lanceolata* showed great variation in its demographic properties which is not astonishing in view of its wide occurrence. When this species was found in a vegetation type together with another plantain species, its adult survival was generally low, whereas in habitats whithout one of the other species, e. g. in open hayfields, it was much higher. In such apparently optimal habitats its longevity equalled that of *P. major*. A characteristic difference between the two species appeared to be the ratio between the numbers of seedlings and adults, which was 2–6 for *P. major* and 0.2–2.5 for *P. lanceolata*; this corresponded with a much higher seedling mortality in *P. major* (95%) than in *P. lanceolata* (40%).

Studies on the genetic properties of the species were mainly focussed on *P. lanceolata*, the two subspecies of *P. major*, and to a lesser extent on *P. coronopus*. In all three species allozyme variation appears to be present in amounts very well comparable with other plant species, though the amount of variation differed between species. The differentiation for allozymes and morphological characters was highly correlated with the reproductive system. In *P. major*, a self-compatible species which is highly self-pollinating, much less allozyme variation was present than in *P. lanceolata*, a self-incompatible species. *P. coro-*

nopus and *P. major* showed a good concordance of the distribution of allozyme and morphological variation, possessing only a slightly higher morphological than electrophoretic differentiation between populations. *P. lanceolata* showed a higher morphological than electrophoretic differentiation between populations.

The comparison of the physiological properties of the species that were studied in conventional water cultures and in water cultures with an exponential addition rate of a limiting nutrient revealed considerable differences. Characteristic differences in response to NaCl existed between *P. maritima* (halophyte) and *P. media* (glycophyte). Tonoplasts of *P. maritima* showed Na^+/H^+ antiport activity when the plants were exposed to NaCl. This activity was absent in controls and in *P. media*. *P. lanceolata* and the two subspecies of *P. major* reacted quite differently to shifts in the nutrient level or in the light conditions. The plasticity of the plant properties of *P. lanceolata*, such as shoot-root ratio, level of cyanide-resistant respiration, levels of activity of glutamate dehydrogenase, glutamine synthase, nitrate reductase and ATPases, was much higher than that of *P. major* ssp. *major*. It is tempting to relate the high degree of plasticity in *P. lanceolata* to its higher degree of heterozygosity, but this hypothesis is falsified by the results with the homozygous *P. major* ssp. *pleiosperma*; the latter possessed the same degree of plasticity as *P. lanceolata*. Also, the hypothesis that a high plasticity can be considered as an adaptation to a high degree of variation under environmental conditions (e.g. changes in soil humidity resulting in shifts in nutrient supply) could not be confirmed in field studies which monitored a number of soil parameter over the course of the year in habitats with populations of *P. lanceolata* and *P. major* ssp. *major*. This lack of evidence for the hypothesis is due to lack of quantitative insight into the nutrient supply of individual plants in the very heterogeneous soil system and in understanding of competition for nutrients between plants. Nevertheless, it is hardly conceivable that a high degree of plasticity, which will involve considerable costs in terms of resources, has no adaptive value and we agree with Einstein that the absence of proof cannot be considered to be proof of absence.

In experiments with an exponential supply of growth-limiting nitrate, plants were obtained with similar characteristics as plants from the field (e.g. no nitrate accumulation). Under these conditions the growth rate and nitrogen productivity of *P. major* were higher than for *P. lanceolata*. At low growth rates the latter species had a more extensive root growth and a lower shoot-root ratio. These results confirm the hypothesis that plants from nutrient-rich environments are adapted to these conditions by means of a high growth rate, whereas the low growth rate of species from nutrient-poor environments will have no adaptive value but rather is the result of a low shoot-root ratio. Similar results were obtained in experiments with both subspecies of *P. major* grown on phosphate. However, the field data collected for *P. major* and *P. lanceolata* (levels of chemical soil factors, nutrient contents of plant material) only partly confirm the above hypothesis.

Summarizing, we conclude that the comparison of related species has yielded important information, particularly for those environmental factors that could be studied in all life stages of the plant, in the field, in experimental plots and in the laboratory. The main discriminating adaptive properties are related to factors such as soil compaction, trampling, grazing and competition, i.e. mainly morphological properties, and to the ability to predict environmental conditions, i.e. characteristics of K- and r-strategy. The species are also differently adapted to sea salt. In laboratory experiments the physiological properties of the species related to the uptake and utilization of nutrients also differed to some extent but no unequivocal relationship with the chemical soil properties could be established. The main problem in comparing these two is the difficulty in translating values of chemical soil factors into a nutrient supply of the roots. Nevertheless, our impression is that factors related to the morphology of *Plantago* species are more important in determining their distribution than chemical soil factors.

9.2 Comparison of Populations

The two subspecies of *P. major* and, particularly *P. lanceolata*, have a wide ecological amplitude and, therefore, they are very suited for a study of the correlation between phenotypic and genetic variation and habitat factors. The correlation between genetic variation and environmental factors might suggest that they are the result of selection processes to which the individual plants are exposed. However, in the field, particularly in a changing environment, the effects of natural selection could not easily be distinguished from gene flow and random drift. The analysis of the relationship between demographic and morphological properties and environmental factors was further complicated by the strong correlation between the variables within each of the two groups of data and by the non-linearity of the relations with fitness and with multiplicative fitness functions. To cope with these problems different approaches were used: comparison of the properties of a large number of natural populations, using a multivariate principal component analysis on uniform experimental plots, and reciprocal transplant experiments at the original sites of the populations. A more precise measurement of the quantitative genetic variability was made in crossing experiments and combining these with data of allozyme variation.

The demographic properties of *P. lanceolata* were most affected by density and height of the vegetation in summer and by the water saturation of the soil in mild winters. Seedlings and small adult plants with few leaves were particularly susceptible to the density of the vegetation. It was not clear, which aspect of vegetation density, e.g. root competition or light interception, was responsible for the low survival. In additional experiments evidence was obtained that light interception was at least partially involved. Notwithstanding the high gene flow, the high plasticity of the individual plants and a considerable amount of independent genetic variation for morphological and fitness parameters, a clustering of these properties was observed as the result of mowing and grazing. This resulted in a pasture ecotype with a large leaf angle and a decumbent rosette, often with daughter rosettes, and a high number of inflorescences with light seeds and a high potential seed production, which is not realized because many inflorescences are eaten. On the other hand, plants from hayfields are characterized by an erect rosette, a small leaf angle, large leaves and scapes, and heavy seeds. In hayfields a relationship between flowering time and mowing date was found. In a common garden experiment the genetic nature of these complexes was established. It was supposed that selection took place on groups of characters rather than on separate ones. Such complexes of characters are possibly governed by apical dominance: an erect hayfield type of growth as a result of a higher apical dominance and a pasture type of growth as a result of a lower one.

Such genetically determined clusters of characters in the outbreeding *P. lanceolata* can only be brought about by high selection pressures. In demonstrating such selection pressures, reciprocal transplant experiments were used. Hayfield and pasture populations had a marked reduction of fitness in each other's habitats. In one population direct evidence was found for selection on growth habit. The local adaptation was manifested in particular in the reproduction, and – to a lesser degree – in the survival of adult plants. The reduction in lifetime seed yield of aliens could be up to 95%, i.e. selection coefficients that are observed in extreme environments, e.g. in case of heavy metal tolerance.

Apart from selection for growth habit other factors may also be involved in local adaptation. Local adaptation was also shown to exist in reciprocal transplant experiments with hayfield populations in which below-ground processes such as root competition may also be involved. Our lack of understanding of the root structure of the populations precludes further speculation in this direction. We should remember that in an experimental plot the root systems of the various *Plantago* species were highly affected by competition, trampling and soil compaction.

In *P. lanceolata* much of the allozyme variation is present within populations whereas the variation between populations is small. Within the variable enzyme loci in this species five linkage groups probably corresponding with five different chromosomes could be distinguished. Linkage of these groups with particular quantitative traits existed. These quantitative trait loci did not appear to be concentrated in major gene complexes. This result confirms Maynard Smith's hypothesis that in species with a high gene flow and which experience a variable environment in successive generations, gene complexes with low recombination rates and fixed allele combinations are less favourable. In *P. lanceolata* a positive correlation exists between the number of heterozygous allozyme loci and fitness characters, which may suggest a relationship between heterozygosity and fitness. However, in a demographic study in which the age of the plants was known, no association between heterozygosity and age could be established. Consequently, the results only partly support the hypothesis that there is a correlation between heterozygosity and fitness.

The high plasticity of metabolic processes as a possible adaptive reaction to changes in the supply of nutrients or light has already been discussed. The individual plant is able to cope with the variation in light conditions and in the availability of water and, consequently, in nutrient supply which it encounters over the growth season in many locations. There was hardly any physiological differentiation between populations. There were some indications of a lower light compensation point in plants from hayfields than in pasture plants. Also, at low light intensity plants from a hayfield increased their shoot to root ratio more than plants from a pasture. As considerable physiological differences must exist between genotypes of *P. lanceolata*, e. g. between hermaphrodites and male steriles (see below), we conclude that there are no population-specific selective forces on physiological processes and that the physiological plasticity needed to cope with variation in environmental conditions in time, is sufficient to meet with the variation in space; alternatively, this differentiation has escaped our measurements.

In the highly self-pollinating species *P. major* with its small neighbourhood size, the gene flow was much lower than in *P. lanceolata*. In the two subspecies a considerable genetically determined, morphological variation was shown to exist between populations, which resulted in a wide array of ecotypes that are considered to be the result of selection: e.g. a lawn type (highly competitive), an erect type from non-trampled roadsides, a path type (trampling-resistant), a coastal meadow type and a riverbank type (adapted to changing environmental conditions) could be discerned. In reciprocal transplant experiments and in competition experiments the local type was best adapted to the specific selection regimes. In general *P. major* ssp. *major* ecotypes had a more perennial character and they were found in environments that were less variable in time than those of the *P. major* ssp. *pleiosperma* ecotypes that have a more annual character.

Many of the loci involved in the genetic variation in morphology of leaves and inflorescence formed a gene complex closely linked to the Pgm-1 locus, which possessed different alleles in both subspecies. This gene complex can exist in at least three different forms, i.e. in ssp. *pleiosperma*, in ssp. *major* lawn type and in ssp. *major* roadside type. The linkage between allozyme loci and fitness-affecting loci can explain the restriction of some enzyme alleles to a particular subspecies. In predominantly selfing species the presence of gene complexes of fitness-related traits may be an advantage as only individuals possessing an integrated gene complex can function adequately in the appropriate well-defined environment. The chance that the complex is transferred as a whole is evidently greater when the complex is more compact.

The variation within populations at the allozyme level was less than in *P. lanceolata*, but between populations there was more variation. The comparison of both species shows that between populations of *P. major* there is more differentiation for both allozyme loci and morphological characters.

In contrast to *P. lanceolata* metabolic differences were found between ecotypes, more particularly between both subspecies. The potential relative growth rate (RGR_{max}) in ssp. *major* was lower than in ssp. *pleiosperma*, which was among others due to a lower specific leaf area, and a higher percentage of dry matter. This resulted in a lower photosynthetic activity per unit leaf weight. These properties, together with the below-ground position of the apex and a higher content of cell wall material, may contribute to the greater tolerance to trampling and soil compaction found in ssp. *major.* Specialization at the subspecies level has also been shown for light intensity and mineral nutrition, and the subspecies also differed in plastic response. The plasticity was particularly high in the riverbank type of ssp. *pleiosperma*, an ecotype with essentially an annual life cycle. In this ecotype the plasticity is likely governed by hormones, particularly cytokinins. Plasticity is apparently related to changes in environmental conditions during its short life cycle. The low plasticity in ssp. *major* ecotypes suggest the occurrence in locations with environmental conditions that are stable in time. However, it has already been mentioned that we were not able to prove the validity of the hypothesis in the field. It was shown that the growth response of *P. major* ssp. *pleiosperma* to a reduction in mineral nutrition was not determined by the actual mineral concentrations in roots and shoots but depended on endogenous cytokinin levels as determined by immunoassay. In a plastic line of this subspecies the growth responded quickly to a decrease in mineral nutrition; growth rate and endogenous cytokinin level decreased simultaneously within 2 to 3 days, whereas mineral contents started to decrease at a much lower rate. Differences in endogenous cytokinin level of *P. major* ssp. *major* and *P. major* ssp. *pleiosperma* at least partly explained the differences in RGR_{max} of the two subspecies.

At the population level much less information on the other species has been collected. The morphological differentiation between populations in *P. coronopus* was low, whereas the individual physiological plasticity was high. In *P. maritima* only two populations were studied that differed considerably in demographic properties. However, in greenhouse experiments these differences disappeared completely when plants of both populations were grown under identical conditions. Therefore, the tentative conclusion is that the differences between both populations are the result of phenotypic plasticity of individual plants. As yet no relevant data are available for *P. media* to allow a comparison of different populations.

Comparison of the differentiation in demographic, genetic and physiological properties between populations shows that considerable differences exist between the species. The predominantly inbreeding species *P. major* is thought to cope with the variation in environmental conditions by means of a high differentiation between populations that are locally adapted and consist of individuals with a high homozygosity and a low plasticity. This would be achieved in the outbreeding species *P. lanceolata* by a genotype with a high plasticity. Apparently, this picture is far too simple. In *P. major* ssp. *pleiosperma* ecotypes with a high plasticity occur in environments with large temporal variation in environmental conditions. *P. lanceolata* is adapted to spatial variation by a high individual plasticity, and a considerable morphological variation between populations has been brought about by high selection pressures such as mowing. In *P. coronopus* this morphological differentiation between populations has not been observed and the species apparently responds to variation in space and time by a high plasticity of the individual plants, although these individuals are phenotypically characterized by a high degree of homeostasis. The few data available for *P. maritima* likewise suggest a strategy with a high degree of individual plasticity, but with a low degree of homeostasis which results in a phenotypic differentiation in demographic and morphological characters of the populations. Clearly, in our plantains adaptation to the habitat has been realized by various combinations of individual plasticity, homeostasis and genetic differentiation.

In evaluating the results obtained at the species and population level, we may conclude that a full appreciation of the relationships between the life-history, genetic and physiological properties of the species and their environmental conditions is only achieved by comparisons in one species populations that are adapted to different environmental conditions. In the *Plantago* species such comparisons confirm the picture that morphological and physiological adaptations together with the genetic structure contribute to the species' strategy of coping with environmental conditions, such as mowing, grazing and trampling.

9.3 Comparison of Individual Plants and Subpopulations

Within populations of the three *Plantago* species that were studied in most detail a considerable genetic variation in allozyme composition and morphological properties was found. The amount of within-population variation was highest in *P. lanceolata*, lowest in *P. major* and intermediate in *P. coronopus*. Populations of *P. major* were composed of genetically homogeneous families. The number of inbred lines in such populations depends on population size and habitat diversity as well as on the selection regime. Within some populations of *P. lanceolata* less morphological differentiation was found than was expected. This is thought to be the result of a strong directional selection in a rather homogeneous habitat. By this selection pressure certain morphological complexes were favoured, causing the loss of morphological variability. In *P. coronopus* a positive relationship was found between outcrossing rates and variability, indicating that higher levels of inbreeding were associated with less variability.

Two studies were conducted to investigate whether or not the genetic differentiation within populations is related to heterogeneity of the environment. One study concerned three different subpopulations of *P. major* ssp. *pleiosperma*, growing in a mosaic environment that differed in nutrient availability and degree of inundation. Considerable phenotypic differences were observed in the field between the plants of these subsites, but they appeared to be mainly the results of a high phenotypic plasticity in fitness components. In the greenhouse small genetic differences in life-history characters between plants of these subsites were established but no indications of specific adaptations to levels of nutrient supply or water availability were found. We suggest that the differences in life-history characteristics are the result of spatial variability in the vegetation structure which in itself is caused by a differential nutrient and water supply.

In a second study genetic differentiation in populations of *P. lanceolata* was studied along a topographical gradient on which subsites were chosen. In a reciprocal transplant experiment it was found that at some sites native plants performed best but plants from different subsites did not differ in important components of plant performance such as growth, survival and seed production. Thus, the differences were by no means comparable to those that occur between populations. The fact that the diversifying selective forces along the gradient were insufficient to achieve much genetic differentiation was due to the levels of gene flow. In addition, the subsite that contributed most to the total seed pool was the site in which no differences between plants from different subsites were found. From the results of both studies we conclude that the absence of much genetic differentiation within populations in relation to environmental heterogeneity shows that the selective forces were too weak to compensate the gene flow.

A special aspect of genetic differentiation within populations is the occurrence of specific genotypes with a different fitness such as male steriles and hermaphrodites in populations of *P. lanceolata*. For one of the male-sterility types such fitness differences were established by planting root cuttings in the field. They appeared to be related to the adult stage, i.e. a higher survival and seed production. The physiological basis of these fitness differences could hither- to not be established. The suggestion of Antonovics' group that they are related to a different level of the alternative respiratory chain was not confirmed by us. The male sterility types occurred in variable proportions in different populations which appeared to be related to some extent to environmental conditions such as soil humidity. This suggests an environment-dependent fitness difference between hermaphrodites and the male-sterility type or between cytoplasmic types. Experiments are in progress, using molecular biological techniques, to localize these cytoplasmic and nuclear genes, and to relate them to physiological properties and fitness.

A particularly interesting situation occurred in sympatric populations of diploids and tetraploids of *P. media*. These genotypes can be considered as different species because they are completely separated genetically (the fitness of the triploid hybrid is zero). The diploids and tetraploids coexist because of a differential selection for flowering time. However, this does not answer the question of the niche differentiation between both karyotypes, which certainly have much in common but in the field cannot be distinguished from another and of which one (the tetraploid) has to compensate the costs of a higher DNA content.

In the preceding years we have collected much information on the adaptations of the *Plantago* species and their populations to environmental conditions, and we have studied how these adaptations in terms of fitness and selection came about and are maintained. We have shown that there is no clear relationship between specific abiotic factors and the occurrence of each species. This is because these factors act mainly through plant-plant and other biotic interactions, such as the vegetation structure, whereas anthropogenic factors (haying, grazing by cattle) are also important in making up the specific set of environment conditions in which each species occurs. The species copes with this set of environmental conditions by the genetic structure of its populations, the genetically determined morphological and physiological characteristics of its individuals and the plastic responses in these characteristics. These plastic responses, i.e. the reaction norms, which in themselves are genetically determined, are often more important than genetic differences that are expressed independently of the environment. From this point of view we consider the niche of these species to some extent as a stochastically composed set of abiotic and biotic environmental factors to which the individuals are adapted with a set of morphological and physiological properties rather than by a unique keyhole relationship in the sense of Hutchinson. However, the reciprocal transplant experiments, in which a high selective advantage of the local genotype was found, show nevertheless that the tuning in of the set of plant properties to the set of environmental properties must be rather accurate.

From the results presented in the preceding chapters we have cut in this chapter a bunch of flowers within the triangle between demography, genetics and physiology. The bunch is a personal selection. The reader is invited to pick his own bunch from the results, which were mainly the outcome of twenty doctoral theses.

References

Adam P, Birks HJB, Huntley B (1977) Plant communities of the Island of Arran, Scotland. New Phytol 79:689–712

Ahmad J, Larker F, Stewart GR (1979) Sorbitol, a compatible osmotic solute in *Plantago maritima*. New Phytol 82:671–678

Ahokas H (1982) Cytoplasmic male sterility in barley: evidence for the involvement of cytokinins fertility restoration. Proc Natl Acad Sci USA 79:7605–7608

Alexander HM (1982) Demography of and intraspecific variation in *Plantago lanceolata* in relation to infection by the fungus *Fusarium moniliforme* var. *subglutinans*. Thesis, Duke Univ, Durham, NC

Alexander HM (1984) Spatial patterns of disease induced by *Fusarium moniliforme* var. *subglutinans* in a population of *Plantago lanceolata*. Oecologia 62:141–143

Alexander HM, Antonovics J, Rausher MP (1984) Relationship of phenotypic and genetic variation in *Plantago lanceolata* to disease caused by *Fusarium moniliforme* var. *subglutinans*. Oceologia 65:89–93

Andary C, Motte-Florac ME, Gargadennec A, Wylde R, Heitz A (1989) Les esters caféïques du genre *Plantago*. Identification et valeur chimiotaxinomique. Plant Med Phytother 22:17–22

Angevine MW, Chabot BF (1979) Seed germination syndromes in higher plants. In: Solbrig OT, Jain S, Johnson GB, Raven PH (eds) Topics in plant population biology. Columbia Univ Press, New York, pp 188–207

Antonovics J (1976) The input from population genetics: The new ecological genetics. Syst Bot 1:233–245

Antonovics J, Primack RB (1982) Experimental ecological genetics in *Plantago*. VI. The demography of seedling transplants of *P. lanceolata*. J Ecol 70:55–75

Ardakani MS, Schulz RK, McLaren AD (1974) A kinetic study of ammonium and nitrite oxidation in a soil field plot. Soil Sci Soc Am Proc 38:273–277

Argyriadis GA, Dijkshoorn W, Lampe JEM (1976) Level and origin of carboxylate in buckwheat. Plant Soil 44:669–681

Assouad MW, Dommee B, Lumaret R, Valdeyron G (1978) Reproductive capacities in the sexual form of the gynodioecious species *Thymus vulgaris* L. Bot J Linn Soc 77:29–39

Atal CK (1958) Cytoplasmic male sterility in psyllium (*Plantago ovata* Forsk.) Curr Sci 27:268

Avid D, Galun E (1980) Restoration of fertility in cytoplasmic male sterile (CMS) *Nicotiana sylvestris* by fusion with X-radiated *N. tabacum* protoplasts. Theor Appl Genet 58:121–128

Baas R (1989) Phosphorus nutrition and vesicular-arbuscular mycorrhizal infection in *Plantago major*. Thesis, State Univ Utrecht, Neth

Bass R, Kuiper D (1989) Effects of vesicular-arbuscular mycorrhizal infection and phosphate on *Plantago major* spp. *pleiosperma* in relation to internal cytokinin concentrations. Physiol Plant 76:211–215

Baas R, Lambers H (1988) Effects of vesicular-arbuscular mycorrhizal infection and phosphate on *Plantago major* ssp. *pleiosperma* in relation to the internal phosphate concentration. Physiol Plant 74:701–707

Baas R, Van Dijk C, Troelstra SR (1989) Effects of rhizosphere soil, vesicular-arbuscular mycorrhizal fungi and phosphate on *Plantago major* ssp. *pleiosperma* Pilger. Plant Soil 113:59–67

Babble GR, Selander RK (1974) Genetic variability in edaphically restricted and widespread plant species. Evolution 28:619–630

Bakker JP (1983) Seed bank and species diversity in moist grassland communities. Acta Bot Neerl 32:246

Bakker JP, Dijkstra M. Russchen PT (1985) Dispersal, germination and early establishment of halophytes and glycophytes on a grazed and abandoned salt-marsh gradient. New Phytol 101:291–308

Bakker K (1964) An analysis of controversies about population theories and their terminologies. Z Angew Entomol 53:187–208

Baldo BA, Chensee QJ, Howden MEH, Sharp PJ (1982) Allergens from plantain (*Plantago lanceolata*): studies with pollen and plant extracts. Int Arch Allerg Appl Immunol 68:295–304

Banwart WL, Pierre WH (1975) Cation-anion balance of field-grown crops. 1. Effect of nitrogen fertilization. Agron J 67:14–19

Barber HN, Halsall GDM, Palmer JH (1968) Genetic, environmental and physiological control of leaf orientation in *Plantago lanceolata*. Aust J Biol Sci 21:641–648

Barret SCH, Helenium K (1981) Floral sex ratios and life history in *Aralia nudicaulus* (Araliaceae). Evolution 35:752–762

Barton NH, Hewitt GM (1985) Analysis of hybrid zones. Annu Rev Ecol Syst 16:113–148

Bassett IJ, Crompton CW (1968) Pollen morphology and chromosome numbers in the family Plantaginaceae in North America. Can J Bot 46:349–361

Bates GH (1935) Vegetation of footpaths, sidewalks, cart tracks and gateways. J Ecol 23:470–487

Bawa KS (1980) Evolution of dioecy in flowering plants. Annu Rev Ecol Syst 11:15–39

Beach JH (1981) Pollinator foraging and the evolution of dioecy. Am Nat 118:572–577

Beck T (1979) Die Nitrifikation in Boden. Z Pflanzenerähr Bodenkd 142:344–364

Beeftink WG (1975) The ecological significance of embankment and drainage with respect to the vegetation of the south-west Netherlands. J Ecol 63:423–458

Beeftink WG (1977) Salt marshes. In: Barnes RSK (ed) The coastline. John Wiley & Sons, New York Chichester, pp 93–123

Belliard G, Vedel F, Pelletier G (1979) Mitochondrial recombination in cytoplasmic hybrids of *Nicotiana tabacum* by protoplast fusion. Nature (London) 281:401–403

Belser LW (1979) Population ecology of nitrifying bacteria. Annu Rev Microbiol 33:309–333

Bennet J (1983) Regulation of photosynthesis by reversible phosphorylation of the light-harvesting chlorophyll a/b protein. Biochem J 212:1–13

Berendse F (1981) Competition and equilibrium in grassland communities. Thesis, State Univ Utrecht, Neth

Berendse F (1983) Interspecific competition and niche differentiation between *Plantago lanceolata* and *Anthoxanthum odoratum* in a natural hayfield. J Ecol 71:379–390

Bestagno BML, Ciferri R, Bestagno G (1958) Ordiamento artificiale delle specie del genere *Coniothyrium* Corda. Sydowia 12:258–320

Bhat KKS, Nye PH, Brereton AJ (1979) The possibility of predicting solute uptake and plant growth response from independently measured soil and plant characteristics. VI. The growth and uptake of rape in solutions of constant nitrate concentration. Plant Soil 53:137–167

Bianchini F, Corbetta F, Pistoia M (1975) La planta della salute. Mondadori, Milano

Bingham ET (1980) Maximizing heterozygosity in autopolyploids. In: Lewis WH (ed) Polyploidy, biological relevance. Plenum, New York, pp 471–489

Bjoerkman Ö (1981) Responses to different quantum flux densities. In: Lange OL, Nobel PS, Osmond CB, Ziegler H (eds) Responses to the physical environment. Encyclopedia of plant physiology, vol 12A. Springer, Berlin Heidelberg New York, pp 57–107

Bjoerkman O, Holmgren P (1963) Adaptability of the photosynthetic apparatus to light intensity in ecotypes from exposed and shaded habitats. Physiol Plant 16:889–914

Bjoerkman O, Holmgren P (1966) Photosynthetic adaptations to light intensity in plants, native to shaded and exposed habitats. Physiol Plant 19:854–859

Blacquière T (1988) Physiology and ecology of ammonium nutrition in *Plantago* and *Alnus*. Thesis, State Univ Groningen, Neth

Blaquière T, Koetsier C (1988) The response of *Plantago lanceolata* L. and *P. major* L. ssp. *major* to nitrate depletion. Plant Soil 107:197–206

Blaquière T, Lambers H (1981) Growth photosynthesis and respiration in *Plantago coronopus* as affected by salinity. Physiol Plant 51:265–268

Blaquière T, Voortman E, Stulen I (1988) Ammonium and nitrate nutrition in *Plantago lanceolata* and *P. major* ssp. *major*. III. Nitrogen metabolism. Plant Soil 106:23–34

Blom CWPM (1976) Effects of trampling and soil compaction on the occurrence of some *Plantago* species in coastal sand dunes I. Soil compaction, soil moisture and seedling emergence. Oecol Plant 11:225–241

Blom CWPM (1977) Effects of trampling and soil compaction on the occurrence of some *Plantago* species in coastal sand dunes II. Trampling and seedling establishment. Oecol Plant 12:363–381

Blom CWPM (1978) Germination, seedling emergence and establishment of some *Plantago* species under laboratory and field conditions. Acta Bot Neerl 27:257–271

Blom CWPM (1979) Effects of trampling and soil compaction on the occurrence of some *Plantago* species in coastal sand dunes. Thesis, Cath Univ Nijmegen, Neth

Blom CWPM (1983) Plasticity of life characteristics in two different populations of *Plantago maritima*. L. Acta Oecol/Oecol Plant 4 (18):377–394

Blom CWPM, Husson LMF, Westhoff V (1979) Effects of trampling and soil compaction on the occurrence of some *Plantago* species in coastal sand dunes. IVa. The vegetation of two dune grasslands in relation to physical soil factors. Proc K Ned Akad Wet, Ser C, Biol Med Sci 82:245–273

Bloss HE (1982) Formation of mycorrhiza in *Plantago ovata*. Mycologia 74:505–508

Böcher TW (1943) Studies on variation and biology in *Plantago lanceolata* L. Dansk Bot Arch 11:1–19

Bolt GH Bruggenwert MGM (eds) (1976) Soil chemistry. A. Basic elements. Elsevier, Amsterdam Oxford New York

Boorman LA (1971) Studies in salt marsh ecology with special reference to the genus *Limonium*. J Ecol 59:103–120

Bos F (1980) Pembrikeshire. Lev Nat 82:69–80 (in Dutch)

Bos M, van der Haring E (1988) Gene flow in *Plantago*. II. Gene flow pattern and population structure, a simulation study. Heredity 61:1–11

Bos M, Steen R, Harmens H (1985) Protogyny in *P. lanceolata* populations: an adaptation to pollination by wind? In: Jacquard P, Heim G, Antovovics J (eds) Genetic differentiation and dispersal in plants. Springer, Berlin Heidelberg New York, pp 327–337

Bos M, Harmens H, Vrieling K (1986) Gene flow in *Plantago* I. Gene flow and neighbourhood in *P. lanceolata*. Heredity 56:43–54

Bowman RA, Cole CV (1978) An exploratory method for fractionation of organic phosphorus from grasslands soils. Soil Sci 125:95–101

Bradshaw AD (1959) Population differentiation in *Agrostis tenuis* Sibth. I. Morphological differentiation. New Phytol 58:208–227

Bradshaw AD (1960) Population differentiation in *Agrostis tenuis* Sibth. III. Populations in varied environments. New Phytol 59:92–103

Bradshaw AD (1965) Evolutionary significance of phenotypic plasticity in plants. Adv Genet 13:115–155

Bradshaw AD (1984) The importance of evolutionary ideas in ecology – and vice versa. In: Shorrocks B (ed) Evolutionary ecology. Blackwell, Oxford, pp 1–25

Braun U (1987) A monograph of the Erisyphales (powdery mildews). Cramer, Berlin

Braun-Blanquet J, Roussine N, Nègre R (1951) Les groupements végétaux de la France méditerranéenne. Centre Natl Rech Sci Serv Carte Group Vég Fr, Paris

Brewbaker JL (1967) The distribution and phylogenetic significance of binucleate and trinucleate pollen grains in the angiosperms. Am J Bot 54:1069–1083

Brockmann I, Bocquet G (1978) Ökologische Einflüsse auf die Geschlechtsverteilung bei *Silene vulgaris* (Moench) Garcke (Caryophyllaceae). Ber Dtsch Bot Ges 91:217–230

Brown AHD (1979) Enzyme polymorphism in plant populations. Theor Popul Biol 15:1–42

Brüggemann W, Janiesch P (1987) Characterization of plasma membrane H^+-ATPase from salt-tolerant and salt-sensitive *Plantago* species. J Plant Physiol 130:395–411

Brüggeman W, Janiesch P (1988) Properties of native and solubilized plasma membrane ATPase from the halophyte *Plantago crassifolia*, grown under saline and non-saline conditions. Physiol Plant 74:615–622

Brunsberg K (1977) Biosystematics of the *Lathyrus pratensis* complex. Opera Bot 42:1–78
Buchsbaum RN, Short FT, Cheney DP (1989) Phenolic-nitrogen interactions in eelgrass, *Zostera marina* L.: possible implications for disease resistance. Aquat Bot 37:291–298
Bünning E (1956) Leaf growth under constant conditions and as influenced by light-dark cycles. In: Milthorpe FL (ed) The growth of leaves. Butterworth, London, pp 119–126
Burdon JJ (1982) The effect of fungal pathogens on plant communities. In: Newman EI (ed) The plant community as a working mechanism. Blackwell Oxford, pp 99–112
Burdon JJ, Harper JL (1980) Relative growth rates of individual members of a plant population. J Ecol 68:953–957
Burdon JJ, Shattock RC (1980) Disease in plant communities. Appl Biol 5:145–219
Burdon JJ (1987) Disease and plant population biology. Univ Press, Cambridge
Campbell R, Newman EI, Lawley RA, Christie P (1982) Relationship between a *Rhizoctonia* species and grassland plants. Trans Br Mycol Soc 79:123–127
Cartier D (1971) Etude biosystématique de quelques espèces du genre *Plantago* (Tourn.) L. (sections Coronopus D.C. et Oreades Decne.) I. Historique, races chromosomiques du *Plantago alpina* L. et du *Plantago serpentina*. Annu Rev Gen Bot 78:493–556
Cavalier-Smith T (ed) (1985) Cell volume and the evolution of eukaryotic genome size. In: The evolution of genome size. John Wiley & Sons, New York, pp 105–184
Cavers PB (1983) Seed demography. Can J Bot 61:3578–3590
Cavers PB, Bassett IJ, Crompton CW (1980) The biology of Canadian weeds 47. *Plantago lanceolata* L. Can J Plant Sci 60:1269–1282
Chandler C (1954) Improvement of *Plantago* for mucilage production and growth in the U.S. Contrib Boyce Thompson Inst 17:495–585
Chapin FS III (1980) The mineral nutrition of wild plants. Annu Rev Ecol Syst 11:232–260
Charlesworth D (1981) A further study of the problem of the maintenance of females in gynodioecious species. Heredity 46:27–39
Charlesworth D, Charlesworth B (1979) The evolutionary genetics of sexual systems in flowering plants. Proc R Soc London SerB 205:513–530
Charlesworth D, Charlesworth B (1987) Inbreeding depression and its evolutionary consequences. Ann Rev Ecol Syst 18:237–268
Cheeseman JM, LaFayette PR, Grunewald JW, Hanson JP (1980) Effect of ATPase inhibitors in cell potential and K^+ influx in corn roots. Plant Physiol 65:1139–1145
Christie P, Newman EI, Campbell R (1978) The influence of neighbouring grassland plants on eachother's endomycorrhizas and root surface microorganisms. Soil Biol Biochem 10:521–527
Chupp C (1953) A monograph of the fungus genus *Cercospora*. Ithaca, New York
Clapham AR, Tutin TG, Warburg EF (1962) Flora of the British Isles. Univ Press, Cambridge
Clark EM, Ihzar S, Hanson MR (1985) Independent segregation of the plastid genome and cytoplasmic male sterility in *Petunis* somatic hybrids. Mol Gen Genet 199:440–445
Clarke B, O'Donald P (1964) Frequency-dependent selection. Heredity 19:201–206
Clement CR, Hopper MY, Jones LHP (1978) The uptake of nitrate by *Lolium perenne* from flowing nutrient solution 1. Effect of NO_3 concentration. J Exp Bot 29:453–464
Clifford HT (1962) Insect pollinators of *Plantago lanceolata* L. Nature (London) 193:196
Clough JM, Teeri JA, Alberte RS (1979) Photosynthetic adaptation of *Solanum dulcamara* L. to sun and shade environments. I. A comparison of sun and shade populations. Oecologia 38:13–21
Cohen D (1966) Optimizing reproduction in a randomly varying environment. J Theor Biol 12:119–129
Cole CV, Innis GS, Stewart JWB (1977) Simulation of phosphorus cycling in semi-arid grasslands. Ecology 58:1–15
Cook R (1980) The biology of seeds in the soil. In: Solbrig OT (ed) Demography and evolution in plant populations. Blackwell, Oxford, pp 107–129
Cooper A (1982) The effects of salinity and waterlogging on the growth and cation uptake of salt marsh plants. New Phytol 90:263–275
Cooper GO (1942) Development of the ovule and the formation of the seed in *Plantago lanceolata* L. Am J Bot 29:577–581

Corré WP (1983) Growth and morphogenesis of sun and shade plants. I. The influence of light intensity. Acta Bot Neerl 32:49–62

Corré WP (1984) Growth and morphogenesis of sun and shade plants. IV. Competition between sun and shade plants in different light environments. Acta Bot Neerl 33:25–38

Cosmides LM, Tooby J (1981) Cytoplasmic inheritance and intragenomic conflict. J Theor Biol 89:83–129

Cowan IR (1978) Water use in higher plants. In: McIntyre AK (ed) Water, planets, plants and people. Aust Acad Sci, Canberra, pp 71–107

Cowan IR (1982) Regulation of water use in relation to carbon gain in higher plants. In: Lange OH, Nobel PS, Osmond CB, Ziegler H (eds) Encyclopedia of plant physiology. New series, vol 12B: Physiological plant ecology II. Springer, Berlin Heidelberg New York, pp 589–614

Crawford TJ (1984) The estimation of neighbourhood parameters for plant populations. Heredity 52:273–283

Crawley MJ (1983) Herbivory. The dynamics of animal-plant interactions. Stud Ecol 10:437

Cronquist A (1968) The evolution and classification of flowering plants. Nelson & Sons Ltd, London

Cruden RW (1977) Pollen-ovule ratios: a conservative indicator of breeding systems in flowering plants. Evolution 31:32–46

Curtis WF (1984) Photosynthetic potential of sun and shade *Viola* species. Can J Bot 62:1273–1278

Darral NM, Wareing PF (1981) The effect of nitrogen nutrition on cytokinin activity and free amino acids in *Betula pendula* Roth and *Acer pseudoplatanus* L. J Exp Bot 32:369–379

Darwin C (1877) Different forms of flowers on plants of the same species. Murray, London

Davies BE (1974) Loss-on-ignition as an estimate of soil organic matter. Soil Sci Soc Am Proc 38:150–151

Davies DD (1973a) Metabolic control in higher plants. In: Milborrow BV (ed) Biosynthesis and its control in plants. Academic Press, New York London, pp 1–20

Davies DD (1973b) Control of and by pH. Symp Soc Exp Biol 27:513–529

Davies DD (1986) The fine control of cytosolic pH. Physiol Plant 67:702–706

Day AG, Stebbins GL (1966) Cytogenic evidence for long continued evolutionary stability in the genus *Plantago*. Science 150:371

De Boer AH (1985) Xylem/symplast ion exchange: mechanism and function in salt-tolerance and growth. Thesis, State Univ Groningen, Neth

De Boer AH, Prins HBA, Zanstra PE (1983) Biphasic composition of trans-root electrical potentials in roots of *Plantago* species: involvement of spatially separated electrogenic bumps. Planta 157:259–266

De Jager A, Posno M (1979) A comparison of the reaction to a localized supply of phosphate in *Plantago major*, *Plantago lanceolata* and *Plantago media*. Acta Bot Neerl 28:479–489

De Nettancourt D (1977) Incompatibility in angiosperms. Monographs on theoretical and applied genetics 3. Springer, Berlin Heidelberg New York

De Nooij MP (1987a) On disease in natural populations of *Plantago lanceolata*. Thesis, State Univ Utrecht, Neth

De Nooij MP (1987b) The effect of pathogens on seedling establishment in *Plantago lanceolata* L. Oecol Plant 8:237–244

De Nooij MP (1988) The role of weevils in the infection process of the fungus *Phomopsis subordinaria* in *Plantago lanceolata*. Oikos 52:51–58

De Nooij MP, Troelstra SR, Wagenaar R (1986) Growth reduction in *Plantago lanceolata* in relation to biotic factors in the soil environment. Oecologia 70:266–272

De Nooij MP, Van Damme JMM (1988a) Variation in host susceptibility among and within populations of *Plantago lanceolata* L. infected by the fungus *Phomopsis subordinaria* (Desm.) Trav Oecol 75:535–538

De Nooij MP, Van Damme JMM (1988b) Variation in pathogenicity among and within populations of the fungus *Phomopsis subordinaria* infecting *Plantago lanceolata*. Evolution 42(6):1166–1171

De Nooij MP, Van der Aa HA (1987) *Phomopsis subordinaria* and associated stalk disease in natural populations of *Plantago lanceolata*. Can J Bot 65:2318–2325

De Nooij MP, Van Gulik WJ (1985) Interaction between soil fungi and seed populations of *Plantago lanceolata*. Prog Rep Inst Ecol Res 84:54–62

De Nooij MP, Van Wingerden YM (1986) The population biological relevance of *Phomopsis* infection in *Plantago lanceolata*. Acta Botanica Neerl 135:107

De Vries DM (1937) Methods used in scientific plant sociology and in agricultural botanical grassland research. Herb Rev 5:187–193

De Wit CT, Dijkshoorn W, Noggle JC (1963) Ionic balance and growth of plants. Agric Res Rep 69–15, Pudoc, Wageningen

Dejaegere R, Neirinckx L, Stassart JM, Delegher V (1981) Mechanism of ion uptake across barley roots. Plant Soil 63:19–24

Delannay X (1978) La gynodioécie ches les angiosperms. Nat Belg 59:223–237

Delannay X, Gouyon PH, Valdeyron G (1981) Mathematical study of the evolution of gynodioecy with cytoplasmic inheritance under the effect of a nuclear restorer gene. Genetics 99:169–181

Dennis WD, Woledge J (1983) The effect of shade during leaf expansion on photosynthesis by white clover leaves. Ann Bot (London) 51:111–118

DeWet JMJ (1980) Origins of polyploids. In: Lewis WH (ed) Polyploidy, biological relevance. Plenum, New York, pp 3–15

Dewey DR (1980) Some applications and misapplications of induced polyploidy in plant breeding. In: Lewis WH (ed) Polyploidy, biological relevance. Plenum, New York, pp 445–470

Dietrich H (1975a) Zum Merkmalsbestand der Plantaginaceae und seiner Bedeutung für die Systematik. Thesis, Univ Jena

Dietrich H (1975b) Cytologische Untersuchungen innerhalb der Familie der Plantaginaceae II. Zusammenstellung bekannter Chromosomenzahlen und der Vergleich zwischen Grundzahl und Lebensdauer. Wiss Z FSU Jena Math Nat R 24:437–461

Dietrich H (1980) Cytologische Untersuchungen innerhalb der Familie der Plantaginaceae. III. Cytotaxonomische Ergebnisse. Wiss Z FSU Jena Math Nat R 29:559–587

Dijkstra P (1990) Physiological aspects of a genotypic difference in relative growth rate within *Plantago major*. Thesis, State Univ Groningen, Neth

Dijkstra P, Kuiper PJC (1989) Effects of exogenously applied growth regulators on shoot growth of inbred lines of *Plantago major* differing in relative growth rate: differential response to gibberellic acid and (2-chloro-ethyl)-trimethyl-ammonium chloride. Physiol Plant 77:512–518

Dijkstra P, Ter Reegen H, Kuiper PJC (1990) Relation between relative growth rate, endogenous gibberellins, and the response to applied gibberellic acid for *Plantago major*. Physiol Plant 79:629–634

Dirzo R (1980) Experimental studies on slug-plant interactions. 1. The acceptability of 30 plant species to the slug *Agriolimax carnuanae*. J Ecol 68:981–998

Doctors van Leeuwen WM (eds) (1984) Het gallenboek, 3rd edn. Thieme, Stuttgart

Doddema H (1978) Uptake of nitrate by chlorate resistant mutants of *Arabidopsis thaliana* (L.) Heynh. Thesis, State Univ Groningen, Neth

Dodds JG (1953) Biological flora of the British Isles. *Plantago coronopus* L. J Ecol 41:467–478

Dommée B, Assouad MW, Valdeyron G (1978) Natural selection and gynodioecy in *Thymus vulgaris* L. Bot J Linn Soc 77:17–28

Douglas TJ, Walker RR (1983) 4-Desmethylsterol composition of citrus rootstocks of different salt exclusion capacity. Physiol Plant 58:69–74

Dowling RE (1933) The reproduction of *Plantago corronopus*: an example of morphological and biological seed dimorphism. Ann Bot (London) 47:861–872

Drew AP (1982) Shoot-root plasticity and episodic growth in red pine seedlings. Ann Bot (London) 49:347–357

Duran R, Fisher GW (1961) The genus *Tilletia*. Washington State Univ, Washington DC

Eavis BW (1972) Soil physical conditons affecting seedling root growth I. Mechanical impedance, aeration and moisture availability as influenced by bulk density and moisture levels in a sandy loam. Plant Soil 36:613–622

Ehrendorfer F (1980) Polyploidy and distribution. In: Lewis WH (ed) Polyploidy, biological relevance. Plenum, New York, pp 45–60

Eichler AW (1875) Blütendiagramme I. Verlag Wilhelm Engelmann, Leipzig

Ellenberg H (1977) Stickstoff als Standortsfaktor, insbesondere für mitteleuropäische Pflanzengesellschaften. Oecol Plant 12:1–22

Ellis WM, Keymer RJ, Jones DA (1977a) The effect of temperature on the polymorphism of cyanogenesis in *Lotus corniculatus* L. Heredity 38:339–347

Ellis WM, Keymer RJ, Jones DA (1977b) The defensive function of cyanogenesis in natural populations. Experientia 33:309–311

Endlicher S (1836) Genera plantarum. Vindobona 1836–1840

Ennos RA (1981) Detection of selection in populations of white clover (*Trifolium repens* L.). Biol J Linn Soc 15:75–82

Ernst WHO, Van Andel J (1985) Autoecologie, oecofysiologische adaptaties. In: Bakker K, Cappenberg ThE, Croin Michielsen M, Freijsen AHJ, Nienhuis PH, Woldendorp JW, Zijlstra JJ (eds) Inleiding tot de oecologie. Bohn, Scheltema & Holtema, Utrecht, pp 70–100

Epstein E (1972) Mineral nutrition of plants: principles and perspectives. John Wiley & Sons, New York, p 412

Erdei L, Kuiper PJC (1979) The effect of salinity on growth, cation content, Na^+-uptake and translocation in salt-sensitive and salt-tolerant *Plantago* species. Physiol Plant 47:95–99

Erdei L, Kuiper PJC (1980) Substrate-dependent modulation of ATPase activity by Na^+ and K^+ in roots of *Plantago* species. Physiol Plant 49:71–77

Erdei L, Tóth I, Zsoldos F (1977) Potassium absorption by rice at different levels of organization. 2. ATPase activity in roots, calluses and re-differentiated roots. Physiol Plant 41:105–108

Erdei L, Stuiver CEE, Kuiper PJC (1980) The effect of salinity on lipid composition and on activity of Ca^{2+}- and Mg^{2+}-stimulated ATPase in salt-sensitive and salt-tolerant *Plantago* species. Physiol Plant 49:315–319

Erdtman G, Berglund B, Praglowski J (1961) An introduction to a Scandinavian pollenflora. Grana Palynologica, vol 2, part 3, 94 pp

Ericson L (1980) The downward migration of plants on a rising Bothnian seashore. Acta Phytogeogr Sued 68:61–72

Faegri K, Van der Pijl L (1979) The principles of pollination ecology. Perg, Oxford

Falconer DS (1981) Introduction to quantitative genetics. 2nd ed. Longman, London New York

Fales FW (1951) The assimilation and degradation of carbohydrates by yeast cells. J Biol Chem 193:113–124

Farquhar GD, Sharkey TD (1982) Stomatal conductance and photosynthesis. Annu Rev Plant Physiol 33:317–345

Favarger C, Vasudevan KN (1972) Etude cytologique et morphologique des populations de *Plantago major* L. et de *Plantago lanceolata* L. de l'Himalaya occidental. Comparison avec des plantes de Suisse. Bull Soc Neuchatel Sci Nat 95:63–74

Feeny P (1976) Plant apparency and chemical defense. Biochemical interaction between plants and insects. Rec Adv Phytochem 10:1–40

Fisher RA (1958) The genetical theory of natural selection, 2nd edn. Dover, New York

Focht DD, Verstraete W (1977) Biochemical ecology of nitrification and denitrification. In: Alexander M (ed) Advances in microbiol ecology. Plenum, New York London, pp 135–214

Fournier P (1946) Les quatre flores de la France. Paul Lechevalier (ed), Paris

Francis SM (1981) *Peronospora alta*. CMI descriptions of pathogenic fungi and bacteria 683:1–3

Frandsen KJ (1946) Studies over *Sclerotinia trifoliorum* Eriksson. PhD thesis, R Vet Agric Coll, Copenhagen

Franke W (1957) Ein Beitrag zur Nährstoffaufnahme durch Blätter. Ber Dtsch Bot Ges 70:297–304

Franke W (1960) Über Beziehungen der Ektodesmen zur Stoffaufnahme durch Blätter. I. Beobachtungen an *Plantago major* L. Planta 55:390–423

Franke W (1964a) Über die Beziehungen der Ektodesmen an der Stoffaufnahme durch Blätter. III. Nachweis der Beteiligung der Ektodesmen an der Stoffaufnahme durch Blätter mittels radioaktiver Stoffe. Planta 61:1–16

Franke W (1964b) Ektodesmenstudien. II. Über die Bedeutung der reduzierenden Substanz beim Ektodesmennachweis. Planta 63:118–132

Frankel OH, Soulé ME (1981) Conservation and evolution. Univ Press, Cambridge

Frankel R, Galun E (1977) Pollination mechanisms, reproduction and plant breeding. Monographs on theoretical and applied genetics 2. Springer, Berlin Heidelberg New York

Frankland B, Poo WK (1980) Phytochrome control of seed germination in relation to natural shading. In: De Greef J (ed) Photoreceptors and plant development. Univ Press, Antwerpen, pp 357–366
Freijsen AHJ, Otten H (1980) The growth and nitrate uptake of *Plantago lanceolata* and *P. major* in culture experiments with intermittently supplied nitrate in low concentrations. Verh K Ned Akad Wetensch, Afd Natuurkd 2 R 75:34–37
Freijsen AHJ, Otten H (1984) The effect of nitrate concentration in a flowing solution system or growth and nitrate uptake of two *Plantago* species. Plant Soil 77:159–169
Fries N (1944) Beobachtungen ueber die thamniscophage Mycorrhiza einiger Halophyten. Bot Notiser 1944:255–264
Fröst S (1959) The cytological behaviour and mode of transmission of accessory chromosomes in *Plantago serraria*. Hereditas 45:191–210
Froud-Williams RJ, Drennan DSH, Chancellor RJ (1984a) The influence of burial and dry storage upon cyclic changes in dormancy, germination and response to light in seeds of various arable weeds. New Phytol 96:473–481
Froud-Williams RJ, Chancellor RJ, Drennan DSH (1984b) The effects of seed burial and soil disturbance on emergence and survival of arable weeds in relation to minimal cultivation. J Appl Ecol 21:629–641
Gadgill M, Solbrig OT (1972) The concept of "r" and "K" selection: evidence from wild flowers and some theoretical considerations. Am Nat 106:14–31
Gäumann E (1959) Die Rostpilze Mitteleuropas. Büchler, Bern
Gasparikovà O, Psenakovà T, Niznanskà A (1976) Influence of various nitrogen sources on the activity and nitrite reductases and glutamate dehydrogenase in *Zea mays* roots. Biologia (Bratislava) 31:527–535
Gauhl E (1976) Photosynthetic response to varying light intensity in ecotypes of *Solanum dulcamara* L from shaded and exposed habitats. Oecologia 22:275–286
Gilbert N (1984) What they didn't tell you about limit cycles. Oecologia 65:112–113
Gimingham CH (1964) Maritime and submaritime communities. In: Burnett JH (ed) Vegetation of Scotland. Oliver & Boyd, Edinburgh, pp 67–143
Givinsh TJ (1982) On the adaptive significance of leaf height in forest herbs. Am Nat 120:353–381
Glass ADM (1976) Regulation of potassium absorption in barley roots. An allosteric model. Physiol Plant 58:33–37
Goethart JWC, Jongmans WJ (1902, 1906) Plantenkaartjes voor Nederland. Univ Leiden, Neth
Gorenflot RR (1960) La polyploidie chez *Plantago coronopus*. Rev Cytol Biol Veg 22:77–108
Gorenflot RR (1961) Etude morphologique et cytologique de quelques triploïdes expérimenteaux chez *Plantago coronopus* L. Bull Soc Bot Fr 108:200–212
Gorenflot RR (1963) Autoallopolyploidie expérimentale chez *Plantago coronopus* L. Bull Soc Bot Fr 108:200–212
Gorenflot RR (1964) Introgression, polymorphisme et taxanomie chez les Plantaginacées. Adansonia 4:393–417
Gorski T, Gorska, Nowicki J (1977) Germination of seeds of various herbacious species under leaf canopy. Flora 166:249–259
Gottlieb LD (1981) Electrophoretic evidence and plant populations. Prog Phytochem 7:1–46
Gottschalk W (1976) Die Bedeutung der Polyploidie für die Evolution der Pflanzen. Fischer, Stuttgart
Gouyon PH, Couvet D (1985) Selfish cytoplasm and adaptation: variations in the reproductive system of thyme. In: Haeck J, Woldendorp JW (eds) Structure and functioning of plant populations. 2. Phenotypic and genotypic variation in plant populations. North Holland, Oxford, pp 299–320
Gouyon PH, Vichot F, Van Damme JMM (1990) Nuclear-cytoplasmic male sterility: single point equilibria versus limit cycles. Am Nat (in press)
Grant V (1981) Plant speciation. Columbia Univ Press, New York
Gravis A (1932) Contribution à l'étude des variations. Nanisme et pédocarpisme de *Plantago coronopus* L. Bull Biol Fr Belg, Suppl 14 In: Recuil de quelques travaux d'anatomie végétale exécutés à Liège de 1929 à 1935 (1936). Hayes, Bruxelles

Greenway H (1965) Plant responses to saline substrates. VII. Growth and ion uptake throughout plant development in two varieties of *Hordeum vulgare*. Aust J Biol Sci 18:763–769

Greenway H, Muns R (1980) Mechanisms of salt tolerance in non-halophytes. Annu Rev Plant Physiol 31:149–190

Gregor JW (1938) Experimental taxonomy II. Initial population differentiation in *Plantago maritima*. L. in Great Britain. New Phytol 37:15–49

Gregor JW (1939) Experimental taxonomy. IV. Population differentiation in North American and European sea plantains allied to *Plantago maritima* L. New Phytol 38:293–322

Gregor JW (1956) Adaptation and ecotypic components. Proc R Soc London Ser B 145:333–337

Gregor JW, Lang JMS (1950) Intra-colonial variation in plant size and habit in sea plantains. New Phytol 49:135–141

Gregory FG, Veale JA (1957) A reassessment of the problem of apical dominance. Symp Soc Exp Biol 2:1–20

Grime JP (1966) Shade avoidance and shade tolerance in flowering plants. In: Bainbridge R, Evans GC, Rackham O (eds) Light as an ecological factor. Blackwell, Oxford, pp 187–209

Grime JP (1977) Evidence for the existence of three primary strategies in plants and its relevance to ecological and evolutionary theories. Am Nat 111:1169–1194

Grime JP (1979) Plant strategies and vegetation processes. John Wiley & Sons, New York Chichester

Grime JP, Hunt R (1975) Relative growth rate, its range and adaptive significance in local flora. J Ecol 63:393–422

Grime JP, MacPherson-Stewart SF, Dearman RS (1968) An investigation of leaf palatibility using the snail *Cepaea nemoralis* L. J Ecol 56:981–998

Grime JP, Mason G, Curtis AV, Rodman J, Band SR, Mowforth M, Neal AM, Shaw S (1981) A coparative study of germination characteristics in a local flora. J Ecol 69:1017–1059

Groot J, Boschhuizen R (1970) A preliminary investigation into the genecology of *Plantago major* L. J Exp Bot 21:835–841

Grove WB (1935) British stem and leaf fungi (Coelomycetes), vol 1. Sphaeropsidales. Univ Press, Cambridge

Haas H, Streibig JC (1981) Status over den vilde flora på danske marker. Ugeskr Jordbrug 7:112–116

Haberlandt G (1981) Physiologische Pflanzenanatomie. Engelmann, Leipzig

Haeck J (1984) A support for the interpretation of the ordination of Dutch plant communities. Verh Kon Ned Akad Wetensch Afd Natuurkd 2 R 82:25–28

Haeck J, Van der Aart PJM, Dorenbosch H, Van der Maarel E, Van Tongeren O (1982) The occurrence of *Plantago* species in ordinated Dutch plant communities. Verh K Ned Akad Wetensch Afd Naturkd 2 R 79:26–33

Haeck J, Van Tongeren O, Van der Maarel E (1985) Phytosociological amplitudes of some Dutch coastal species and their ecological interpretation. Vegetatio 61:77–85

Halm BJ, Stewart JWB, Halstead RL (1972) The phosphorus cycle in a native grassland ecosystem. In: Isotopes and radiation in soil-plant relationships including forestry. Proc Symp IAEA, pp 571–586

Hammer K (1978) Evolutionary trends concerning pollination ecology in *Plantago*. Flora 167:41–56 (in German)

Hammond J (1982) *Plantago* as a host of economically important viruses. Adv Virus Res 27:103–140

Hamrick JL, Linhart YB, Mitton JP (1979) Relationships between life history characteristics and electrophoretically detectable genetic variation in plants. Annu Rev Ecol Syst 10:173–200

Hansson G, Kylin A (1969) ATPase activities in homogenates from sugar beet roots, relation to Mg^{2+} and ($Na^{+}+K^{+}$)-stimulation. Z Pflanzenphysiol 60:270–275

Harper JL (1977) Population biology of plants. Academic Press, New York London San Francisco

Harper JL, Benton RA (1966) The behaviour of seeds in soil, part 2. The germination of seeds on the surface of a water supplying surface. J Ecol 54:151–166

Harper JL, Ogden J (1970) The reproductive strategy of higher plants. I. The concept of strategy with special reference to *Senecio vulgaris* L. J Ecol 58:681–698

Harper JL, Lovell PH, Moore KG (1970) The shapes and sizes of seeds. Ann Rev Ecol Syst 1:327–356

Havill DC, Lee JA, DeFelice J (1977) Some factors limiting nitrate utilization in acidic and calcareous grasslands. New Phytol 78:649–659

Hawthorn WR (1974) The biology of Canadian weeds: *Plantago major* and *P. rugelii*. Can J Plant Sci 54:383–396

Hawthorn WR, Cavers PB (1978) Resource allocation in young plants of two perennial species of *Plantago*. Can J Bot 56:2533–2537

Haynes RJ, Goh KM (1978) Ammonium and nitrate nutrition of plants. Biol Rev 53:465–510

Heale JB, Isaac I (1963) Wilt of lucerne caused by species of *Verticillium* IV. Ann Appl Biol 52:439–451

Heap AJ (1979) Root microorganisms and their significance for grassland plants. PhD Thesis, Univ Bristol

Heap AJ, Newman EI (1980a) Links between roots and hyphae of vesicular-arbuscular mycorrhizas. New Phytol 85:169–171

Heap AJ, Newman EI (1980b) The influence of vesicular-arbuscular mycorrhizas on phosphorus transfer between plants. New Phytol 85:173–179

Hedrick PH (1980) Hitchhiking: a comparison of linkage and partial selfing. Genetics 94:791–808

Hess HE, Landolt E, Hirzel R (1972) Flora der Schweiz. Birkhäuser, Basel

Heukels H (1910) De flora van Nederland. Noordhoff Groningen

Heukels H, van Ooststroom SJ (1975) Flora van Nederland. Noordhoff Groningen

Heukels H, Van der Meijden R, Weeda EJ, Adema FACB, De Jonckheere GJ (1983) Flora van Nederland, 20 edn. Wolters-Noordhof, Groningen

Hewitt GM (1989) The subdivision of species by hybrid zones. In: Otte D, Endler JA (eds) Speciation and its consequences. Sinauer, Sunderland, MA, pp 85–110

Heydecker W, Orphanos PJ (1968) The effect of excess moisture on the germination of *Spinacia oleracea* L. Planta 83:237–247

Hiatt AJ (1978) Critique of 'Absorption and utilization of ammonium nitrogen by plants'. In: Nielsen DR, MacDonald JG (eds) Nitrogen in the environment, soil-plant-nitrogen relationships, vol. 2. Academic Press, New York London San Francisco, pp 191–199

Hiatt AJ, Leggett JE (1974) Ionic interactions and antagonisms in plants. In: Carson EW (ed) The plant root and its environment. Univ Press Virginia, Charlottesville, pp 101–134

Hickey DA, McNeilly T (1975) Competition between metal tolerant and normal plant populations: A field experiment on normal soil. Evolution 29:458–464

Hickman JC (1979) The basic biology of plant numbers. In: Solbrig OT, Jain S, Johnson GB (eds) Topics in plant population biology. Columbia Univ Press, New York, pp 232–263

Hill MO (1973) Reciprocal averaging: an eigenvector method of ordination. J Ecol 61:237–249

Hill MO (1974) Correspondence analysis: a neglected multivariate method. Appl Statist 23:340–354

Hitz WD, Rhodes D, Hanson AD (1981) Radiotracer evidence implicating phosphoryl and phosphatidyl bases as intermediates in betaine synthesis by water-stressed barley leaves. Plant Physiol 68:814–822

Hoagland DR, Snijder WC (1933) Nutrition of strawberry plant under controlled conditions. Proc Am Soc Hortic Sci 30:288–296

Hoflacher H, Bauer H (1982) Light acclimation in leaves of the juvenile and adult life phases of ivy (*Hedera helix*). Physiol Plant 56:177–182

Holm L (1957) Etudes taxonomiques sur les Pléosporacées. Lundequistska, Uppsala

Holmes FO (1954) Inheritance of resistance of viral diseases in plants. Adv Virus Res 2:1–30

Horvath I, Vigh L, Farkas T (1981) The manipulation of polar head group composition of phospholipids in the wheat Miranovskaja 808 affects frost tolerance. Planta 151:103–108

Houba VJG, Van Egmond F, Wittich EM (1971), Changes in production of organic nitrogen and carboxylates (C-A) in young sugar-beet plants grown in nutrient solutions of different nitrogen composition. Neth J Agric Sci 19:39–47

Hsiao TC (1973) Plant responses to waterstress. Annu Rev Plant Physiol 24:519–570

Huffaker RC, Rains DW (1978) Factors influencing nitrate acquisition by plants: assimilation and fate of reduced nitrogen. In: Nielsen DR, MacDonald JG (eds) Nitrogen of the environment. 2. Soil-plant-nitrogen relationships. Academic Press, New York London, pp 1–43

Hutchings MJ, Waite S (1986) Cohort behaviour and environmental determination of life histories within a natural population of *Plantago coronopus* L. In: Haeck J, Woldendorp JW (eds) Structure and functioning of plant populations 2. North Holland, Amsterdam, pp 171–184

Hütter R (1958) Untersuchungen über die Gattung *Pyrenopeziza* Fuck. Phytopathol Z 33:1–54

Hyde HA, Williams DA (1946) Studies in atmospheric pollen. III. Pollen production and pollen incidence in ribwort plantain (*Plantago lanceolata* L.). New Phytol 45:271–277

Irmisch TH (1857) Über die Keimung und die Erneuerungsweise von *Convolvulus sepium* and *C. arvensis*, sowie über hypokotylische Adventifknospen bei phanerogamen Pflanzen. Bot Z 15:464–474

Israel DW, Jackson WA (1978) The influence of nitrogen nutrition on ion uptake and translocation by leguminous plants. In: Andrew CS, Kamprath EJ (eds) Mineral nutrition of legumes in tropical and subtropical soils. CSIRO, Melbourne, pp 113–129

Israel DJ, Jackson WA (1982) Ion balance, uptake, and transport processes in N_2-fixing and nitrate- and urea-dependent soybean plants. Plant Physiol 69:171–178

Jackson MB, Fenning TM, Drew MC, Saker LR (1985) Stimulation of ethylene production and gas-space (aerenchyma) formation in adventitious roots of *Zea mays* L. by small partial pressures of oxygen. Planta 165:486–492

Jaworski EG (1971) Nitrate reductase assay in intact plant tissues. Biochem Biophys Res Commun 43:1274–1279

Jefferies RL (1977) Growth responses of coastal halophytes to inorganic nitrogen. J Ecol 65:847–865

Jefferies RL, Perkins N (1977) The effects on the vegetation of the additions of inorganic nutrients to salt marsh soils at Stiffkey, Norfolk. J Ecol 65:867–882

Jefferies RL, Rudmik T, Dillon EM (1979) Responses of halophytes to high salinities and low water potentials. Plant Physiol 64:989–994

Jehle RA, Walker EA, Preston PA, Jenkins AE (1947) Additional records on distribution of plantain scab. Plant Disease Rep 31:162–166

Jerling L (1981) Effects of microtopography on the summer survival of *Plantago maritima* seedlings. Holarct Ecol 4:120–126

Jerling L (1982) Population dynamics of a perennial herb (*Plantago maritima* L.) along a distributional gradient: a demographical study. Thesis, Univ Stockholm, Swed

Jerling L (1983) Composition and viability of the seed bank along a successional gradient on a Baltic seashore meadow. Holarct Ecol 6:150–156

Jerling L (1985a) Population dynamics of *Plantago maritima* along a distributional gradient on a Baltic seashore meadow. Vegetatio 61:155–161

Jerling L (1985b) Are plants and animals alike? A note on evolutionary plant population ecology. Oikos 45:150–153

Jerling L, Andersson M (1982) Effects of selective grazing by cattle on the reproduction of *Plantago maritima*. Holarct Ecol 5:405–411

Jerling L, Liljelund LE (1984) Dynamics of *Plantago maritima* along a distribution gradient: a demographic study. Holarct Ecol 7:280–288

Jones DA (1972) Cyanogenic glycosides and their function. In: Harborne JB (ed) Phytochemical ecology. Academic Press, New York London, pp 103–124

Jones DA (1973) Co-evolution and cyanogenesis. In: Heywood VH (ed) Taxonomy and ecology. Academic Press, New York London, pp 213–242

Jones DA (1977) On the polymorphism of cyanogenesis in *Lotus corniculatus* L. VII. The distribution of the cyanigenic form in Western Europe. Heredity 39:27–44

Kanamori T, Konishi S, Takahashi E (1972) Inducible formation of glutamate dehydrogenase in rice plant roots by addition of ammonia to the roots. Physiol Plant 26:1–6

Kang BG (1979) Epinasty. In: Haupt W, Feinleib ME (eds) Physiology of movements. Springer, Berlin Heidelberg New York, pp 646–667 (Encyclopedia of Plant Physiology, New Ser, vol 7 eds: Pirson A, Zimmermann MH)

Karling JS (1961) *Synchytrium erieum* sp. nov. Sydowia 15:100

Karssen CM (1970a) The light promoted germination of the seeds of *Chenopodium album* L. III. Effects of the photoperiod during growth and development of the plants on the dormancy of the produced seeds. Acta Bot Neerl 19:81–94

Karssen CM (1970b) The light promoted germination of the seeds of *Chenopodium album* L. IV. Effects of red, far-red and white light on non-photoblastic seeds incubated in mannitol. Acta Bot Neerl 19:95–108

Karssen CM (1970c) The light-promoted germination of *Chenopodium album* L. VI. Pfr requirements during different stages of the germination process. Acta Bot Neerl 19:297–312

Karssen CM (1980/81) Environmental conditions and endogenous mechanisms involved in secondary dormancy of seeds. Isr J Bot 29:45–64

Kaul MLH (1988) Male sterility in higher plants. Springer, Berlin Heidelberg New York

Kawase M (1982) Anatomical and morphological adaptation of plants to waterlogging. Hortsci 16:30–34

Kays S, Harper JL (1974) The regulation of plant and tiller density in a grass sward. J Ecol 62:97–105

Keyworth WG (1962) Plant pathology. Rep Nat Veg Res Stn Warwick 13:56–58

Kirkby EA (1981) Plant growth in relation to nitrogen supply. In: Clark FE, Roswall T (eds) Terrestrial nitrogen cycles. Ecol Bull 33:249–267

Kirkby EA, Knight AH (1977) Influence of the level of nitrate nutrition on ion uptake and assimilation, organic acid accumulation, and cation-anion balance in whole tomato plants. Plant Physiol 60:349–353

Kivilaan A, Bandurski RS (1973) The 90 year period for Dr. Beal's seed viability experiment. Am J Bot 60:140–145

Kloepper JW, Garrot DG (1981) Acquisition by aphids of spiroplasmas from *Plantago major*. Phytopathology 71:886

Knuth P (1905) Handbuch der Blütenbiologie. Wilhelm Engelmann, Leipzig

Knuth P (1906) Handbook of flower pollination. Clarendon, Oxford

Königshofer H (1981) Stoffwechselphysiologische Untersuchungen an *Plantago*-Arten unter besonderer Berücksichtigung ökologischer Aspekte. Thesis, Univ Vienna

Königshofer H, Kinzel H (1986) Zum Ionenhaushalt einiger *Plantago*-Arten. Flora 178:167–176

Konings H (1982) Ethylene-promoted formation of aerenchyma in seedling roots of *Zea mays* L. under aerated and non-aerated conditions. Physiol Plant 54:119–124

Konings H, De Wolf A (1984) Promotion and inhibition by plant growth regulators of aerenchyma formation in seedling roots of *Zea mays* L. Physiol Plant 60:309–314

Krüger W (1970) Über den Einfluß unterschiedlicher Bodenverdichtung auf einige bodenphysikalische Eigenschaften und das Pflanzenwachstum. Albrecht-Thaer-Archiv 14:613–624

Kruijne AA, De Vries DM, Mooi H (1967) Contribution to the ecology of the Dutch grassland plants. Agric Res Rep 696. Pudoc, Wageningen (in Dutch, with English summary)

Kuiper D (1982) Genetic differentiation in *Plantago major*: Ca^{2+}- and Mg^{2+}-stimulated ATPases from roots and their role in phenotypic adaptation. Physiol Plant 56:436–443

Kuiper D (1983a) Genetic differentiation in *Plantago major*: growth and root respiration and their role in phenotypic adaptation. Physiol Plant 57:222–230

Kuiper D (1983b) Ca^{2+}- and Mg^{2+}-stimulated ATPases from roots of *Plantago* species: response to different light energy fluxes. New Phytol 94:39–45

Kuiper D (1984) Genetic differentiation and phenotypic plasticity in *Plantago* species. Thesis, State Univ Groningen, Neth

Kuiper D (1985) Genetic differentiation and phenotypic plasticity in populations of *Plantago lanceolata* in response to nutrient levels. In: Proc Table Ronde CNRS, Population biology of plants, Montpellier

Kuiper D, Kuiper PJC (1979a) Ca^{2+}- and Mg^{2+}-stimulated ATPases from roots of *Plantago major* and *Plantago maritima*: response to alterations of the level of mineral nutrition and ecological significance. Physiol Plant 45:1–6

Kuiper D, Kuiper PJC (1979b) Ca^{2+}- and Mg^{2+}-stimulated ATPases from roots of *Plantago lanceolata, Plantago media* and *Plantago coronopus:* response to alterations of the level of mineral nutrition and ecological significance. Physiol Plant 45:240–244

Kuiper D, Kuiper PJC (1988) Phenotypic plasticity in a physiological perspective. Acta Oecologica Oecol Plant 9:43–59

Kuiper D, Smid A (1985) Genetic differentiation and phenotypic plasticity in *Plantago major* ssp. *major*: I The effect of differences in level of irradiance on growth, photosynthesis, respiration and chlorophyll content. Physiol Plant 65:520–528

Kuiper D, Staal M (1986) Genetic differentiation and phenotypic plasticity in *Plantago major* ssp. *major*. II. The effect of differences in level of irradiance on Mg^{2+}- and Ca^{2+}-stimulated ATPase activity from roots and shoot. Physiol Plant 66:674–678

Kuiper PJC (1984) Functioning of plant cell membranes under saline conditions: membrane lipid composition and ATPases. In: Staples RC (ed) Salinity tolerance in plants: strategies for crop improvement. John Wiley & Sons, New York, pp 77–92

Kylin A, Gee R (1970) Adenosine triphosphase activities in leaves of the mangrove *Avicennia nitida* Jacq. Plant Physiol 45:169–172

Lambers H, Blacquière T, Stuiver CEE (1981 a) Photosynthesis and respiration as affected by salinity. II. Interactions between osmoregulation and the alternative pathway. Physiol Plant 51:63–68

Lambers H, Posthumus F, Stulen I, Lanting L, Van de Dijk S, Hofstra R (1981 b) Energy metabolism of *Plantago lanceolata* as dependent on the supply of mineral nutrients. Physiol Plant 51:85–92

Lambers H, Posthumus F, Stulen I, Lanting L, Van de Dijk S, Hofstra R (1981 c) Energy metabolism of *Plantago major* ssp. *major* as dependent on the supply of mineral nutrients. Physiol Plant 51:245–252

Lange OL, Schulze E-D, Kappen L, Buschbom U, Evenari M (1975) Photosynthesis of desert plants as influenced by internal and external factors. In: Gates DM, Smerl RB (eds) Ecological Studies vol 12: Perspectives of biophysical ecology. Springer, Berlin Heidelberg New York, pp 121–143

Larcher W (1981) Effects of low temperature stress and frost injury on plant productivity. In: Johnson CB (ed) Physiological processes limiting plant productivity. Butterworth's, London

Laura RD (1977) Salinity and nitrogen mineralization in soil. Soil Biol Biochem 9:333–336

Leaver CJ, Gray MW (1982) Mitochondrial genome organisation and expression in higher plants. Annu Rev Plant Phys 33:373–402

Leclerc MC, Robin P (1983) Etude in situ de l'influence de l'humidité et de la teneur en nitrate d'un sol dunaire sur l'accumulation et la réduction du nitrate chez l'oyat (*Ammophila arenaria* L.). Plant Soil 70:229–241

Leclerc MC, Robin P (1985) Variations saisonnières de la disponibilité du nitrate et de son utilisation par la végétation des dunes littorales de la Mediterrannée. Oecol Plant 6:87–106

Ledig FT, Guries RP, Bonefeld BA (1983) The relation of growth to heretozygosity in pitch pine. Evolution 37:1127–1138

Lee SM, Cavers PB (1981) The effects of shade on growth, development and resource allocation patterns of three species of foxtail. Can J Bot 59:1776–1786

Leereveld H, Meeuse ADJ, Stelleman P (1976) Anthecological relations between reputedly anemphilous flowers and syrphid flies. II. *Plantago media* L. Acta Bot Neerl 25:205–211

Lerner IM (1954) Genetic homeostatis. Oliver & Boyd, Edinburgh

Letham DS (1978) Cytokinins. In: Letham DS, Goodwin PB, Higgins TJV (eds) Phytohormones and related compounds a comprehensive treatise, Vol 1. Elsevier/North Holland Biomedical Press, Amsterdam, pp 295–348

Levin DA (1975) Minority cytotype exclusion in local plant populations. Taxon 24:35–43

Levin DA (1978) The origin of isolation mechanisms in flowering plants. In: Hecht HM, Steere WC, Wallace B (eds) Evolutionary biology II. Plenum, New York, pp 185–317

Levin DA (1983) Polyploidy and novelty in flowering plants. Am Nat 122:1–25

Levin DA, Kerster HW (1974) Gene flow in seed plants. Evol Biol 7:139–220

Levitt J (1980) Factors related to freezing resistance. In: Levitt J (ed) Responses of plants to environmental stresses, vol. 1. Academic Press, New York London, pp 163–179

Lewis D (1941) Male sterility in natural populations of hermaphrodite plants. New Phytol 40:56–63

Lewis D (1949) Incompatibility in flowering plants. Biol Rev 24:472–496

Lewis WH (1980) Polyploidy in species populations. In: Lewis WH (ed) Polyploidy, biological relevance. Plenum, New York, pp 104–143

Li HL (1946) On the identity of *Kokonaria*. J Washington Acad Sci 35:384–385
Liddle MJ, Greig-Smith P (1975) A survey of tracks and paths in a sand dune ecosystem II Vegetation. J Appl Ecol 12:909–930
Liddle MJ, Moore KG (1974) The microclimate of sand dune tracks: the relative contribution of vegetation removal and soil compression. J Appl Ecol 11:1057–1068
Lincoln DE, Langenheim JH (1979) Variation of *Satureja douglasii* Monoterpenoids in relation to light intensity and herbivory. Biochem Syst Ecol 7:289–298
Lindley J (1836) Naturae systema, p 267
Lineberger RD, Steponkus P (1980) Effects of freezing on the release and inactivation of chloroplast coupling factor. Cryobiology 17:486–494
Linn MB (1940) The yellows disease of lettuce and andive. Bull Cornell Univ Agric Exp Sta 742:1–33
Livshits G, Kobyliansky E (1985) Lerner's concept of developmental homeostatis and the problem of heterozygosity level in natural populations. Heredity 55:341–353
Lloyd DG (1974) Theoretical sex ratios of dioecious and gynodioecious angiosperms. Heredity 32:11–34
Lloyd DG (1975) The maintenance of gynodioecy and androdioecy in angiosperms. Genetica 45:325–339
Lloyd DG (1976) Transmission of genes via pollen and ovules in gynodioecious populations. Theor Popul Biol 9:299–316
Lloyd DG (1982) Selection of combined versus separate sexes in seed plants. Am Nat 120:571–585
Lona F (1959) Notes on the effect of gibberellic acid and other growth substances in relation to light reactions. In: Coll int photo-thermopériodism II, Farma 1957, Bull 34, UISB, Ser B pp 141:147
Lotz LAP (1989) Variation in life-history characteristics between and within populations of *Plantago major* L. Thesis, Catholic Univ Nijmegen, Neth
Lotz LAP (1990) The relation between age and sizeat first flowering of *Plantago major* in various habitats. J Ecol 70:55–75
Lotz LAP, Blom CWPM (1986) Plasticity in life-history traits of *Plantago major* ssp. *pleiosperma* Pilger. Oecologia 69:25–30
Lotz LAP, Spoormakers LDH (1988) Differentiation in reproductive characteristics within a population of *Plantago major* L. ssp. *pleiosperma* Pilger. Oecol Plant 9:11–18
Lotz LAP, Olff H, Van Tienderen PH (1990) Within-population variability in morphology and life-history of *Plantago major* L. ssp. *pleiosperma* Pilger in relation to environmental heterogeneity. Oecologia 84:404–410
Louis JP, Durand B (1978) Studies with the dioecious angiosperm *Mercurialis annua* L. (2n = 16). Correlation between genic and cytoplasmic male sterility, sex segregation and feminizing hormones (cytokinins). Mol Gen Genet 165:309–322
Lumaret R (1988) Cytology, genetics and evolution in the genus *Dactylis*. CRC Crit Rev Plant Sci 7:55–91
Lumaret R, Guillerm JL, Delay J, Ait Lhaj Loutfi A, Jay M (1987) Polyploidy and habitat differentiation. Oecologia 73:436–446
Maathuis FJM (1991) Salt-tolerance in *Plantago* and application of the patch-clamp technique in plant cell membranes. Thesis, State Univ Groningen, Neth
Maathuis FJM, Prins HBA (1990) Patch clamp studies on cell vacuoles of a salt tolerant and a salt sensitive *Plantago* species. Plant Physiol 92:23–28
Mahall BE, Park RB (1976) The ecotone between *Spartina foliosa* Trin and *Salicornia virginica* L. in salt marshes of northern San Francisco Bay I. Biomass and production. J Ecol 64:421–433
Mattei GL (1909) Penzime naturale delle Plantagineae. Malpighii 23:309–379
Matthews REF (1970) Plant virology. Academic Press, New York
Mayer AM, Poljakoff-Mayber A (1982) The germination of seeds. In: Wareing PF, Galston AY (eds) Series of monographs on pure and applied biology. Pergamon, Oxford London
McDonald JF, Ayala FJ (1974) Genetic response to environmental heterogeneity. Nature (London) 250:572–574
McIntyre DS (1958) Soil splash and the formation of surface crusts by raindrop impact. Soil Sci 85:261–266

McIntyre GI (1977) The role of nutrition in apical dominance. Symp Soc Exp Biol 31:251–273
McNeilly T (1979) Studies on the ecological genetics of heavy metal tolerant plant populations. Aquilo Ser Zool 20:17–25
McNeilly T, Bradshaw AD (1968) Evolutionary processes in populations of copper tolerant *Agrostis tenuis* Sibth. Evolution 22:108–118
Meeuse ADJ (1984) Rate of dependence of *Plantago media* L. on entomophilous reproduction-preliminary report. Acta Bot Neerl 33:129–130
Melnik VA (1977) Opredelitel gribov roda *Ascochyta* Lib. Nauka, Leningrad
Mengel K (1974) Plant ionic status. In: Carson EW (ed) The plant root and its environment. Univ Press Virginia, Charlottesville, pp 63–81
Mengel K, Kirkby EA (1978) Principles of plant nutrition. Int Potash Inst, Bern
Mikkelsen VM, Laurensen F (1966) Markukrudtet i Danmark omkring 1960. Bot Tidsskr 62:1–26
Mital SP, Issar SC (1970) Fertility restorer for cytoplasmic male sterility in isaphgul (*Plantago ovata* Forsk.). Sci Cult Bangalore 36:550–551
Mitton JB, Grant MC (1984) Association among protein heterozygosity, growth rate, and developmental homeostasis. Annu Rev Ecol Syst 15:479–499
Mølgaard P (1976) *Plantago major* ssp. *major* and ssp. *pleiosperma*. Morphology, biology and ecology in Denmark. Bot Tidsskr 71:31–56
Mølgaard P (1986a) Population genetics and geographical distribution of caffeic acid esters in leaves of *Plantago major* in Denmark. J Ecol 74:1127–1137
Mølgaard P (1986b) Food plant preferences by slugs and snails: a simple method to evaluate the relative palatability of the food plants. Biochem Syst Ecol 14:113–121
Mølgaard P, Ravn H (1988) Evolutionary aspects of caffeoyl ester distribution in dicotyledons. Phytochemistry 27:2411–2421
Mølgaard P, Skjøth L, Kaufmann U (1980) A genetic investigation of the esters of caffeic acid with rhamnose or glucose in leaves of *Plantago major*. Biochem Syst Ecol 8:277–278
Mook JH, Haeck J, Van der Toorn J (1981) Survival of *Plantago lanceolata* in vegetations of varying structure. Verh K Ned Adak Wetensch, Afd Natuurkd, 2 R 77:32–35
Mook JH, Haeck J, Van der Toorn J, Van Tienderen PH (1989) Comparative demography of *Plantago*. I. Observations on eight populations of *Plantago lanceolata*. Acta Bot Neerl 38:67–78
Mueller M, Santarius KA (1978) Changes in chloroplast membrane lipids during adaptation of barley to extreme salinity. Plant Physiol 62:326–329
Müller I (1947) Die pollenanalytischen Nachweise der menschlichen Besiedlung im Federsee- und Bodenseegebiet. Planta 35:70–87
Munns R, Greenway H, Delane R, Gibbs J (1982) Ion concentration and carbohydrate status of the elongation leaf tissues of *Hordeum vulgare* growing at high NaCl. J Exp Bot 33:574–583
Musgrave ME, Antonovics J, Siedow JN (1986) Is male sterility in plants related to lack of cyanide-resistant respiration in tissues? Plant Sci 44:7–11
Nannenga-Bremenkamp NE (1974) De Nederlandse Myxomyceten. K. N. N. V. Thieme, Stuttgart, Zutphen
Naylor REL (1980) Effects of seed size and emergence time on subsequent growth of perennial ryegrass. New Phytol 84:313–318
Neales TF, Sharkey PJ (1982) Effect of salinity on growth and on mineral and organic constituents of the halophyte *Disphyma australe* (SoLand.) IM Black. Aust J Plant Physiol 8:165–179
Nelson PV, Kuiper PJC (1975) Properties of (sodium/potassium)-stimulated ATPase in English ryegrass roots and implications in ion transport. Physiol Plant 35:263–268
Nevo E, Beiles A, Ben-Shlomo R (1984) The evolutionary significance of genetic diversity: ecological, demographic and life history correlates. In: Mani GS (ed) Lecture notes in biomathematics, vol 53. Evolutionary dynamics of genetic diversity. Springer, Berlin Heidelberg New York, pp 13–213
Newman EI (1978) Root microorganisms: their significance in the ecosystem. Biol Rev 53:511–554
Newman EI, Heap AJ, Lawley RA (1981) Abundance of mycorrhizas and root surface microorganisms of *Plantago lanceolata* in relation to soil and vegetation: a multivariate approach. New Phytol 89:95–108
Noë R, Blom CWPM (1982) Occurrence of three *Plantago* species in coastal dune grasslands in relation to pore-volume and organic matter content of the soil. J Appl Ecol 19:177–182

Oberdorfer E (1979) Pflanzensoziologische Exkursionsflora. Ulmer, Stuttgart

Ochman H, Jones JS, Selander RK (1983) Molecular area effects in *Cepaea*. Proc Natl Acad Sci USA 80:4189–4193

Oelmüller F, Mohr H (1984) Responsivity amplification by light in phytochrome-mediated induction of chloroplast glyceraldehyde-3-phosphate dehydrogenase (NADP-dependent) in the shoot of milo (*Sorghum vulgare* Pers). Plant Cell Environ 7:29–37

Okamoto H, Ichino K, Katou K (1978) Radial electrogenic activity in the stem of *Vigna sesquipedales*: involvement of spatially separate pumps. Plant, Cell and Environment 1:279–284

Olff H, Kuiper D, van Damme JMM, Kuiper PJC (1989) Gynodioecy in *Plantago lanceolata*. VI. Functions of cytokinins in growth, development and reproduction of two sex types. Can J Bot 67:2765–2769

Olsen SR, Cole CV, Watanabe FS, Dean LA (1954) Estimation of available phosphorus in soils by extraction with sodium bicarbonate. USDA Circ 939

Ondok JP (1971) Indirect estimation of primary values used in growth analysis. In: Sestak Z, Catsky J, Jarvis PG (eds) Plant photosynthetic production. Mannual of methods. Junk, The Hague, pp 392–410

Ondok JP, Gloser J (1978) Net photosynthesis and dark respiration in a stand of *Phragmites communis* Trin. calculated by means of a model. I Description of the model. II Results. Photosynthetica 12:328–343

Oomes MJH, Elberse WT (1976) Germination of six grassland herbs in microsites with different water contents. J Ecol 64:745–755

Öquist G, Brunes L, Hällgren JE (1982) Photosynthetic efficiency during ontogenesis of leaves of *Betula pendula*. Plant Cell Environ 5:17–21

Osmond CG (1976) Ion absorption and carbon metabolism in cells of higher plants. In: Lüttge U, Pitman MG (eds) Encyclopedia of plant physiology, New Ser, vol 2. Transport in plants, pt A: Cells. Springer, Berlin Heidelberg New York, pp 347–372

Oudemans CAJA (1923) Enumeratio systematica fungorum, vol 4. Hagae comitum. Nyhoff, Den Haag

Palmer CE (1981) Influence of abscisic acid on nitrate accumulation and nitrate reductase activity in potato tubes slices. Plant Cell Physiol 22:1541–1551

Parker JJ, Taylor HM (1965) Soil strength and seedling emergence relations I. Soil type, moisture tension, temperature and planting depth effects. Agron J 57:289–291

Pate JS (1983) Distribution of metabolites. In: Steward FC, Bidwell RGS (eds) Plant physiology: a treatise, vol 8. Nitrogen metabolism. Academic Press, New York, pp 335–403

Paul EA, Juma NG (1981) Mineralization and immobilization of soil nitrogen by microorganisms. In: Clark FE, Rosswall T (eds) Terrestrial nitrogen cycles. Ecol Bull 33:179–195

Perrot V, Dommee B, Jacquard P (1982) Etude expérimentale de la concurrence entre individus issus d'autofécondation et d'allofécondation chez le thym (*Thymus vulgaris* L.). Oecol Plant 3:171–184

Pick H (1982) Über den Einfluß des Lichtes auf die Gestalt und Orientierung der Zellen des Assimilationsgewebes. Bot Centralbl 11:400–406

Pilger R (1898) Vergleichende Anatomie der Gattung *Plantago*, mit Rücksicht auf die Existenzbedingungen. Bot Jahrb Syst Pflanzenges Pflanzengeogr 25:296–351

Pilger R (1937) Plantaginaceae. In: Engler A, Diels L (eds) Das Pflanzenreich, vol 4, H 102. Engelman, Leipzig, p 269

Pitman MG, Cram WJ (1977) Regulation of ion content in whole plants. Symp Soc Exp Biol 31:391–424

Pöhl F (1929) Beziehungen zwischen Pollenbeschaffenheit, Bestäubungsart und Fruchtknotenbau. Beitr Bot Centralbl 46:247–285

Poljakoff-Mayber A (1975) Morphological and anatomical changes in plants as a response to salinity stress. In. Poljakoff-Mayber A, Gale J (eds) Plants in saline environments. Springer, Berlin Heidelberg New York, p 97

Pons TL (1983) Significance of inhibition of seed germination under the leaf canopy in ash coppice. Plant Cell Environ 6:385–392

Pons TL (1984) Possible significance of changes in the light requirement of *Cirsium palustre* seeds after dispersal in ash coppice. Plant Cell Environ 7:263–268

Pons TL (1986) Response of *Plantago major* seeds to the red/far-red ratio as influenced by other environmental factors. Physiol Plant 68:252–258

Pons TL (1989) Breaking of seed dormancy by nitrate as a gap detection mechanism. Ann Bot (London) 63:139–145

Pons TL, Van der Toorn J (1988) Establishment of *Plantago lanceolata* L. and *Plantago major* L. among grass. I. Significance of light for germination. Oecologia 75:394–399

Poorter H, Lambers H (1986) Growth and competitive ability of a highly plastic and a marginally plastic genotype of *Plantago major* in a fluctuating environment. Physiol Plant 67:217–222

Powell JR (1971) Genetic polymorphisms in varied environments. Science 174:1035–1036

Powell JR, Wistrand H (1978) The effect of heterogeneous environments and a competitor on genetic variation in *Drosophila*. Am Nat 112:935–947

Price MW, Waser NM (1979) Pollen dispersal and optimal outbreeding in *Delphinium nelsoni.* Nature (London) 277:294–298

Primack RB (1976) The evolutionary basis of population dynamics in the genus *Plantago.* Thesis, Duke Univ Durham, NC

Primack RB (1978) Evolutionary aspects of wind pollination in the genus *Plantago* (Plantaginaceae). New Phytol 81:449–458

Primack RB, Antonovics J (1981) Experimental ecological genetics in *Plantago.* V. Components of seed yield in the ribwort plantain *Plantago lanceolata.* Evolution 35(6):1069–1079

Prins HHT, Ydenberg RC, Drent RH (1980) The interaction of Brent Geese *Branta bernicla* and sea plantain *Plantago maritima* during spring staging: field observations and experiments. Acta Bot Neerl 29:585–596

Rabenhorst L (1907) Die Pilze Deutschlands, Österreichs und der Schweiz. VIII. *Fungi imperfecti.* Kummer, Leipzig

Rahn K (1954) Experimental and cytological studies in *Plantago media.* Bot Tidsskr 51:300–307

Rahn K (1957) Chromosome number in *Plantago.* Bot Tideskr 53:369–378

Ranachandran SR, Venkataraman K (1968) The seed husk of *Plantago ovata* in printing and finishing. J Soc Dyers Color 54:462–464

Ranwell DS (1972) Ecology of salt marshes and sand dunes. Chapman & Hall, London

Rauh W, Rappert F (1954) Über das Vorkommen und die Histogenese von Scheitelgruben bei krautigen Dikotylien, mit besonderer Berücksichtigung der Ganz- und Halbrosettenpflanzen. Planta 43:325–360

Rauh W, Senghas K (1976) Flora von Deutschland, 86th edn. Schmeil O, Fitschen J (eds) Quelle & Meyer, Heidelberg, p 50

Raven JA (1985) pH regulation in plants. Sci Progr 69:495–509

Raven JA, Smith FA (1973) The regulation of intracellular pH as a fundamental biological process. In: Anderson WP (ed) Ion transport in plants. Academic Press, New York London, pp 271–278

Raven JA, Smith FA (1976) Nitrogen assimilation and transport in vascular land plants in relation to intracellular pH regulation. New Phytol 76:415–431

Ravn H, Brimer L (1988) Structure and antibacterial activity of plantamajoside, a caffeic acid sugar ester from *Plantago major* ssp. *major.* Phytochemistry 27:3433–3437

Reid DA (1970) *Thanatphorus langlei-regis* Reid, sp. nov. Trans Br Mycol Soc 52:22

Reisenauer HM (1978) Absorption and utilization of ammonium nitrogen by plants. In: Nielsen DR, MacDonald JG (eds) Nitrogen in the environment, soil-plant-nitrogen relationships, vol 2. Academic Press, New York London San Francisco, pp 157–170

Rezk MR (1974) Sensitivity to light in *Plantago* seeds as related to seed coat structure. Acta Biol N Ser 20:89–93

Rhoades FD, Cates RG (1976) Toward a general theory of plant antiherbivore chemistry. In: Recent advances in phytochemistry, vol 10. Biochemical interaction between plants and insects. Plenum, New York London, pp 168–213

Rhodes D, Rendon GA, Stewart GR (1976) The regulation of ammonia assimilating enzymes in *Lemna minor.* Planta 129:203–210

Richardson B, Whittaker JB (1982) The effect of varying the reference material on ranking of acceptability indices of plant species to a polyphagous herbivore, *Agriolimax reticulatus.* Oikos 39:237–240

Richter DD, Switzer GL (1982) A technique for determining quantitative expressions of dormancy in seeds. Ann Bot (London) 50:459–463

Rimbach A (1898) Die kontraktielen Wurzeln und ihre Thätigkeit. Beitr Wiss Bot 2:1–28

Rimbach A (1899) Beiträge zur Physiologie der Wurzeln. Ber Dtsch Bot Ges 17:18–35

Rimbach A (1929) Die Verbreitung der Wurzelverkürzung im Pflanzenreich. Ber Dtsch Bot Ges 47:22–31

Roberts EH (1972) Dormancy: a factor affecting seed survival in soil. In: Roberts EH (ed) Viability of seeds. Univ Press, Syracuse, pp 321–359

Robinson D, Rorison IH (1983) Relationships between root morphology and nitrogen availability in a recent theoretical model describing nitrogen uptake from soil. Plant Cell Environ 6:641–647

Ross MD (1970) Breeding systems in *Plantago*. Heredity 25:129–133

Ross MD (1973) Inheritance of self incompatibility in *Plantago lanceolata*. Heredity 30:169–176

Ross MD (1978) The evolution of gynodioecy and subdioecy. Evolution 32:174–188

Ross MD, Gregorius HR (1985) Selection with gene-cytoplasm interactions. II. Maintenance of gynodioecy. Genetics 109:427–439

Rothera SL, Davy AJ (1986) Polyploidy and habitat differentiation in *Deschampsia cespitosa*. New Phytol 102:449–467

Roughgarden J (1979) Theory of population genetics and evolutionary ecology: an introduction. MacMillian, New York

Rouwendal GJA, Van Damme JMM, Wessels JGH (1987) Cytoplasmic male sterility in *Plantago lanceolata* L.: differences between male-sterile cytoplasms at the DNA- and RNA-level. Theor Appl Gen 75:59–65

Rush DW, Epstein E (1976) Genotypic responses to salinity: differences between salt-sensitive and salt-tolerant genotypes of tomato. Plant Physiol 57:162–166

Rush C (1970) Transmission of Tobacco Ringspot Virus from native hosts to *Cucumis sativus* by *Xiphinema americanum*. Phytopathology 60:917–918

Rushton BS (1983) Male sterility in natural populations of *Plantago coronopus* L. in Ireland. J Life Sci 4:231–327

Sagar GR (1959) The biology of some sympatric species of grassland. D Phil Thesis Univ of Oxford

Sagar GR, Harper JL (1960) Factors affecting the germination and early establishment of plantains (*Plantago lanceolata*, *P. media* and *P. major*). In: Harper JL (ed) Br Ecol Soc Symp, vol 1. The biology of weeds. Blackwell, Oxford, pp 236–245 (Brit Ecol Soc Symp 1:236–245)

Sagar GR, Harper JL (1961) Controlled interference with natural populations of *Plantago lanceolata*, *P. major* and *P. media*. Weed Res 1:163–176

Sagar GR, Harper JL (1964) Biological flora of the british isles. *Plantago major* L., *Plantago media* L. and *Plantago lanceolata* L. J Ecol 52:189–221

Salisburgy EJ (1961) Weeds and aliens. Collins new naturalistic series. Collins, London

Salt GA (1979) The increasing interest in 'minor pathogens'. In: Schippers B, Gams W (eds) Soil borne plant pathogens. Academic Press, New York, pp 289–312

Sawney VK (1974) Morphogenesis of the stamenless-2 mutant in tomato III Relative levels of gibberellins in the normal and mutant plants. J Exp Bot 25:1004–1009

Schaal BA, Levin DA (1976) The demographic genetics of *Liatris cylindracea* Michx. (Compositae). Am Nat 110:192–206

Schat H (1981) Seed polymorphism and germination ecology of *Plantago coronopus* L. Acta Oecol 2(16):367–380

Schat H (1982) On the ecology of some Dutch dune slack plants. Thesis, Free Univ Amsterdam, Neth

Scheibe J, Lang A (1965) Lettuce seed germination: evidence for a reversible light-induced increase in growth potential and for phytochrome mediation of the low temperature effect. Plant Physiol 40:485–492

Scheiner SM, Goodnight CJ (1984) The comparison of phenotypic plasticity and genetic variation in populations of the grass *Danthonia spicata*. Evolution 38:845–855

Schmidt EL (1982) Nitrogen in soil. In: Stevenson FJ (ed) Nitrogen in agricultural soils. Agronomy 22:253–288. Am Soc Agron, Madison, WI

Schulze E-D, Hall AE (1982) Stomatal responses, waterloss and CO_2 assimilation rates of plants in contrasting environments. In: Lange OL, Novel PS, Osmund CB, Ziegler H (eds) Encyclopedia of plant physiology, New Ser, vol 12B. Physiological plant ecology II. Springer, Berlin Heidelberg New York, pp 181–230

Scott Russell R, Goss MM (1974) Physical aspects of soil fertility. The response of roots to mechanical impedance. Neth J Agric Sci 22:305–318

Seibert B, Koch W, Eller BM (1980) Leaf temperatures in a gas exchange chamber and in the open air. J Exp Bot 31:863–871

Shaw DV, Kahler AL, Allard RW (1981) A multilocus estimator of mating system parameters in plant populations. Proc Natl Acad Sci USA 78:1298–1302

Sheldon JC (1974) The behaviour of seed in soil. III The influence of seed morphology and the behaviour of seedlings on the establishment of plants from surface-lying seeds. J Ecol 62:47–66

Shone MGT, Gale J (1983) Effect of sodium chloride stress and nitrogen source on respiration, growth and photosynthesis in lucerne. J Exp Bot 34:1117–1125

Silvertown J (1981) Seed size, life span, and germination date as coadapted features of plant life history. Am Nat 118:860–864

Silvertown J (1982) Introduction to plant population ecology. Longman, London

Silvertown J (1989) The paradox of seed size and adaptation. TREE 4:24–26

Slim P, Van der Toorn J (1983) Variability in morphological characteristics of *Plantago lanceolata.* Verh K Ned Akad Wetensch Afd Natuurkd 2R81:36–41

Smillie RM, Hetherington SE (1983) A screening method for chilling tolerance using chlorophyll in vivo. In: Sybesma C (ed) Advances in photosynthesis research, vol 4. Nijhoff/Junk, The Hague Boston Lancaster, pp 471–474

Smit AJ, Woldendorp JW (1981) Nitrate production in the rhizosphere of *Plantago* species. Plant Soil 61:43–52

Smith AJ, Hoare DS (1968) Acetate assimilation by *Nitrobacter agilis* in relation to its "obligate autotrophy". J Bacteriol 95:844–855

Smith FA, Raven JA (1979) Intracellular pH and its regulation. Annu Rev Plant Physiol 30:289–311

Smith H (1982) Light quality, photoperception, and plant strategy. Annu Rev Plant Physiol 33:481–518

Smouter H, de Boer AH, Lambers JT (1984) Proton pump activity in a calcicole and a calcifuge *Plantago* species. In: 2nd Int Symp Structure and functioning of plant populations – phenotypic and genotypic variation within and between plant populations, Wageningen. Poster Abstr

Snyder WE (1948) Mechanism of the photoperiodic response of *Plantago lanceolata* L., a long-day plant. Am J Bot 35:520–525

Soekarjo R (1979) Regeneration in *Plantago* species. Verh K Ned Akad Wetensch Afd Natuurkd 2R73:344–346

Soekarjo R (1980) Vegetative regeneration in *Plantago* species and the relation between vegetative and generative regeneration. Verh K Ned Akad Wetensch Afd Natuurkd 2R75:19–23

Soekarjo R (1981) Effect of the supply of water and nutrients on vegetative regeneration in *Plantago* species. Verh K Ned Akad Wetensch Afd Natuurkd 2R77:40–45

Soekarjo R (1982a) Effects of low temperature on the germination of seeds of *Plantago major.* Verh K Ned Akad Wetensch Afd Natuurkd 2R79:51–55

Soerkarjo R (1982b) Effects of low temperatures on the germination of seeds of *Plantago major.* Verh K Ned Akad Wetensch Afd Natuurkd 2R81:51–54

Solbrig OT (1980) Principles of demographic analysis applied to natural polyploid populations. In: Lewis WH (ed) Polyploidy, biological relevance. Plenum, New York, pp 537–538

Soltis DE (1984) Autopolyploidy in *Tolmiea menziesii* (Saxifragaceae). Am J Bot 71:1171–1174

Soltis DE, Rieseberg LH (1986) Autopolyploidy in *Tolmiea menziesii* (Saxifragaceae): genetic insights from enzyme electrophoresis. Am J Bot 73:310–318

Souèges R (1921) Embryogénie des Scrophulariacées. Développement de l'embryon chez le *Veronica arvensis.* CR Acad Sci Paris 172:703

Souèges R (1923) Embryogénie des Plantaginacées. Développement de l'embryon chez le *Plantago lanceolata* L. CR Acad Sci Paris 177:864–887

Soulé M (1971) The variation problem: the gene flow-variation hypothesis. Taxon 20:37–50
Staal M, Maathuis FJM, Elzenga JTM, Overbeek JHM, Prins HBA (1991) Na^+/H^+ antiport activity in tonoplast vesicles in roots from the salt-tolerant *Plantago maritima* L. and the salt-sensitive *Plantago media* L. Physiol Plant 82:179–184
Stålfelt MG (1955) The stomata as a hydrophotic regulator of the water deficit of the plant. Physiol Plant 8:572–593
Stebbins GL (1971) Chromosomal evolution in higher plants. Arnold, London
Stebbins GL (1980) Polyploidy in plants: Unsolved problems and prospects. In: Lewis LH (ed) Polyploidy, biological relevance. Plenum, New York, pp 495–520
Stebbins GL (1985) Polyploidy, hybridization and the invasion of new habitats. Ann MO Bot Garden 72:824–832
Stebbins GL, Dawe JC (1987) Polyploidy and distribution in the European flora: a reappraisal. Bot Jahrb Syst Pflanzenges Pflanzengeogr 108:343–354
Steinbauer GP, Grigsby B (1975) Dormancy and germination characteristics of the seeds of *Plantago*. Proc Assoc Off Seed Anal 47:158–164
Steiner WWM (1977) Niche width and genetic variation in Hawaiian *Drosophila*. Am Nat 111:1037–1045
Stelleman P (1981) Anthecological relations between reputedly anemophilous flowers and syrphid flies. V. Some special aspects of the visiting of *Plantago media* and *P. lanceolata* by insects. Beitrag Biol Pflanzen 55:157–167
Stelleman P (1982) De betekenis van biotische bestuiving bij *Plantago lanceolata*. Thesis, State Univ Amsterdam, Neth
Stelleman P (1984a) The significance of biotic pollination in a nominally anemophilous plant: *Plantago lanceolata*. Proc K Ned Akad Wet C87:95–119
Stelleman P (1984b) Reflections on the transition from wind pollination to ambophyly. Acta Bot Neerl 33:497–508
Stelleman P, Meeuse ADJ (1976) Anthecological relations between reputedly anemophilous and syrphid flies I The possible role of syrphid flies as pollinators of *Plantago*. Tijdschr Entomol 119:15–31
Stelzer R, Läuchli A (1978) Salt- and flooding tolerance of *Puccinellia personas*. III. Distribution and localization of ions in plants. J Pflanzenphysiol 88:437–448
Steubing L (1949) Beiträge zur Tauwasseraufnahme höherer Pflanzen. Biol Zentralbl 68:252–259
Stevens DP, Van Damme JMM (1988) The evolution and maintenance of gynodioecy in sexually and vegetatively reproducing plants. Heredity 61:329–338
Stocker O (1976) The water-photosynthesis syndromė and the geographical plant distribution in the saharan desert. In: Lange OC, Kappen L, Schulze E-D (eds) Water and plant life. Ecole Stud vol 19. Springer, Berlin Heidelberg New York, pp 506–522
Stoutjesdijk P (1980) A small thermo-electric psychometer and its performance. Verh K Ned Akad Wetensch Afd Natuurkd 2R75:23–24
Stuiver CEE, de Kok LJ, Hendrik AE, Kuiper PJC (1982) The effect of salinity on phospholipid content and composition of two *Plantago* species differing in salt-tolerance. In: Wintermans JFGM, Kuiper PJC (eds) Biochemistry and metabolism of plant lipids. Elsevier, Amsterdam, pp 455–458
Stulen I, Lanting L, Lambers H, Posthumus F, Van de Dijk SJ, Hofstra R (1981 a) Nitrogen metabolism of *Plantago lanceolata* as dependent on the supply of mineral nutrients. Physiol Plant 51:93–98
Stulen I, Lanting L, Lambers H, Posthumus F, Van de Dijk SJ, Hofstra R (1981 b) Nitrogen metabolism of *Plantago major* spp. *major* as dependent on the supply of mineral nutrients. Physiol Plant 52:108–114
Sun M, Ganders FR (1986) Female frequencies in gynodioecious populations correlated with selfing rates in hermaphrodites. Am J Bot 73:1645–1648
Takhtajan A (1953) Die Evolution der Angiospermen. VEB Fischer Verlag, Jena
Tanaka T, Mizuno M, Namba T (1982) Pharmacognostical studies of Plantaginis Herba; (2) On the leaves of Japanese *Plantago*. Shoyakugaku Zasshi 36(2):107–118
Tánczos O, Erdei L, Snijder J (1981) Uptake and translocation of sodium in salt-sensitive and salt-tolerant *Plantago* species. Plant Soil 63:27–32

Taylor AA, Havill DC (1981) The effect of inorganic nitrogen on the major enzymes of ammonium assimilation in grassland plants. New Phytol 87:53–62

Taylor HM (1962) Seedling emergence of wheat, grain sorghum and quar as affected by rigidity and thickness of surface crusts. Soil Sci Soc Proc 26:431–432

Taylor HM (1971) Effects of soil strength on seedling emergence, root growth and crop yield. In: Compaction of agricultural soils. ASAE Monogr, Am Soc Agric Eng, Michigan

Taylor HM, Parker JJ, Roberson GM (1966) Soil strength and seedling emergence relations III. A generalized relation of Gramineae. Agron J 58:393–395

Taylorson RB, Hendricks SB (1971) Changes in phytochrome expressed by germination of *Amaranthus retroflexus* L. seeds. Plant Physiol 47:619–622

Teramura A (1978) Localized ecotypic differentiation in three contrasting populations of *Plantago lanceolata* L. Thesis, Duke Univ Durham, NC

Teramura AH (1983) Experimental ecological genetics in *Plantago*. IX. Differences in growth and vegetative reproduction in *Plantago lanceolata* (Plantaginaceae) from adjacent habitats. Am J Bot 70(1):53–58

Teramura AH, Strain BR (1979) Localized populational differences in the photosynthetic response to temperature and irradiance in *Plantago lanceolata*. Can J Bot 57:2559–2563

Teramura AH, Antonovics J, Strain BR (1981) Experimental ecological genetics in *Plantago*. IV. Effects of temperature on growth rates and reproduction in three populations of *Plantago lanceolata* L. Am J Bot 68(3):425–434

Thompson K, Grime JP (1979) Seasonal variation in the seed banks of herbaceous species in ten contrasting habitats. J Ecol 67:893–921

Thompson K, Grime JP, Mason G (1977) Seed germination in relation to diurnal fluctuations of temperature. Nature (London) 267:147–149

Thornberry HH, Anderson WH (1937) Some bacterial diseases of plants in Illinois. Phytopathology 27:946–949

Tilman D (1985) The resource ratio hypothesis of succession. Am Nat 125:827–852

Tomilin BA (1979) Opredelitel gribov roda *Mycosphaerella* Johans. Nauka, Leningrad

Tomlinson JA, Garrett RG (1964) Studies on lettuce big-vein virus and its vector *Olpidium brassicae* (Wor) Dang. Ann Appl Biol 54:45–61

Tonsor SJ (1985a) Intrapopulational variation in pollen-mediated gene flow in *Plantago lanceolata* L. Evolution 39:775–782

Tonsor SJ (1985b) Leptokurtic pollen-flow, non-leptokurtic gene flow in a wind pollinated herb. *Plantago lanceolata* L. Oecologia 67:442–446

Toole VK (1973) Effects of light and temperature and their interaction on the germination of seeds. Seed Sci Technol 1:339–396

Torrey JG (1976) Root hormones and plant growth. Annu Rev Plant Physiol 27:435–459

Troelstra SR (1978) Spatial variability of soil chemical properties in an older dune area (Westduinen) of Goeree. Verh K Ned Akad Wetensch Afd Natuurkd 2R71:321–327

Troelstra SR (1983) Growth of *Plantago lanceolata* and *Plantago major* on a NO_3/NH_4 medium and the estimation of the utilization of nitrate and ammonium from ionic balance aspects. Plant Soil 70:183–197

Troelstra SR, Blacquière T (1985) Growth, ionic balance, proton excretion, and NRA in *Alnus* and *Hippophaë* supplied with different sources of nitrogen. Plant Soil 91:381–384

Troelstra SR, Smant W (1979) The ionic balance of some plant species from natural vegetations and its relation to nitrogen uptake and salt tolerance. Verh K Ned Akad Wetensch Afd Natuurkd 2R73:381–389

Troelstra SR, Smant W (1980) The ionic balance of some plant species from natural vegetations: comparison of plants grown in the greenhouse and in the field. Verh K Ned Akad Wetensch Afd Natuurkd 2R75:46–51

Troelstra SR, Wagenaar R (1980) Seasonal patterns in the availability of mineral nitrogen in a relatively old dune area (Westduinen) on the island of Goeree. Verh K Ned Akad Wetensch Afd Natuurkd 2R75:37–41

Troelstra SR, Sluimer L, Smant W, Wagenaar R, Van der Meulen MA (1981) On the soil chemistry of natural habitats of *Plantago* species and *Hypochaeris radicata* in various parts of The

Netherlands in relation to chemical composition of the plants. Verh K Ned Akad Wetensch Afd Natuurkd 2R77:20–32

Troelstra SR, Smant W, Wagenaar R (1983a) Different K and P requirements for *Plantago major* spp. *major* and *Plantago lanceolata*? Verh K Ned Akad Wetensch Afd Natuurkd 2R81:43–54

Troelstra SR, Freijsen AHJ, Smant W (1983b) The ionic balance of *Plantago* species and its relation to nitrogen nutrition. In: Lee, JA, McNeill S, Rorison IH (eds) Nitrogen as an ecological factor. Blackwell, Oxford, pp 437–438

Troelstra SR, Smant W, Van der Meulen MA (1984a) The chemical composition of *Plantago lanceolata* and *Plantago major* spp. *major* at some selected sites. Verh K Ned Akad Wetensch Afd Natuurkd 2R82:28–36

Troelstra SR, Wagenaar R, Sluimer L, Van der Meulen MA, Smant W (1984b) Soil chemical aspects of some selected *Plantago* sites. Verh K Ned Akad Wetensch Afd Natuurkd 2R83: 62–72

Troelstra SR, Van Dijk C, Blacquière T (1985) Effects of N source on proton excretion, ionic balance and growth of *Alnus glutinosa* (L.) Gaertner: comparison of N_2 fixation with single and mixed sources of NO_3 and NH_4. Plant Soil 84:361–385

Troll W (1937) Vergleichende Morphologie der höheren Pflanzen, vol 1. Vegetationsorgane, pt 1. Bornträger, Berlin, p 714

Troll W (1939) Vergleichende Morphologie der höheren Pflanzen, vol 1. Vegetationsorgane, pt 2. Bornträger, Berlin, pp 1193; 1643

Troll W, Rauh W (1950) Das Erstarkungswachstum krautiger Dikotylen, mit besonderer Berücksichtigung der primären Verdickungsvorgänge. Sitzungsber Heidelberger Akad Wiss Math Naturwiss Kl Abh 1:1–86

Turesson G (1922) The species and the variety as ecological units. Hereditas 3:100–113

Turesson G (1930) The selective effect of climate upon the plant species. Hereditas 14:99–152

Turner ME, Stephens JC, Anderson WW (1982) Homozygosity and patch structure in plant populations as a result of nearest-neighbour pollination. Proc Nat Acad Sci 79:203–207

Tutin TG, Heywood VH, Burgers NA, Moore DM, Valentine DH, Walters SM, Webb DA (eds) (1964–1980), Flora Europaea I–V. Cambridge Univ Press, London

USDA. (1960) Index of plant diseases in the United States. USDA Handbook 165, US Govt Printing Office Washington

Vakhmistrov DB, Tikthaya NJ, Mishustina NE (1982) Characterization and comparison of membrane-bound Na, K, Mg-ATPase from tissues of *Hordeum vulgare* L. and *Halocnemum strobilaceum* L. Physiol Plant 55:155–160

Van Altena SC, Minderhoud JW (1972) Keimfähige Samen von Gräsern und Kräutern in der Narbenschicht der niederländischen Weiden. Z Acker Pflanzenb 136:95–109

Van Damme JMM (1982) Gynodioecy in *Plantago lanceolata* L. I. Polymorphism for plasmon type. Heredity 49:303–318

Van Damme JMM (1983a) On the gynodioecy of *Plantago lanceolata*. Thesis, State Univ Groningen, Neth

Van Damme JMM (1983b) Gynodioecy in *Plantago lanceolata* L. II. Inheritance of three male sterility types. Heredity 50:253–273

Van Damme JMM (1984) Gynodioecy in *Plantago lanceolata* L. III. Sexual reproduction and the maintenance of male steriles. Heredity 52:77–93

Van Damme JMM (1985) Why are so many *Plantago* species gynodioecious? In: Haeck J, Woldendorp JW (eds) Structure and functioning of plant populations. 2. Phenotypic and genotypic variation in plant population. North Holland, Oxford, pp 241–251

Van Damme JMM (1986) Gynodioecy in *Plantago lanceolata* L. V. Frequencies and spatial distribution of nuclear and cytoplasmic genes. Heredity 56:355–365

Van Damme JMM (1991) A restover gene in gynodioecious *Plantago coronopus* subject to selection in the gametophytic and seedling stage. Heridity 66: in press

Van Damme JMM, Graveland J (1985) Does cytoplasmic variation in *Plantago lanceolata* contribute to ecological differentiation? In: Jacquard P, Helm G, Antonovics J (eds) Genetic differentiation and dispersal in plants. Springer, Berlin, pp 67–81

Van Damme JMM; Van Damme R (1986) On the maintenance of gynodioecy: Lewis' result extended. J Theor Biol 121:339–350

Van Damme JMM, Van Delden W (1982) Gynodioecy in *Plantago lanceolata* L. I. Polymorphism for plasmon type. Heredity 49:303–318

Van Damme JMM, Van Delden W (1984) Gynodioecy in *Plantago lanceolata* L. IV. Fitness components of sex types in different life cycle stages. Evolution 38:1326–1336

Van de Dijk SJ (1981) Nitrogen metabolism in higher plants in different environments. Thesis, State Univ Groningen, Neth

Van de Dijk SJ, Lanting L, Lambers H, Posthumus F, Stulen I, Hofstra R (1982) Kinetics of nitrate uptake by different species from nutrient-rich and nutrient-poor habitats as affected by the nutrient supply. Physiol Plant 55:103–110

Van Delden W (1985) The significance of genetic variation in plants as illustrated by *Plantago* populations. In: Haeck J, Woldendorp JW (eds) Structure and functioning of plant populations 2. Phenotypic and genotypic variation in plant populations. North Holland, Amsterdam, pp 219–239

Van Delden W (1988) Multigenic selection in *Plantago* and *Drosophila*, two different approaches. In: de Jong C (ed) Population genetics and evolution. Springer, Berlin Heidelberg New York, pp 173–186

Van den Bergh JP (1979) Changes in the composition of mixed populations of grassland species. In: Werger MJA (ed) The study of vegetation. Jung, The Hague, pp 59–80

Van der Aart PJM (1985) Demographic, genetic and ecophysiological variation in *Plantago major* and *Plantago lanceolata* in relation to habitat type. In: White J (ed) Handbook of vegetation science, vol 3. Population structure of vegetation. Junk, Dordrecht, pp 441–462

Van der Goot VS, Grabandt RAJ (1970) Some species of the genera *Melanostoma*, *Platycheirus* and *Pyrophaena* (Diptera, Syrphidae) and their relations to flowers. Ent Ber 30:135–143

Van der Paauw F (1971) An effective water extraction method for the determination of plant-available soil phosphorus. Plant Soil 34:467–481

Van der Paauw F (1978) Die Stellung der P-Wassermethode zur Erfassung des P-Angebotes des Bodens. Landw Forsch Sonderh 34:109–120

Van der Paauw F, Sissingh HA, Ris J (1971) An improved method of water extraction for the assessment of availability of soil phosphate: Pw value. Agric Res Rep 749. Pudoc, Wageningen (in Dutch, with English summary)

Van der Pijl (1972) Principles of dispersal in higher plants. Springer, Berlin Heidelberg New York

Van der Toorn J, Haeck J, Mook JH (1980) Some remarks on the demography of *Plantago* species. Verh K Ned Akad Wetensch 2R75:16–19

Van der Toorn J, Pons TL (1988) Establishment of *Plantago lanceolata* L. and *Plantago major* L. among grass. II. Shade tolerance of seedlings and selection on time of germination. Oecologia 76:341–347

Van der Veen R (1970) The importance of the red-far red antagonism in photoblastic seeds. Acta Bot Neerl 19:809–812

Van der Veen R, Meijer G (1958) Licht und Pflanzen. Philips' Technische Bibliotheek, Eindhoven, p 38

VanDerWoude WJ, Toole VK (1980) Studies of the mechanism of enhancement of phytochrome-dependent lettuce seed germination by prechilling. Plant Physiol 66:220–224

Van Dijk H (1984) Genetic variability in *Plantago* species in relation to their ecology. 2. Quantitative characters and allozyme loci in *P. major*. Theor Appl Genet 68:43–52

Van Dijk H (1985a) Genetic variability in *Plantago* species in relation to their ecology. Thesis, State Univ Groningen, Neth

Van Dijk H (1985b) Allozyme genetics, self-incompatibility and male sterility in *Plantago lanceolata*. Heredity 54:53–64

Van Dijk H (1987) A method for the estimation of gene flow parameters from a population structure caused by restricted gene flow and genetic drift. Theor Appl Genet 73:724–736

Van Dijk H (1989) Genetic variability in *Plantago* species in relation to their ecology. 4. Ecotypic differentiation in *P. major*. Theor Appl Genet 77:749–759

Van Dijk H, Kuiper PJC (1989) No evidence for a pleiotropic relationship between male sterility and cyanide resistant respiration in *Plantago lanceolata*. Physiol Plant 77:579–586

Van Dijk H, Van Delden W (1981) Genetic variability in *Plantago* species in relation to their ecology. I. Genetic analysis of the allozyme variation in *P. major* subspecies. Theor Appl Genet 60:285–290

Van Dijk H, Wolff K, De Vries A (1988) Genetic variability in *Plantago* species in relation to their ecology. 3. Genetic structure of populations of *P. major, P. lanceolata* and *P. coronopus.* Theor Appl Genet 75:518–528

Van Dijk PJ, Van Delden W (1990) Evidence for autotetraploidy in *Plantago media* and comparisons between natural and artificial cytotypes concerning cell size and fertility. Heredity 65:349–357

Van Dijk PJ, Hartog MV, Van Delden W (1991) Single cytotype areas in *Plantago media* L. Biol J Linn Soc (in press)

Van Egmond F (1975) The ionic balance of the sugar-beet plant. Thesis, Agric Univ Wageningen, Neth

Van Egmond F (1978) Nitrogen nutritional aspects of the ionic balance of plants. In: Nielsen DR, MacDonald JG (eds) Nitrogen in the environment, soil-plant-nitrogen relationship, vol 2. Academic Press, New York London San Francisco, pp 169–189

Van Groenendael JM (1985 a) Selection for different life histories in *Plantago lanceolata* L. Thesis, Cath Univ Nijmegen, Neth

Van Groenendael JM (1985 b) Life history characteristics of two ecotypes of *Plantago lanceolata* L. In: White J (ed) Studies on plant demography: a "Festschrift" for John L Harper. Academic Press, New York London, pp 51–67

Van Groenendael JM (1985 c) Teratology and metameric plant construction. New Phytol 99:171–178

Van Groenendael JM (1986) Life history characteristics of two ecotypes of *Plantago lanceolata* L. Acta Bot Neerl 32:71–86

Van Groenendael JM, Slim P (1988) The contrasting dynamics of two populations of *Plantago lanceolata* classified by age and size. J Ecol 76:585–599

Van Tienderen PH (1989) On the morphology of *Plantago lanceolata*: selection, adaptation, constraints. Thesis, State Univ Utrecht, Neth

Van Tienderen PH, Van der Toorn J (1991 a) Genetic differentiation between populations of *Plantago lanceolata.* I Local adaptation in three contrasting habitats. J Ecol 79:27–42

Van Tienderen PH, Van der Toorn J (1991 b) Genetic differentiation between populations of *Plantago lanceolata.* II Phenotypic selection in a transplant experiment in three contrasting habitats. J Ecol 79:43–59

Van Tuil HDW (1965) Organic salts in plants in relation to nutrition and growth. Agric Res Rep 657, Pudoc, Wageningen

Van Valen L (1965) Morphological variation and width of ecological niche. Am Nat 99:377–390

Vasudevan KN (1976) Contribution to the cytotaxonomy and cytogeography of the flora of the western Himalayas. part III. Ber Schweiz Bot Ges 86:152–203

Venable DL (1984) Using intraspecific variation to study the ecological significance and evolution of plant life histories. In: Dirzo A, Sarukhán J (eds) Perspective on plant population ecology. Sinauer, Sunderland, pp 166–187

Vernet P, Guillem JL, Gouyon PH (1977) Le polymorphisme chimique de *Thymus vulgaris.* I Répartition des formes chimiques en relation avec certains facteurs écologiques. Oecol Plant 12:159–179

Von Gaisberg E (1922) Zur Deutung der Monokotylenblätter als Phyllodien. Flora NF 15:177–188

Von Wettstein F (1937) Experimentelle Untersuchungen zum Artbildungsproblem I. Zellgrossen-Regulation und Fertilwerden einer polyploiden Bryum-Sippe. Z Ind Abstamm Vererb 74:34–53

Waddington KD (1983) Pollen flow and optimal outcrossing distance. Am Nat 122:147–151

Waite S (1984) Changes in the demography of *Plantago coronopus* at two coastal sites. J Ecol 72:809–826

Waite S, Hutchings MJ (1982) Plastic energy allocation patterns in *Plantago coronopus.* Oikos 38:333–342

Wallsgrove RM, Keys AJ, Lea PJ, Miflin BJ (1983) Photosynthesis, photorespiration and nitrogen metabolism. Plant Cell Environ 6:301–309

Walmsley D, Cornforth IS (1973) Methods of measuring available nutrients in West Indian soils. II. Phosphorus, Plant Soil 39:93–101

Walter H (1968) Die Vegetation der Erde, vol 2. Fischer, Stuttgart

Warwick SI (1980) The genecology of lawn weeds. VII. The response of different growth forms in *Plantago major* L. and *Poa annua* L. to simulated trampling. New Phytol 85:461–469

Warwick SI, Briggs D (1978) The genecology of lawn weeds. II. Evidence for disruptive selection in *Poa annua* L. in a mosaic environment of bowling green lawns and flower beds. New Phytol 81:725–737

Warwick SI, Briggs D (1979) The genecology of lawn weeds. III. Cultivation experiments with *Achillea millefolium* L., *Bellis perennis* L.; *Plantago lanceolata* L., *Plantago major* L. and *Prunella vulgaris* L. collected from lawns and contrasting grassland habitats. New Phytol 83:509–536

Warwick SI, Briggs D (1980) The genecology of lawn weeds. V. The adaptive significance of different growth habit in lawn and roadside populations of *Plantago major.* L. New Phytol 85:289–300

Waser MM, Price MV (1983) Optimal and actual outcrossing in plants, and the nature of plant-pollinator interaction. In: Jones CE, Little RJ (eds) Handbook of experimental pollination biology. Sci and Acad Ed, New York, pp 341–359

Watson RJ (1969) Evolution in closely adjacent plant populations. VII. An entomophilous species, *Potentilla erecta,* in two contrasting habitats. Heredity 24:407–422

Weaver SE, Cavers PB (1979) Dynamics of seed populations of *Rumex crispus* and *Rumex obtusifolius* (Polygonaceae) in disturbed and undisturbed soil. J Appl Ecol 16:909–917

Weberbauer A (1911) Die Vegetation der Erde. XII. Die Pflanzenwelt der peruanischen Anden. Leipzig

Wehmeyer LE (1961) A world monograph of the genus *Pleospora* and its segregates. Univ Michigan Press, Ann Arbor

Weidner M, Mathee C, Schmitz FK (1982) Phenotypical temperature adaptation of protein synthesis in wheat seedlings. Plant Physiol 69:1281–1288

Weissenböck G (1969) Einfluß des Bodensalzgehaltes auf Morphologie und Ionenspeicherung von Halophyten. Flora Ser B 158:369–389

Werner PA (1975) A seed trap for determining patterns of seed deposition in terrestrial plants. Can J Bot 53:810–813

Westhoff V, den Held AJ (1969) Plantengemeenschappen in Nederland. Thieme, Zutphen

Westhoff V, van der Maarel (1973) The Braun-Blanquet approach. In: Whittaker RH (ed) Handbook of vegetation science 5. Junk, The Hague, pp 691–726

Westhoff V, van Leeuwen CG, Adriani MJ (1961) Vegetatie en fauna van Goeree. Jaarboek natuurmon. De Beer, Den Haag

Westhoff V, Bakker PA, van Leeuwen CG, van der Voo EE (1970) Wilde planten; flora en vegetatie in onze natuurgebieden. Veren Behoud Natuurmonum Ned, 's Gravenland

Wiesner J (1891) Formänderungen von Pflanzen bei Cultur im absolut feuchten Raume und im Dunkeln. Ber Dtsch Bot Ges 9:46–53

Wild A (1981) Mass flow and diffusion. In: Greenland DJ, Hayes MHB (eds) The chemistry of soil processes. John Wiley and Sons, New York Chichester Brisbane Toronto, pp 37–80

Williams ED (1983) Germinability and enforced dormancy in seeds of species of indigenous grasslands. Ann Appl Biol 102:557–566

Williams RF (1948) The effect of phosphorus supply on the rates of intake of phosphorus and nitrogen and upon certain aspects of phosphorus metabolism in gramineous plants. Aust J Sci Res 1:333–361

Williamson M (1972) The analysis of biological populations. Arnold, London

Willson MF (1982) Sexual selection and dicliny in angiosperms. Am Natur 119:579–583

Whittingham J, Read DJ (1982) Vesicular arbuscular mycorrhiza in natural vegetation systems II. Nutrient transfer between plants with mycorrhizal interconnections. New Phytol 90:277–284

Woldendorp JW (1983) The relationship between the nitrogen metabolism of *Plantago* species and the characteristics of the environment. In: Lee JA, McNeill S, Rorison IH (eds) Nitrogen as an ecological factor. Blackwell, Oxford, pp 137–166

Wolff K (1987) Genetic analysis of ecologically relevant morphological variability in *Plantago lanceolata* L. 2. Localisation and organisation of quantitative trait loci. Theor Appl Genet 73:903–914

Wolff K (1988) Genetic analysis of ecologically relevant morphological variability in *Plantago lanceolata* L. 3. Natural selection in an F_2 population. Theor Appl Genet 75:772–778

Wolff K (1990) Genetic analysis of ecologically relevant morphological variability in *Plantago lanceolata* L. 5. Diallel analysis in two natural populations. Theor Appl Genet 79:481–489

Wolff K (1991 a) Genetic analysis of morphological variability in three *Plantago* species with different mating systems. Theor Appl Genet 81:111–119

Wolff K (1991 b) Analysis of allozyme variability in three *Plantago* species and a comparison to morphological variability. Theor Appl Genet 81:119–127

Wolff K, Haeck J (1990) Genetical analysis of ecologically relevant morphological variability in *Plantago lanceolata* L. 6. Heterozygote superiority in a natural population. J, Evol Biol 3:243–255

Wolff K, Van Delden W (1987) Genetic analysis and ecological relevance of morphological variability in *Plantago lanceolata* L. I. Population characteristics. Heredity 58:183–192

Wolff K, Van Delden W (1989) Genetic analysis of ecological relevant morphological variability in *Plantago lanceolata* L. IV. Response and correlated response to bidirectional selection for leaf angle. Heredity 62:153–160

Wolff K, Friso B, Van Damme JMM (1988) Outcrossing rates and male sterility in natural populations of *Plantago coronopus.* Theor Appl Genet 76:190–196

Wollenweber B, Kinzel H (1988) Role of carboxylate in the nitrogen metabolism of plants from different natural habitats. Physiol Plant 72:321–328

Woodhouse RP (1954) Hayfever plants. Chemica Botanica, Waltham, MA Mass

Wu L, Antonovics J (1975) Experimental ecological genetics in *Plantago.* I. Induction of roots and shoots on leaves for large-scale vegetative propagation and metal tolerance testing in *P. lanceolata.* New Phytol 75:277–282

Wu L, Antonovics J (1976) Experimental ecological genetics in *Plantago.* II. Lead tolerance in *Plantago lanceolata* and *Cynodon dactylon* from a road side. Ecology 57:205–208

Zoller T (1968) Die Vegetation vom ausgehenden Miozän bis ins Holozän. In: Ur- und Frühgeschichtliche Archeologie der Schweiz I. Die Ältere und Mittlere Steinzeit, pp 146, Basel

Subject Index

MIX
Papier aus verantwortungsvollen Quellen
Paper from responsible sources
FSC® C105338

If you have any concerns about our products,
you can contact us on
ProductSafety@springernature.com

In case Publisher is established outside the EU,
the EU authorized representative is:
Springer Nature Customer Service Center GmbH
Europaplatz 3, 69115 Heidelberg, Germany

Printed by Libri Plureos GmbH
in Hamburg, Germany